The Nature of Natural Philosophy
in the Late Middle Ages

Studies in Philosophy and the
History of Philosophy

General Editor: Jude P. Dougherty

Volume 52

The Nature of Natural Philosophy in the Late Middle Ages

Edward Grant

The Catholic University of America Press
Washington, D.C.

The paper used in this publication meets the minimum re-
quirements of American National Standards for Information
Science—Permanence of Paper for Printed Library Materials,
ANSI Z39.48-1984

∞

Library of Congress Cataloging-in-Publication Data

Grant, Edward, 1926–

The nature of natural philosophy in the late Middle Ages /
    Edward Grant.

      p. cm. — (Studies in philosophy and the history of
      philosophy ; v. 52)

    Includes bibliographical references and index.

    ISBN 978-0-8132-1738-3 (cloth : alk. paper)   1. Science,

    Medieval—Philosophy.   I. Title.   II. Series.

    Q174.8.G725 2010

    509.4′0902—dc22

             2009036242

Contents

Preface

Until the early twentieth century, the expression "medieval science" would have been regarded as an oxymoron. This changed when Pierre Duhem, a famous French physicist, became interested in early science and subsequently devoted himself to investigating what might have passed for scientific activity during the Middle Ages. In his examination of numerous medieval manuscripts by a variety of Scholastic natural philosophers, Duhem became convinced that there was indeed science in the late Middle Ages. Between 1902 and 1916, he published some seventeen volumes on medieval science and made the study of medieval science a respectable research activity.[1] He believed that the scientific activity he had discovered in the Middle Ages played an essential role in producing the Scientific Revolution of the seventeenth century. At the same time, he also struck a blow for the continuity of the history of science, demonstrating that the period 1200 and 1500 in Western Europe was not a barren intellectual wasteland, as was almost universally assumed, but a fertile period that flowed naturally into the Scientific Revolution.

There is no doubt that Duhem sometimes made extravagant claims, but his overall results were impressive. Later historians of medieval science, however, were not as certain as Duhem that medieval natural philosophy and science played a significant role in generating the Scientific Revolution, but they were convinced that it probably played a positive role. But historians of the Scientific Revolution would have none of Duhem's claims, or those made by subsequent historians of medieval science. They accused medievalists of "whiggism," or "presentism," by which they meant that medievalists viewed the science and natural philosophy of the Middle Ages with modern eyes, focusing on ideas and achievements that sounded modern and could be interpreted as anticipa-

---

1. His most prodigious work was his ten-volume *Le Système du monde. Histoire des doctrines cosmologiques de Platon a Copernic*, 10 vols. (Paris, 1913–1959).

tions of scientific ideas and theories that were actually proclaimed considerably later.

"Medieval claims were further subverted by Alexandre Koyré, a pre-eminent historian of the Scientific Revolution, who insisted that the classical science of the seventeenth century was in no way a continuation of medieval physics, even when medieval ideas and concepts were strikingly similar to ideas proposed in the Scientific Revolution. It was, he argued a 'decisive mutation' *(mutation decisive)*. The ideas and concepts were embedded in radically different intellectual contexts. Or, to use the language made famous by Thomas Kuhn in *The Structure of Scientific Revolutions,* the respective *normal paradigms* of medieval and seventeenth-century physics were *incommensurable.* The physics and cosmology of the Middle Ages, it was argued, were based wholly upon Aristotelian natural philosophy, which was incompatible with the new science that emerged in the seventeenth century. Indeed, Aristotelian natural philosophy was viewed as the major obstacle to the birth of the new science. Only by its repudiation could the Scientific Revolution have succeeded."[2] Many medievalists—including this author—accepted Koyré's judgment about the relationship of the Middle Ages and the advent of early modern science in the seventeenth century. Although we argued that some interesting contributions had been made in the Middle Ages, we regarded them as playing little or no role in generating the Scientific Revolution. The difficulty of demonstrating that medieval scientific ideas had exerted any direct influence on seventeenth-century natural philosophers led most medievalists to cease making such claims. This attitude shaped my ideas about medieval physical thought in my early book *Physical Science in the Middle Ages.*[3]

But some fifteen to twenty years later, a stunningly dramatic question occurred to me, one that changed my whole attitude about the relationship between medieval natural philosophy and early modern science: could a Scientific Revolution have occurred in the seventeenth century if the level of science and natural philosophy had remained what it was in the first half of the twelfth century?

2. From my book, *The Foundations of Modern Science in the Middle Ages* (Cambridge: Cambridge University Press, 1996), xii.

3. New York: John Wiley and Sons, 1971. Responsibility for the book was assumed by Cambridge University Press in 1977.

That is, could a scientific revolution have occurred in the seventeenth century if the massive translations of Greco-Arabic science and natural philosophy into Latin had never taken place? The response seemed obvious: no, it could not. Without the translations, many centuries would have been required before Western Europe could have reached the level of Greco-Arabic science, thus delaying any possibility of a transformation of science. But the translations did occur and so did the Scientific Revolution. It follows that something happened between approximately 1200 and 1600 that proved conducive for the production of a scientific revolution.[4]

This momentous question is equivalent to asking whether modern physics could have occurred in the nineteenth and twentieth centuries if the level of physics was the same as it was in the days of Isaac Newton in the seventeenth and early eighteenth centuries. The leap to modern physics was a steady and gradual development. It could not have occurred by ignoring the many achievements and advances that had occurred between Newton and the late nineteenth and twentieth centuries. The same could be said of modern medicine and biology, and chemistry, and so on.

If the Middle Ages contributed to the revolution of early modern science, it did not do so by advances in the exact sciences, although some significant contributions in physics and mathematics were made. Indeed, advances were also made in medicine and alchemy. Whether early modern scientists and natural philosophers knew of these advances is largely unknown. In truth, however, all this is largely irrelevant. I would argue that in the Latin Middle Ages of Western Europe an intellectual environment was established that proved conducive to the emergence of early modern science. The new intellectual environment was generated and shaped by "certain attitudes and institutions that were generated in Western society from approximately 1175 to 1500. These attitudes and institutions were directed toward learning as a whole and toward science and natural philosophy in particular. Together they coalesced into what may be appropriately called 'the foundations of modern science.' They were new to Europe and unique to the world. Because there is nothing to which we can compare this extraordinary process, no one can say whether it was fast or slow."[5]

4. Grant, *Foundations of Modern Science*, xiii.
5. Ibid., 170–71.

Three of the most important preconditions that laid the foundations of a new medieval intellectual world and made the Scientific Revolution possible are: "(1) the translation of Greco-Arabic works on science and natural philosophy into Latin, (2) the formation of the medieval university, and (3) the emergence of theologian-natural philosophers."[6] The translations furnished scholars in the Latin West with Aristotle's very substantial body of natural philosophy and Averroes's commentaries on those works.

Aristotle's works—especially his logic and treatises on natural philosophy—quickly formed the basis of an undergraduate education in the newly formed universities of Paris and Oxford. By 1500 there were approximately sixty-five universities spread across Western Europe, virtually all of which taught logic and natural philosophy as the basic undergraduate discipline. Never before had a scientific subject such as natural philosophy been disseminated so widely and deeply in any society. By virtue of its emphasis on reason, natural philosophy made the use of reason commonplace in Western society, thus emphasizing one of the most significant tools for the development of science.

Reason became of great significance in Western thought in part because Scholastics approached Aristotelian natural philosophy by framing questions. Commentaries on any of Aristotle's books on natural philosophy were in the form of a series of successive questions from beginning to end. The questions were answered by use of reason, and in the fourteenth century, as a consequence of various articles condemned in 1277, use of the imagination was emphasized in counterfactual questions about cosmic conditions and circumstances that were regarded as naturally impossible in Aristotle's world.

I have used the expression "theologian-natural philosophers" because virtually all university-trained medieval theologians studied logic and natural philosophy at whatever university they attended. They were expected to be very familiar with natural philosophy. The importance of the theologian-natural philosophers

cannot be overestimated. If theologians at the universities had decided to oppose Aristotelian learning as dangerous to the faith, it could not have become

6. Ibid., 171. For the present context, I have eliminated the word "the" before "theologian-natural philosophers."

the focus of study in European universities. Without the approval and sanction of these scholars, Greco-Arabic science and Aristotelian natural philosophy could not have become the official curriculum of the universities.

To emphasize the great significance of the theologian-natural philosophers, I shall continue on from the passage just cited:

The development within the universities of Western Europe of a class of theologian-natural philosophers was extraordinary. Not only did they endorse a secular arts curriculum, but most believed that natural philosophy was essential for a proper elucidation of theology. Schools of theology expected their entering students to have attained a high level of competence in natural philosophy. As evidence of this, students who wished to matriculate for a theology degree were usually required to have acquired a master of arts degree. Because of the intimate relationship between theology and natural philosophy during the Middle Ages and because arts masters had been forbidden by oath (since 1272) to treat theological problems, it fell to the theologians to apply natural philosophy to theology and theology to natural philosophy. Their training in both disciplines enabled them to do so with relative ease and confidence, whether this involved, for example, the application of science and natural philosophy to scriptural exegesis, the application of the concept of God's absolute power to hypothetical possibilities in the natural world, or the invocation of scriptural texts to support or oppose scientific ideas and theories. Theologians had a remarkable degree of intellectual freedom to cope with such problems and rarely allowed theology to hinder their inquiries into the physical world. If there was any temptation to produce a "Christian science," medieval theologians successfully resisted it. Biblical texts were not employed to "demonstrate" scientific truths by appeal to divine authority.[7]

The points I have made in this preface are illustrated and discussed in the essays that have been included in this book. In a certain sense, these essays are all about natural philosophy and the role it played in shaping medieval thought. The titles are sometimes good indicators of what the essay discusses, as is true of "Science and the Medieval University," "Medieval Departures from Aristotelian Natural Philosophy," "Scientific Imagination in the Middle Ages," and "Science and Theology in the Middle Ages." Two articles—"When Did Modern Science Begin?" and "What Was Natural Philosophy in the Late Middle Ages?"—describe

7. Ibid., 174, 175.

how natural philosophy played a significant role in producing the counterfactual questions that were proposed as a consequence of the articles
condemned in 1277. In "God, Science, and Natural Philosophy in the Late
Middle Ages," I argue with those who believe that natural philosophy is
not in any way associated with science, because, as they argue, natural
philosophy is always about God and modern science is never about God.

"God and the Medieval Cosmos" and "Medieval Natural Philosophy:
Empricism without Observation" belong to the category of essays concerned with the substantive character of medieval natural philosophy as
manifested in its details. "The Fate of Ancient Greek Philosophy in the
Middle Ages: Islam and Western Christianity" attempts to show why natural philosophy in the West flourished while, in Islam, it lost its vitality
and gradually faded away. The final essay in this book—"Aristotelianism
and the Longevity of the Medieval Worldview"—seeks to explain how
medieval Aristotelian natural philosophy, and the view of the cosmos it
upheld for approximately five centuries, managed to remain the dominant way to interpret the world for so long.

I shall conclude my preface with a rather lengthy quotation from
the "Conclusion" of "What Was Natural Philosophy in the Late Middle
Ages?"[8] It argues for the monumental importance of natural philosophy
for the history of science.

What was the legacy of medieval natural philosophy to the modern world? Before 1500, the exact sciences in Islam had reached lofty heights, greater than
they achieved in medieval Western Europe, but they did so without a vibrant
natural philosophy. By contrast, in Western Europe natural philosophy was
highly developed, whereas the exact sciences were merely absorbed (from the
body of Greco-Arabic scientific literature) and maintained at a modest level.
After 1500, Islamic science effectively ceased to advance, but Western science
entered upon a revolution that would culminate in the seventeenth century.
What can we learn from this state of affairs? Let me propose the following: that
the exact sciences were [changed from "are"] unlikely to flourish in isolation
from a well-developed natural philosophy, whereas natural philosophy is apparently sustainable at a high level even in the absence of significant achievements
in the exact sciences. One or more of the exact sciences, especially mathematics,
was practiced in a number of societies that never had a fully developed, broad-

8. *History of Universities* 20, no. 2 (2005): 39–40.

ly disseminated natural philosophy. In none of these societies had scientists attained as high a level of competence and achievement as they had in Islam. Was the subsequent decline of science in Islam perhaps connected with the relatively diminished role of natural philosophy in that society and to the fact that it was never institutionalized in higher education? This is a distinct possibility. In Islamic society, where religion was so fundamental, the absence of support for natural philosophy from theologians, and, more often, their open hostility toward that discipline might have proved fatal to it and, eventually, to the exact sciences as well.

Acknowledgments

The essays in this book were lightly edited for inclusion here. The work was originally presented or published as follows:

1. "When Did Modern Science Begin?" *The American Scholar* (Winter 1997): 105–13. Reprinted in *Frame and Focus: An Anthology for Investigative Reading,* edited by Paul Winters and James Schneider, 4th ed., 211–23. Boston: Pearson Custom Publishing, 2004.

2. "Science and the Medieval University." In *Rebirth, Reform and Resilience: Universities in Transition 1300–1700,* edited by James M. Kittelson and Pamela J. Transue, 68–102. Columbus: Ohio State University Press, 1984.

3. "The Condemnation of 1277, God's Absolute Power, and Physical Thought in the Late Middle Ages." *Viator* 10 (1979): 211–44.

4. "God, Science, and Natural Philosophy in the Late Middle Ages." In *Between Demonstration and Imagination: Essays in the History of Science and Philosophy Presented to John D. North,* edited by Lodi Nauta and Arjo Vanderjagt, 243–67. Leiden: Brill, 1999.

5. "Medieval Departures from Aristotelian Natural Philosophy." In *Studies in Medieval Natural Philosophy,* edited by Stefano Caroti, 237–56. Biblioteca di Nuncius, Studi e Testi. Firenze: Leo S. Olschki, 1989.

6. "God and the Medieval Cosmos." Public lecture at Harris Manchester College Chapel, Oxford University, July 29, 1999.

7. "Scientific Imagination in the Middle Ages." *Perspectives on Science* 12, no. 4 (December 2004): 394–423. Copyright of *Perspectives on Science* is the property of MIT Press, and its content may not be copied or emailed to multiple sites or posted to a listserv without the copyright holder's express written permission. However, users may print, download, or email articles for individual use.

8. "Medieval Natural Philosophy: Empiricism without Observation." In *The Dynamics of Aristotelian Natural Philosophy from Antiquity to the Seventeenth Century,* edited by Cees Leijenhorst, Christoph Luthy, and Johannes M. M. H. Thijssen, 141–68. Leiden: Brill, 2002.

9. "Science and Theology in the Middle Ages." In *God and Nature: Historical Essays on the Encounter between Christianity and Science,* edited by David

C. Lindberg and Ronald L. Numbers, 49–75. Berkeley and Los Angeles: University of California Press, 1986.

10. "The Fate of Ancient Greek Natural Philosophy in the Middle Ages: Islam and Western Christianity." *The Review of Metaphysics* 61 (March 2008): 503–26. Originally presented as a lecture at a conference titled "Children of Abraham: Judaism, Christianity, and Islam," Pennsylvania State University, March 31–April 1, 2000.

11. "What Was Natural Philosophy in the Late Middle Ages?" *History of Universities* 20, no. 2 (2005): 12–46.

12. "Aristotelianism and the Longevity of the Medieval World View." *History of Science* 16 (1978): 93–106. Although considerably, and perhaps even radically, altered, preliminary versions of this essay were delivered at the banquet of the eighteenth annual meeting of the Midwest Junto of the History of Science Society, March 26, 1976, University of Notre Dame; and again on November 4, 1976, at the Medieval Studies Conference, "Medium Aevum Transdisciplinale: Approaches to the Middle Ages," Indiana University, Bloomington. I am grateful to my colleague, Professor J. Alberto Coffa, who made valuable suggestions for the improvement of this essay.

The Nature of Natural Philosophy
in the Late Middle Ages

1 ⤎ When Did Modern Science Begin?

Although science has a long history with roots in ancient Egypt and Mesopotamia, it is indisputable that modern science emerged in Western Europe and nowhere else. The reasons for this momentous occurrence must therefore be sought in some unique set of circumstances that differentiate Western society from other contemporary and earlier civilizations. The establishment of science as a basic enterprise within a society depends on more than expertise in technical scientific subjects, experiments, and disciplined observations. After all, science can be found in many early societies. In Islam, until approximately 1500, mathematics, astronomy, geometric optics, and medicine were more highly developed than in the West. But science was not institutionalized in Islamic society. Nor was it institutionalized in ancient and medieval China, despite significant achievements. Similar arguments apply to all other societies and civilizations. Science can be found in many of them but was institutionalized and perpetuated in none.

Why did science as we know it today materialize only in Western society? What made it possible for science to acquire prestige and influence and to become a powerful force in Western Europe by the seventeenth century? The answer, I believe, lies in certain fundamental events that occurred in Western Europe during the period from approximately 1175 to 1500. Those events, taken together, should be viewed as forming the foundations of modern science, a judgment that runs counter to prevailing scholarly opinion, which holds that modern science emerged in the seventeenth century by repudiating and abandoning medieval science and natural philosophy, the latter based on the works of Aristotle.

The Scientific Revolution appeared first in astronomy, cosmology, and physics in the course of the sixteenth and seventeenth centuries. Whether or not the achievements of medieval science exercised any influence on these developments is irrelevant. What must be emphasized, however, is that the momentous changes in the exact sciences of physics and astronomy that epitomized the Scientific Revolution did not develop from a vacuum. They could not have occurred without certain foundational events that were unique products of the late Middle Ages. To realize this, we must inquire whether a scientific revolution could have occurred in the seventeenth century if the level of science in Western Europe had remained much as it was in the first half of the twelfth century, before the transformation that occurred as a consequence of a great wave of translations from the Greek and Arabic languages into Latin that began around 1150 and continued on to the end of the thirteenth century. Could a scientific revolution have occurred in the seventeenth century if the immense translations of Greco-Arabic (or Greco-Islamic) science and natural philosophy into Latin had never taken place? Obviously not. Without those translations, many centuries would have been required before Western Europe could have reached the level of Greco-Arabic science. Instead of the Scientific Revolution of the seventeenth century, our descendants might look back upon a "Scientific Revolution of the Twenty-first Century." But the translations did occur in the twelfth and thirteenth centuries, and so did a scientific revolution in the seventeenth century. It follows that something happened between, say, 1175 and 1500 that paved the way for that Scientific Revolution. What that "something" was is my subject here.

To describe how the late Middle Ages in Western Europe played a role in producing the Scientific Revolution in the physical sciences during the seventeenth century, two aspects of science need to be distinguished: the contextual and the substantive. The first—the contextual—involves changes that created an atmosphere conducive to the establishment of science, made it feasible to pursue science and natural philosophy on a permanent basis, and made those pursuits laudable activities within Western society. The second aspect—the substantive—pertains to certain features of medieval science and natural philosophy that were instrumental in bringing about the Scientific Revolution.

The creation of an environment in the Middle Ages that eventually made a scientific revolution possible involved at least three crucial preconditions. The first of these was the translation of Greco-Arabic science and natural philosophy into Latin during the twelfth and thirteenth centuries. Without this initial, indispensable precondition, the other two might not have occurred. With the transfer of this large body of learning to the Western world, the old science of the early Middle Ages was overwhelmed and superseded. Although modern science might eventually have developed in the West without the introduction of Greco-Arabic science, its advent would have been delayed by centuries.

The second precondition was the formation of the medieval university, with its corporate structure and control over its varied activities. The universities that emerged by the thirteenth century in Paris, Oxford, and Bologna were different from anything the world had ever seen. From these beginnings, the medieval university took root and has endured as an institution for some eight hundred years, being transformed in time into a worldwide phenomenon. Nothing in Islam, or China, or India, or in the ancient civilizations of South America is comparable to the medieval university. It is in this remarkable institution, and its unusual activities, that the foundations of modern science must be sought.

The university was possible in the Middle Ages because the evolution of medieval Latin society allowed for the separate existence of church and state, each of which, in turn, recognized the independence of corporate entities, the university among them. The first universities, of Paris, Oxford, and Bologna, were in existence by approximately 1200, shortly after most of the translations had been completed. The translations furnished a ready-made curriculum to the emerging universities, a curriculum that was overwhelmingly composed of the exact sciences, logic, and natural philosophy.

The curriculum of science, logic, and natural philosophy established in the medieval universities of Western Europe was a permanent fixture for approximately four hundred and fifty to five hundred years. It was the curriculum of the arts faculty, which was the largest of the traditional four faculties of a typical major university, the others being law, medicine, and theology. Courses in logic, natural philosophy, geometry, and astronomy formed the core curriculum for the baccalaureate and master of arts degrees and were taught on a regular basis for centuries. These two arts degrees were virtual prerequisites for entry into the higher disciplines of law, medicine, and theology.

For the first time in the history of the world, an institution had been created for teaching science, natural philosophy, and logic. An extensive four- to six-year course in higher education was based on those subjects, with natural philosophy as the most important component. As universities multiplied during the thirteenth to fifteenth centuries, the same science-natural philosophy-logic curriculum was disseminated throughout Europe, extending as far east as Poland.

The science curriculum could not have been implemented without the explicit approval of church and state. To a remarkable extent, both granted to the universities corporate powers to regulate themselves: universities had the legal right to determine their own curricula, to establish criteria for the degrees of their students, and to determine the teaching fitness of their faculty members.

Despite some difficulties and tensions between natural philosophy and theology—between, essentially, reason and revelation—arts masters and theologians at the universities welcomed the arrival of Aristotle's natural philosophy as evidenced by the central role they gave it in higher education. Why did they do this? Why did a Christian society at the height of the Catholic Church's power readily adopt a pagan natural philosophy as the basis of a four- to six-year education? Why didn't Christians fear and resist such pagan fare rather than embrace it?

Because Christians had long ago come to terms with pagan thought and were agreed, for the most part, that they had little or nothing to fear from it. The rapprochement between Christianity and pagan literature, especially philosophy, may have been made feasible by the slowness with which Christianity was disseminated. The spread of Christianity beyond the Holy Land and its surrounding region began in earnest after St. Paul proselytized the Gentile world, especially Greece, during the middle of the first century. In retrospect—and by comparison with the spread of Islam—the pace of the dissemination of Christianity appears quite slow. Not until 300 A.D. was Christianity effectively represented throughout the Roman Empire. And not until 313, in the reign of Constantine, was the Edict of Milan (or Edict of Toleration) issued, which conferred on Christianity full legal equality with all other religions in the empire. In 392, Christianity was made the state religion of the Roman Empire. In that year, Emperor Theodosius ordered all pagan temples closed, and also prohibited pagan worship, thereafter classified as treason. Thus it was not until 392 that Christianity became the exclusive religion supported by the state. After almost four centuries of existence, Christianity was triumphant.

By contrast, Islam, following the death of Mohammad in 632, was carried over an enormous geographical area in a remarkably short time.

In less than one hundred years, it was the dominant religion from the Arabian peninsula westward to the Straits of Gibraltar, northward to Spain and eastward to Persia, and beyond. But where Islam was largely spread by conquest during its first hundred years, Christianity spread slowly and, with the exception of certain periods of persecution, relatively peacefully. It was this slow percolation of Christianity that enabled it to come to terms with the pagan world and thus prepare itself for a role that could not have been envisioned by its early members.

The time it took before Christianity became the state religion enabled Christianity to adjust to the pagan society around it. In the second half of the third century, Christian apologists concluded that Christianity could profitably utilize pagan Greek philosophy and learning. In a momentous move, Clement of Alexandria (ca. 150–ca. 215) and his disciple Origen of Alexandria (ca. 185–ca. 254) laid down the basic approach that others would follow. Greek philosophy, they argued, was not inherently good or bad, but one or the other depending on how it was used by Christians. Although the Greek poets and philosophers had not received direct revelation from God, they did receive natural reason and were therefore pointed toward truth. Philosophy—and secular learning in general—could thus be used to interpret Christian wisdom, which was the fruit of revelation. They were agreed that philosophy and science could be used as "handmaidens to theology"—that is, as aids to understanding Holy Scripture—an attitude that had already been advocated by Philo Judaeus, a resident of the Jewish community of Alexandria, early in the first century A.D.

The "handmaiden" concept of Greek learning became the standard Christian attitude toward secular learning by the middle of the fourth century. That Christians chose to accept pagan learning within limits was a momentous decision. They might have heeded the words of Tertullian (ca. 150–ca. 225), who asked pointedly: "What indeed has Athens to do with Jerusalem? What concord is there between the Academy and the Church?" With the total triumph of Christianity at the end of the fourth century, the church might have reacted adversely toward Greek pagan learning in general, and Greek philosophy in particular, since there was much in the latter that was offensive to the church. The Catholic Church might even have launched a major effort to suppress pagan thought as a danger to the church and its doctrines. But it did not.

The handmaiden theory was obviously a compromise between the rejection of traditional pagan learning and its full acceptance. By approaching secular learning with caution, Christians could utilize Greek philosophy—especially metaphysics and logic—to better understand and explicate Holy Scripture and to cope with the difficulties generated by the assumption of the doctrine of the Trinity and other esoteric dogmas. Ordinary daily life also required use of the mundane sciences such as astronomy and mathematics. Christians came to realize that they could not turn away from Greek learning.

When Christians in Western Europe became aware of Greco-Arabic scientific literature and were finally prepared to receive it in the twelfth century, they did so eagerly. They did not view it as a body of subversive knowledge. Despite a degree of resistance that was more intense at some times than at others, Aristotle's works were made the basis of the university curriculum by 1255 in Paris, and long before that at Oxford.

The emergence of a class of theologian-natural philosophers was the third essential precondition for the Scientific Revolution. Their major contribution was to sanction the introduction and use of Aristotelian natural philosophy in the curriculum of the new universities. Without that approval, natural philosophy and science could not have become the curriculum of the medieval universities. The development of a class of theologian-natural philosophers must be regarded as extraordinary. Not only did most theologians approve of an essentially secular arts curriculum, but they were convinced that natural philosophy was essential for the elucidation of theology. Students entering schools of theology were expected to have achieved a high level of competence in natural philosophy. Since a master of arts degree, or the equivalent thereof, signified a thorough background in Aristotelian natural philosophy, and since a master's degree in the arts was usually a prerequisite for admittance to the higher faculty of theology, almost all theologians can be said to have acquired extensive knowledge of natural philosophy. Many undoubtedly regarded it as worthy of study in itself and not merely because of its traditional role as the handmaiden of theology.

If theologians at the universities had chosen to oppose Aristotelian learning as dangerous to the faith, it could not have become the center of study at the university. But medieval theologians interrelated natural philosophy and theology with relative ease and confidence, whether this involved the application of science and natural philosophy to scriptural exegesis, the application of the concept of God's absolute power to hypothetical possibilities in the natural world, or the frequent invocation of scriptural texts to support or oppose scientific ideas and theories. Theologians rarely permitted theology to hinder their inquiries into the physical world. If there was any temptation to produce a "Christian science," they successfully resisted it. Although biblical texts were often cited in natural philosophy, they were not used to demonstrate scientific truths by appeal to divine authority.

The relatively small degree of trauma that accompanied Greco-Arabic science and natural philosophy into Western Europe, and the subsequent high status that science and natural philosophy achieved in Western thought, is attributable in no small measure to theologian-natural philosophers of this kind. Some of the most significant contributors to science and mathematics came from their ranks: Albertus Magnus, Robert Grosseteste, John Pecham, Theodoric of Freiberg, Thomas Bradwardine, Nicole Oresme, and Henry of Langenstein. Theologians used natural philosophy so extensively in their theological treatises that, from time to time, the church had to admonish them to refrain from frivolously using natural philosophy to resolve theological problems. Although there were occasional theological reactions against natural philosophy—as in the early thirteenth century when Aristotle's works were banned for some years at Paris, and in the later thirteenth century when the bishop of Paris issued the Condemnation of 1277—they were relatively minor aberrations when viewed against the grand sweep and scope of the history of Western Christianity.

To appreciate the importance of a class of theologian-natural philosophers for the development of science and natural philosophy in the Latin West, one has only to compare the Western reception of natural philosophy with its treatment in the civilization of Islam, where religious authorities regarded the study of natural philosophy as potentially dangerous to the faith. Despite the fact that for many centuries—say, from the ninth to the end of the fifteenth—the level of science in the civilization of Islam, especially the exact sciences and medicine, far exceeded that of Western Europe, Aristotelian natural philosophy encountered many obstacles. Because of fears that natural philosophy might subvert the faith, and perhaps for other reasons as well, natural philosophy and also the exact sciences were never institutionalized in Islam and thus never made a regular part of the educational process.

By contrast, the universities that were founded in the West European Middle Ages preserved and enhanced natural philosophy. The university as we know it today was invented in the late Middle Ages. Universities were powerful and highly regarded institutions, corporate entities with numerous privileges that increased century by century. They were always there, dispensing natural philosophy and thereby keeping alive a tradition of scientific inquiry. Despite plagues, wars, and revolutions, they carried on, giving natural philosophy and science a sense of permanence. They could do so because the church and its theologians, who were the guardians of dogma and doctrine, had acquiesced in the major role accorded to Aristotelian natural philosophy. For the first time in history, science and natural philosophy had a permanent institutional base. No longer was the preservation of natural philosophy left to the whims of fortune and to isolated teachers and students.

Without the development of these three preconditions, it is difficult to imagine how a scientific revolution could have occurred in the seventeenth century. Although these preconditions, permanent features of medieval society, were vital for the emergence of early modern science, and therefore qualify as foundational elements, they were not in themselves sufficient. The reasons why science took root in Western society must ultimately be sought in the nature of the science and natural philosophy that were developed.

If we leave medicine aside, science in the Middle Ages is appropriately divisible into two parts: the exact sciences (primarily mathematics, astronomy, and optics) and natural philosophy. Although the Latin Middle Ages preserved the major texts of the exact sciences in mathematics, astronomy, and optics, and even added to their sum total, I am unaware of any methodological or technical changes that proved to be significant for the Scientific Revolution. Preserving the texts, as well as studying them, and even writing new treatises on these subjects, was itself a major achievement. Not only did these activities keep the exact sciences alive, but they reveal the existence of a group of individuals who, during the medieval centuries, were competent in dealing with these sciences. At the very least, expertise in these sciences was maintained, so that the Copernicuses, Galileos, and Keplers of the new science had something to study, something to which they might react and alter for the better. Because the late Middle Ages is not highly regarded for its contributions to the exact sciences, let us concentrate on natural philosophy, where there were significant achievements.

The role of natural philosophy during the Middle Ages differed radically from that of the exact sciences. With natural philosophy, we are not concerned with the mere preservation of Greco-Arabic knowledge, but rather with the transformation of an inheritance into something ultimately beneficial for the development of early modern science. Natural philosophers in the arts faculties of the universities converted Aristotle's natural philosophy into a large number of questions that were put to nature on a range of subjects that eventually crystallized into specific sciences, among them physics, geology, meteorology, and others. To each of these questions, a yes or no response was usually required.

Within the format of a yes or no reply, however, Scholastic authors presented numerous arguments and conclusions in defense of their different positions. Revolutionary changes occurred when the responses that were acceptable to natural philosophers in the Middle Ages were found inadequate by scholars in the sixteenth and seventeenth centuries. By the end of the seventeenth century, new conceptions of physics, and of the cosmos as a whole, drastically altered natural philosophy. Aristotle's cosmology and physics were largely abandoned, though his ideas about many other aspects of nature—including material change, zoology, and psychology—were still found useful. In biology, Aristotle's influence continued into the nineteenth century.

During the fourteenth century, Aristotelian natural philosophy was significantly transformed. This transformation played a role in the revolution to come. But it was not because of any particular achievements in science, important though these were. Medieval natural philosophers emphasized ways of knowing and approaching nature—that is, they became interested in what we might characterize as scientific method. They sought to explain how we come to understand nature, even though they rarely pursued the consequences of their own methodological insights.

A few of these methodological changes were relevant to mathematics. The mathematical treatment of the variation of qualities was characteristic of medieval natural philosophy.

The problems were usually imaginary and hypothetical, but the application of mathematics to resolve them was commonplace. In treating such problems, Scholastic authors frequently introduced infinites and infinitesimals. By the sixteenth and seventeenth centuries, mathematical ways of thinking, if not mathematics itself, had been incorporated into natural philosophy. The stage was set for the consistent application of natural philosophy to real physical problems, rather than to imaginary variations of qualities.

Most of the methodological contributions to science were, however, philosophical. Scholastic natural philosophers formulated sound interpretations of concepts such as causality, necessity, and contingency. Some—and John Buridan, an eminent arts master at the University of Paris in the fourteenth century, was one of them—concluded that final causes were superfluous and unnecessary. For them, efficient causes were sufficient to determine the agent of a change. John Buridan was also involved in another major methodological development when he insisted that scientific truth is not absolute, like mathematical truth, but has degrees of certitude. The kind of certainty Buridan had in mind consisted of undemonstrable principles that formed the basis of natural science— as, for example, that all fire is warm and that the heaven moves. For Buridan, these principles are not absolute, but are derivable from inductive generalization; or, as he put it, "they are accepted because they have been observed to be true in many instances, and to be false in none."

Moreover, Buridan regarded these inductively generalized principles as conditional because their truth is predicated on the assumption of the "common course of nature." This was a profound assumption that effectively eliminated the effect on science of unpredictable, divine interventions. In short, it eliminated the need to worry about miracles in the pursuit of natural philosophy. Miracles could no longer affect the validity of natural science. Nor indeed could chance occurrences that might occasionally impede or prevent the natural effects of natural causes. Just because individuals are occasionally born with eleven fingers does not negate the fact that in the common course of nature we can confidently expect ten fingers. On this basis Buridan proclaimed that "for us the comprehension of truth with certitude is possible." Using reason, experience, and inductive generalizations, he sought to "save the phenomena" in accordance with the principle of Occam's razor—that is, by the simplest explanation that fits the evidence. Buridan had only made explicit what was implied by his Scholastic colleagues. The widespread use of the principle of simplicity was a feature typical of medieval natural philosophy. It was also characteristic of science in the seventeenth century, as when Johannes Kepler declared that "it is the most widely accepted axiom in the natural sciences that Nature makes use of the fewest possible means."

Medieval natural philosophers investigated the "common course of nature," not its uncommon, or miraculous, path. They characterized this approach, admirably, by the phrase "speaking naturally" *(loquendo naturaliter)*—that is, speaking by means of natural science, and not by means of faith or theology. That such an expression should have emerged, and come into common usage in medieval natural philosophy, is a tribute to the scholars who took as their primary mission the explanation of the structure and operation of the world in purely rational and secular terms.

The widespread assumption of "natural impossibilities" or counterfactuals—or, as they are sometimes called, "thought-experiments"—was a significant aspect of medieval methodology. An occurrence would have been considered "naturally impossible" if it was thought inconceivable for it to occur within the accepted framework of Aristotelian physics and cosmology. The frequent use of natural impossibilities derived largely from the powerful medieval concept of God's absolute power, in which it was conceded that God could do anything whatever short of a logical contradiction. In the Middle Ages, such thinking resulted in conclusions that challenged certain aspects of Aristotle's physics. Where Aristotle had shown that other worlds were impossible, medieval Scholastics showed not only that the existence of other worlds was possible, but that they would be compatible with our world.

The novel replies that emerged from the physics and cosmology of counterfactuals did not cause the overthrow of the Aristotelian worldview, but they did challenge some of its fundamental principles. They made many aware that things could be quite different from what was dreamt of in Aristotle's philosophy. But they accomplished more than that. Not only did some of the problems and solutions continue to influence Scholastic authors in the sixteenth and seventeenth centuries, but this characteristically medieval approach also influenced significant non-Scholastics, who reveal an awareness of the topics debated by Scholastics.

One of the most fruitful ideas that passed from the Middle Ages to the seventeenth century is the concept of God annihilating matter and leaving behind a vacuum—a concept used effectively by John Locke, Pierre Gassendi, and Thomas Hobbes in their discussions of space.

A famous natural impossibility derived from a proposition condemned in 1277. As a consequence, it was mandatory after 1277 to concede that God could move our spherical world rectilinearly, despite the vacuum that might be left behind. More than an echo of this imaginary manifestation of God's absolute power reverberated through the seventeenth century, when Pierre Gassendi and Samuel Clarke (in his famous dispute with Leibniz) found it useful to appeal to God's movement of the world. In medieval intellectual culture, where observation and experiment played negligible roles, counterfactuals were a powerful tool because they emphasized metaphysics, logic, theology, and the imagination—the very areas in which medieval natural philosophers excelled.

The scientific methodologies described here produced new conceptualizations and assumptions about the world. Ideas about nature's simplicity, its common course, as well as the use of counterfactuals, emphasized new and important ways to think about nature. Galileo and his fellow scientific revolutionaries inherited these attitudes, and most would have subscribed to them.

Another legacy from the Middle Ages to early modern science was an extensive and sophisticated body of terms that formed the basis of later scientific discourse—such terms as *potential, actual, substance, property, accident, cause, analogy, matter, form, essence, genus, species, relation, quantity, quality, place, vacuum, infinite,* and many others. These Aristotelian terms formed a significant component of Scholastic natural philosophy. The language of medieval natural philosophy, however, did not consist solely of translated Aristotelian terms. New concepts, terms, and definitions were added in the fourteenth century, most notably in the domains of change and motion. Definitions of uniform motion, uniformly accelerated motion, and instantaneous motion were added to the lexicon of natural philosophy. By the seventeenth century, these terms, concepts, and definitions were embedded in the language and thought of European natural philosophers.

Medieval natural philosophy played another momentous role in the transition to early modern science. It furnished some—if, it is true, not many—of the basic problems that exercised the minds of non-Scholastic natural philosophers in the sixteenth and seventeenth centuries. Medieval natural philosophers produced hundreds of specific questions about nature, the answers to which included a vast amount of scientific information. Most of the questions had multiple answers, with no genuine way of choosing between them. In the sixteenth and seventeenth centuries, new solutions were proposed by scholars who found Aristotelian answers unacceptable, or, at best, inadequate. The changes they made, however, were mostly in the answers, not in the questions. The Scientific Revolution was not the result of new questions put to nature in place of medieval questions. It was, at least initially, more a matter of finding new answers to old questions, answers that came, more and more, to include experiments, which were exceptional occurrences in the Middle Ages. Although the solutions differed, many fundamental problems were common to both groups. Beginning around 1200, medieval natural philosophers, largely located at European universities, exhibited an unprecedented concern for the nature and structure of the physical world. The contributors to the Scientific Revolution continued the same tradition, because by then these matters had become an integral part of intellectual life in Western society.

The Middle Ages did not just transmit a great deal of significantly modified, traditional, natural philosophy, much of it in the form of questions; it also conveyed a remarkable tradition of relatively free, rational inquiry. The medieval philosophical tradition was fashioned in the faculties of arts of medieval universities. Natural philosophy was their domain, and almost from the outset masters of arts struggled to establish as much academic freedom as possible. They sought to preserve and expand the study of philosophy. Arts masters regarded themselves as the guardians of natural philosophy and fought for the right to apply reason to all problems about the physical world. By virtue of their independent status as a faculty, with numerous rights and privileges, they achieved a surprisingly large degree of freedom during the Middle Ages.

Theology was always a potential obstacle, true, but in practice theologians offered little opposition, largely because they too were heavily imbued with natural philosophy. By the end of the thirteenth century, the arts faculty had attained virtual independence from the theological faculty. By then, philosophy and its major subdivision, natural philosophy, had emerged as an independent discipline based in the arts faculties of European universities. True, arts masters were always subject to restraints with regard to religious dogma, but the subject areas where such issues arose were limited. During the thirteenth century, arts masters had learned how to cope with the problematic aspects of Aristotle's thought. They treated those problems hypothetically, or announced that they were merely repeating Aristotle's opinions, even as they offered elaborations of his arguments. During the Middle Ages, natural philosophy remained what Aristotle had made it: an essentially secular and rational discipline. It remained so only because the arts faculty struggled to preserve it. In doing so, they transformed natural philosophy into an independent discipline that embraced as well as glorified the rational investigation of all problems relevant to the physical world. In the 1330s, William of Ockham expressed the sentiments of most arts masters and many theologians when he declared:

Assertions . . . concerning natural philosophy, which do not pertain to theology, should not be solemnly condemned or forbidden to anyone, since in such matters everyone should be free to say freely whatever he pleases.[1]

Everyone who did natural philosophy in the sixteenth and seventeenth centuries was the beneficiary of these remarkable developments. The spirit of free inquiry nourished by medieval natural philosophers formed part of the intellectual heritage of all who engaged in scientific investigation. Most, of course, were unaware of their legacy and would probably have denied its existence, preferring to heap ridicule and scorn on Aristotelian Scholastics and Scholasticism. That ridicule was not without justification. It was time to alter the course of medieval natural philosophy.

Some Aristotelian natural philosophers tried to accommodate the new heliocentric astronomy that had emerged from the brilliant efforts of Copernicus, Tycho Brahe, and Galileo. By then, accommodation was no longer sufficient. Medieval natural philosophy was destined to vanish by the end of the seventeenth century. The medieval Scholastic legacy, however, remained—namely, the spirit of free inquiry, the emphasis on reason, a variety of approaches to nature, and the core of legitimate prob-

---

1. Translated by Mary Martin McLaughlin, *Intellectual Freedom and Its Limitations in the University of Paris in the Thirteenth and Fourteenth Centuries* (New York: Arno Press, 1977), 96.

lems that would occupy the attention of the new science. Inherited from the Middle Ages, too, was the profound sense that all of these activities were legitimate and important, that discovering the way the world operated was a laudable undertaking. These enormous achievements were accomplished in the late Middle Ages, between 1175 and 1500.

To illustrate how medieval contributions to the new science ought to be viewed, let me draw upon an analogy from the Middle Ages. In the late thirteenth century in Italy, the course of the history of medicine was altered significantly when human dissection was allowed for postmortems and was shortly afterward introduced into medical schools, where it soon became institutionalized as part of the anatomical training of medical students. Except in ancient Egypt, human dissection had been forbidden in the ancient world. By the second century A.D., it was also banned in Egypt. It was never permitted in the Islamic world. Its introduction into the Latin West marked a new beginning, made without serious objection from the church. It was a momentous event. Dissection of cadavers was used primarily in teaching, albeit irregularly until the end of the fifteenth century. Rarely, if at all, was it employed to enhance scientific knowledge of the human body. The revival of human dissection and its incorporation into medical training throughout the Middle Ages laid a foundation for what was to come.

Without it, we cannot imagine the significant anatomical progress that was made by such keen anatomists as Leonardo da Vinci (1452–1519), Bartolommeo Eustachio (1520–1574), Andreas Vesalius (1514–1564), and many others.

What human dissection did for medicine, the translations, the universities, the theologian-natural philosophers, and the medieval version of Aristotelian natural philosophy did collectively for the Scientific Revolution of the seventeenth century. These vital features of medieval science formed a foundation that made possible a continuous, uninterrupted eight hundred years of scientific development, a development that began in Western Europe and spread around the world.

2   Science and the Medieval University

Prior to the monumental research on medieval science by Pierre Duhem in the first two decades of this century,[1] the title of this essay would have evoked laughter and/or scorn. Any juxtaposition of the terms "science" and "medieval" would have been thought a contradiction in terms. Since Duhem's time, however, and largely because of him and a series of brilliant successors, we have grown accustomed to the concept of medieval science, which has even developed into a significant research field. But now that historians of science have grown accustomed to the idea that there was indeed science in the Middle Ages, the time has come to risk laughter and/or scorn once again by proposing the prima faciae outrageous claim that the medieval university laid far greater emphasis on science than does its modern counterpart and direct descendant. It is no exaggeration or distortion to claim that the curriculum of the medieval university was founded on science and largely devoted to teaching about the nature and operation of the physical world.[2] For better or worse, this

1. An emiment physicist, Duhem not only published hundreds of papers in physics, but also wrote fifteen volumes on medieval science embraced within three works: *Les Origines de la Statique,* 2 vols. (Paris, 1905–1906); *Etudes sur Léonard de Vinci ceux qu'il a lus et ceux qui l'ont lu,* 3 vols. (Paris, 1906–1913); and *Le Systéme du monde: Histoire des doctrines cosmologiques de Platon à Copernic,* 10 vols. (1913–1959), the last five volumes of which were published posthumously. For many topics, these works still form an indispensable point of departure. A brief biographical sketch of Duhem (with primary and secondary bibliography) by Donald G. Miller appears in the *Dictionary of Scientific Biography,* edited by Charles C. Gillispie, 16 vols., 4: 225–33 (New York, 1970–1980).

2. Walter J. Ong has perceptively observed that "because of the university curriculum, a distinctive feature of late medieval civilization was an organized and protracted study of physics which was more intense and widespread than ever before. Greek or Roman civilizations had seen nothing on this scale" (*Ramus: Method and the Decay of Dialogue from the Art of Discourse to the Art of Reason* [Cambridge, Mass., 1958], 144). By "physics" Ong means "natural science" (142) or natural philosophy. He notes further (144–45) that the medieval study of Aristotle polarized around a "logic-and-physics" context rather than one of "metaphysics-and-theology."

is surely not true today. This essay will attempt to describe not only the origins of this incredible development, but to present the details that will substantiate the claim that the medieval university provided to all an education that was essentially based on science.

That science became the foundation and core of a medieval university education is directly attributable to the unprecedented translation activity of the twelfth and early thirteenth centuries.[3] From approximately 1125 to around 1230, a large portion of Greco-Arabic science had been translated from Arabic and Greek into Latin. Prior to this activity, only a miniscule portion of Greek science had ever been made available in Latin. From the Roman Empire period to the twelfth century, Western Europe subsisted on a meager scientific fare that had been absorbed into handbooks and encyclopedic treatises associated with the names of Chalcidius, Macrobius, Martianus Capella, Boethius, Isidore of Seville, Cassiodorus, and Venerable Bede. When not merely repetitive, the sum total of science embedded in these treatises was frequently inaccurate, contradictory, and largely superficial. Nothing illustrates the sorry state of affairs better than the virtual absence of Euclid's *Elements*. Without the most basic text of geometry, the physical sciences of astronomy, optics, and mechanics were impossible. Although a cosmological picture of the world was available in Chalcidius's partial translation of Plato's *Timaeus*, the latter treatise in and of itself did not provide a detailed natural philosophy with adequate physical and metaphysical principles. Despite the lack of geometry and technical science and an inadequate natural philosophy, twelfth-century scholars at Chartres, such as Adelard of Bath, Bernard Silvester, Thierry of Chartres, William of Conches, and Clarenbaldus of Arras, had begun to interpret natural phenomena, and even biblical texts, with critical objectivity.[4] Whether, if given sufficient time,

3. For a recent valuable and informative article on the translations, see David C. Lindberg, "The Transmission of Greek and Arabic Learning to the West," in *Science in the Middle Ages*, edited by David C. Lindberg, 52–90 (Chicago, 1978). Although translations of scientific works from Arabic to Latin actually began during the tenth and eleventh centuries, the works that would be fundamental to the university arts curriculum, especially those of Aristotle, were only translated during the twelfth and thirteenth centuries.

4. See M. D. Chenu, "The Discovery of Nature," in *Nature, Man, and Society in the Twelfth Century: Essays on New Theological Perspectives in the Latin West*, selected, edited, and translated by Jerome Taylor and Lester K. Little, 4–18 (Chicago, 1968; original French version 1957), 33; Brian Stock, *Myth and Science in the Twelfth Century: A Study of Bernard Silvester* (Princeton, 1972), 271–73. For the strongest claims on the critical objectivity of twelfth-century scholars,

this bold intellectual venture would have generated new insights and theories about the physical world will never be known. For the influx of Greco-Arabic science into Western Europe had already begun and would soon overwhelm the incipient rational science that had been evolving within the context of the old learning.

The achievements of the international brigade of translators that labored in Spain, Sicily, and northern Italy were truly monumental. Within a period of approximately one hundred years, they made available in Latin the works of Aristotle and the commentaries of Averroes, which together would dominate scientific thought for the next four hundred years; Euclid's *Elements;* Ptolemy's *Almagest,* the greatest astronomical treatise until the *De revolutionibus* of Copernicus; Alhazen's *Optics;* the *Algebra* of al-Khwarizmi; and the medical works of Galen, Hippocrates, and Avicenna.[5] Many lesser scientific works were also rendered into Latin. And if we push into the 1260s and 1270s, we must add the approximately forty-nine translations from Greek into Latin by William of Moerbeke, which included the works of Archimedes and his commentator Eutochius, Proclus, and the Greek Aristotelian commentators, Simplicius, Themistius, Alexander of Aphrodisias, and John Philoponus, as well as works by Hero of Alexandria and Ptolemy.[6] To improve the quality of the texts of Aristotle, Moerbeke translated almost the whole of the Aristotelian corpus from Greek to Latin.

When compared to the paucity of scientific texts prior to the age of translation, the achievements of the translators of the twelfth and thirteenth centuries are truly staggering. It enabled two things to occur that might not otherwise have happened. First, it laid the true foundation for the continuous development of science to the present day, and second, it provided a powerful and comprehensive subject matter that enabled the university to emerge as a fundamental intellectual force in medieval society.

---

see two articles by Tina Stiefel: "Science, Reason and Faith in the Twelfth Century: The Cosmologists' Attack on Tradition," *Journal of European Studies* 6 (1976): 1–16, and "The Heresy of Science: A Twelfth-Century Conceptual Revolution," *Isis* 68 (1977): 347–62.

5. Many of these were translated by a single prolific translator, Gerard of Cremona, whose translations are listed and discussed by Michael McVaugh in *A Source Book in Medieval Science,* edited by Edward Grant, 35–38 (Cambridge, Mass., 1974); see also Richard Lemay, "Gerard of Cremona," *Dictionary of Scientific Biography* (Supplement 1), 15:173–92.

6. For a list of these translations, see Grant, *Source Book in Medieval Science,* 39–41.

The first of these consequences of the translations of the twelfth and thirteenth centuries is not the subject matter of this essay, but will be mentioned again, since its importance cannot be overestimated. It is, however, the second momentous consequence of the translations that shall be the primary concern. With the introduction of Aristotelian science and philosophy and the numerous other works that came along with it, the basis for an extensive curriculum became available and it is hardly surprising that by 1200 two of the three greatest universities of Christendom, Oxford and Paris, were already in existence with curricula based on the new science. To substantiate the claim that the medieval universities taught an essentially science curriculum, it is necessary to distinguish two aspects of medieval science. The first, and most important, was natural philosophy, or natural science, which consisted of the "natural books" *(libri naturales)* of Aristotle and formed one of the major subdivisions under what was usually called the "Three Philosophies," which also embraced moral philosophy and metaphysics.[7] Along with Aristotelian logic,[8] natural philosophy constituted the most significant part of the arts curriculum of every medieval university and will receive emphasis here.

Before turning to it, however, we must describe and discuss the second aspect of medieval science, which was concerned with the exact sciences of arithmetic, geometry, astronomy, and music. Here indeed you will recognize the old *quadrivium* of the venerable seven liberal arts. When compared with the quadrivium as represented in the curriculum of the monastic and cathedral schools prior to the translations, it is read-

---

7. The *libri morales* consisted of Aristotle's *Nichomachean Ethics, Politics,* and *Economics;* the metaphysics consisted primarily of the books of Aristotle's *Metaphysics.* The *libri naturales* will be detailed below. For the lists of books studied under the three philosophies, see James A. Weisheipl, O.P., "Curriculum of the Faculty of Arts at Oxford in the Early Fourteenth Century," *Mediaeval Studies* 26 (1964): 173–76, hereafter cited as "Curriculum at Oxford"; for more on the *libri morales,* see Nancy G. Siraisi, "The *libri morales* in the Faculty of Arts and Medicine at Bologna: Bartolomeo de Varignana and the Pseudo-Aristotelian Economics," *Science, Medicine and the University: 1200–1500: Essays in Honor of Pearl Kibre, Part 1, Manuscripta* 20 (1976): 105–18.

8. Logic was one of the subjects of the *trivium.* For a list of the works studied in logic during the thirteenth and early fourteenth centuries at Oxford, see Weisheipl, "Curriculum at Oxford," 169–70. Although Weisheipl observed that logic at Oxford "occupied about half of the actual curriculum," it will not be considered further here since it was a tool of analysis rather than a science in its own right. Its importance in medieval university education was, however, enormous.

ily apparent that the exact sciences as taught in the medieval universities shared little more than the name "quadrivium" with what was dispensed under that rubric in the early Middle Ages.[9] The emphasis on the exact sciences was not, however, of equal breadth and scope in all medieval universities. Although they formed an integral part of the curriculum at Oxford from the thirteenth century onward, they received much less emphasis at Paris and other places. For example, mathematics was not regularly taught at Paris in the thirteenth century and only sporadically in the fourteenth. At Paris it was more usual for masters to offer mathematical instruction privately during feast days. Mathematics and the other quadrivial sciences were thus rarely part of the regular course of instruction. Such courses were offered by interested masters to students who probably had special interests in the exact sciences and were presumably well motivated.[10]

It was Oxford that served as the model for regular instruction in the exact sciences. From lists compiled by Father James Weisheipl, we can obtain a good sense of the books used in the quadrivial courses.[11] At the

9. The term "quadrivium" was rarely used in university statutes (see Pearl Kibre, "The *Quadrivium* in the Thirteenth Century Universities [with Special Reference to Paris]," in *Arts libéreaux et philosophie au moyen âge: Actes du quatriéme congrés international de philosophie médiévale,* Université de Montréal, Canada, 27 août–2 septembre 1967 [Montreal: Institut d'études mediévalés; Paris, Librairie philosophique J. Vrin, 1969], 175; hereafter cited as *Arts libéraux et philosophie au moyen âge*) and does not seem to occur in curriculum lists. The explanation may lie in the fact that the four traditional quadrivial sciences were not conceived as part of a liberal arts education. Indeed, the seven liberal arts, though transmitted to the Middle Ages by the Latin Encyclopedists (Martianus Capella, Isidore of Seville, Boethius) were not taught as such in the medieval universities. Thus, although all the subjects of the seven liberal arts were usually represented in the university curriculum, they were absorbed into a larger whole in which natural philosophy, metaphysics, and moral philosophy were the major components (See Philippe Delhaye, "La place des arts libéraux dans les programmes scolaires du xiiie siècle," in *Arts libéraux et philosophie au moyen âge,* 169, 172). Moreover, the disciplines of the traditional quadrivium had undergone a transformation. Arithmetic, geometry, and astronomy, which were theoretical subjects in the liberal arts tradition, were enlarged in scope during the late Middle Ages to embrace practical and applied knowledge.

10. For the contrast between Oxford and Paris in the study of mathematics and the exact sciences in general, see Guy Beaujouan, "Motives and Opportunities for Science in the Medieval Universities," in *Scientific Change: Historical Studies in the Intellectual, Social and Technical Conditions for Scientific Discovery and Technical Invention, from Antiquity to the Present,* Symposium on the History of Science, University of Oxford, 9–15 July 1961, edited by A. C. Crombie, 221–22 (New York, 1963). Arithmetic was also taught outside the university at Oxford and in Italy.

11. Weisheipl, "Curriculum at Oxford," 170–73. For additional curriculum information on the arts and sciences as taught at Bologna, Paris, and Oxford, see Hastings Rashdall, *The Uni-*

heart of the exact science curriculum was geometry and Euclid's *Elements*. Of the thirteen genuine and two spurious books of the medieval Latin version of the *Elements*, only the first six were formally required.[12] Practical, or applied, geometry was also stressed.[13] In this category, use was made of the *Treatise on the Quadrant (Tractatus quadrantis)* of Robertus Anglicus, which described the use of an astronomical instrument known as the quadrant; the *Treatise on Weights (Tractatus de ponderibus)* associated with the name of Jordanus de Nemore and concerned with the subject of statics;[14] and treatises on perspective or optics drawn from works by Ptolemy, Alhazen, John Pecham, Roger Bacon, and others.[15]

Although medieval technical astronomy was based on the famous *Almagest* of Ptolemy, which appears on curriculum lists, it is implausible to suppose that anything more than the descriptive sections of the first book could have served as a text. Since the objective of astronomical instruction was "to enable students to understand the position of the planets and to calculate the variable feast days of the ecclesiastical year,"[16]

versities of Europe in the Middle Ages, a new edition by F. M. Powicke and A. B. Emden, 3 vols., 1.233–53 (Bologna); 1.433–96 (Paris); 3.140–68 (Oxford), (Oxford, 1936).

12. These books were probably in one of the versions attributed to Adelard of Bath in the twelfth century. For the history of the translations of Euclid's *Elements* in the Latin Middle Ages, see John E. Murdoch, "Euclid: The Transmission of the Elements," *Dictionary of Scientific Biography* (New York, 1971): 4:443–48.

13. "Treatises titled *Practica geometriae* (Applied or Practical Geometry) were written by Hugh of St. Victor, Leonardo Fibonacci, and Dominicus de Clavasio, while others, under different titles or anonymously, wrote similar treatises with substantially the same content." In these works, geometry was applied to height measurement *(altimetria)*, surface measurement *(planimetria)*, and the measurement of solids *(cosimetria* or *stereometria)*. "In each of these parts geometry was applied to determine various measurements in astronomy and optics, as well as to measure heights of mountains, depths of valleys, and in general, lengths, areas, and volumes" (Grant, *Source Book in Medieval Science,* 180). Although such works do not appear on the Oxford lists supplied by Weisheipl, they represent the most general treatises on applied geometry and would have been more appropriate than any of the works in this genre cited below. For typical problems translated from the *Practica geometriae* of Dominicus de Clavasio, see Grant, *Source Book in Medieval Science,* 181–87.

14. According to Weisheipl, the text generally used was the *Elementa Jordani de ponderibus,* which, however, does not fit the titles identified by E. A. Moody and Marshall Clagett in their edition of *The Medieval Science of Weights (Scientia de ponderibus), Treatises Ascribed to Euclid, Archimedes, Thabit ibn Qurra, Jordanus de Nemore, and Blasius of Parma,* with English Introductions, English Translations, and Notes (Madison, Wis., 1959).

15. For the manuscripts, printed editions, and translations of the optical works of these authors, see David C. Lindberg, *A Catalogue of Medieval and Renaissance Optical Manuscripts, Subsidia Mediaevalia* 4 (Toronto, 1975).

16. Weisheipl, "Curriculum at Oxford," 172.

two elementary thirteenth-century texts came to serve the first of these goals, namely, the understanding of the planetary positions. The most famous of these is surely the *Sphere (De sphaera)* of John of Sacrobosco, which provided a general cosmological and astronomical sketch of the different components of the finite, spherical universe accepted by all during the Middle Ages.[17] From Sacrobosco's introduction we learn that he has divided the treatise into four chapters, "telling first, what a sphere is, what its center is, what the axis of a sphere is, what the pole of the world is, how many spheres there are, and what the shape of the world is. In the second we give information concerning the circles of which this material sphere is composed and that supercelestial one, of which this is the image.... In the third we talk about the rising and setting of the signs, and the diversity of days and nights which happens to those inhabiting diverse localities, and the division into climes. In the fourth the matter concerns the circles and motions of planets, and the causes of eclipses."[18] The treatment of the planets in the fourth book was, however, so meager that an unknown teacher of astronomy composed another treatise, *The Theory of the Planets (Theorica planetarum)*,[19] that consisted of numerous definitions describing all aspects of planetary motion. Along with Sacrobosco's *Sphere*, the anonymous *Theory of the Planets* served to introduce generations of students to the basic elements of planetary astronomy and to provide them with a skeletal frame of the cosmos.

To achieve the second objective and enable students to compute the variable feast days in the ecclesiastical calendar, *compotus* treatises, representing practical astronomy, were employed, most notably those written in the thirteenth century by Robert Grosseteste and John of Sacrobosco.[20]

17. For the Latin text and English translations, see Lynn Thorndike, ed. and trans., *The Sphere of Sacrobosco and Its Commentators* (Chicago, 1949), 76–142.

18. Ibid., 118.

19. Olaf Pedersen estimates at least two hundred extant manuscripts of the *Theorica planetarum* (see his "The Theorica Planetarum—Literature of the Middle Ages," *Classica et Mediaevalia: Revue Danoise de Philologie et d'Histoire* 23 [1962]: 225–26). For Pedersen's introduction to, and translation of, the *Theorica*, see Grant, *Source Book in Medieval Science*, 451–65.

20. Weisheipl, "Curriculum at Oxford," 172–73. University professors wrote numerous treatises on the quadrivial sciences, some, or even many, of which may have been used as texts at some time or other. Mere absence from a curriculum list is not an accurate guide as to whether or not a particular work may have served as an actual text. This is true not only because extant curriculum lists are rare in themselves, but even if they were abundant it is probable that many

Only in arithmetic and music was there a continuation with the quadrivial tradition of the early Middle Ages. In these subjects, Boethius's *Arithmetica* and *Musica*[21] served as the basic links. But even here treatises translated in the twelfth century or newly composed in the thirteenth and fourteenth augmented the Boethian texts. Arithmetic, which in its Boethian tradition was of a largely theoretical nature, was supplemented by books seven to nine of Euclid's *Elements,* which treated of number theory.[22] To this was added a strong practical component in the form of treatises that described and exemplified the four arithmetic operations for whole numbers, as, for example, Sacrobosco's enormously popular *Algorismus vulgaris,*[23] and fractions, the latter usually under titles such as *Al-*

---

texts would not have appeared on official curriculum lists because they were assigned and required by the professor himself without official university sanction (Weisheipl, "Curriculum at Oxford," 168). It does, however, seem plausible to assume that a given treatise served as a text if a large number of manuscripts of it have been preserved. For mention of numerous quadrivial works composed by faculty at the University of Paris, see Kibre, "The *Quadrivium* in the Thirteenth Century Universities," 175–91.

21. Both treatises have been edited by G. Friedlein, *Boetii De institutione arithmetica libri duo; De institutione musica libri quinque* (Leipzig, 1867). An English translation of parts of the *Arithmetica* appears in Grant, *Source Book in Medieval Science,* 17–24. Boethius's treatise is actually a paraphrase and near translation of Nichomachus of Gerasa's Greek treatise on arithmetic composed around 100 A.D. (For a translation of the latter, see *Nichomachus of Gerasa: Introduction to Arithmetic,* translated by Martin Luther D'Ooge, with studies in Greek arithmetic by Frank E. Robbins and Louis C. Karpinski [New York, 1926].)

22. Also more advanced than Boethius was the rather widely used *Arithmetica* by Jordanus de Nemore in the thirteenth century (for translation of a few of its propositions, see Grant, *Source Book in Medieval Science,* 102–6). In the fourteenth century, Thomas Bradwardine composed an *Arithmetica speculativa,* which has been described as "little more than the extraction of the barest essentials of Boethian arithmetic" intended, it seems, "for arts students who may have wished to learn something of the quadrivium, but with a minimal exposure to mathematical niceties" (cited in John E. Murdoch, "Bradwardine, Thomas," *Dictionary of Scientific Biography* [New York, 1970], 2:395; according to Murdoch, "the *Arithmetica speculativa* was first printed in Paris, 1495 and reprinted many times during the fifteenth and sixteenth centuries" [396]).

23. Sacrobosco's treatise, also known by the title *De arte numerandi,* was based on al-Khwarizmi's ninth-century Arabic treatise, which had been translated into Latin before the middle of the twelfth century with the title *De numero indorum* (for the Latin text of this translation, see Kurt Vogel, *Mohammed ibn Musa Alchwarizmi's Algorismus das früheste Lehrbuch zum Rechnen mit indischen Ziffern* [hereafter cited as *Mohammed*] Nach der einzigen [lateinischen] Handschrift [Cambridge Un. Lib. Ms. Ii. 6.5] in Faksimile mit Transkription und Kommentar herausgegeben [Aalen, 1963]). Although other practical arithmetic works describing the basic operations with Arabic numerals and containing the term "algorismus" (an obvious corruption of al-Khwarizmi's name) in their titles were written during the Middle Ages (Vogel, *Mohammed,* 42; the popular *Carmen de algorismo,* written around 1200 by Alexandre de Villedieu, was

gorismus minutiarum or *Algorismus de minutiis.*[24] In music, the tradition-
al treatises of Boethius and St. Augustine *(De musica)* were supplemented
by the early fourteenth-century treatises of Johannis de Muris (John of
Murs). Of some four or five musical treatises, most significant were his
Musica speculativa secundum Boetium, a commentary on the *Musica* of
Boethius, and his *Ars nove musice (The Art of the New Music).*[25]

---

in the form of a poem in 284 Latin hexameters; for an analysis of it and Sacrobosco's treatise,
see Guy Beaujouan, "L'enseignement de l'arithmetique elementaire a l'universite de Paris aux
xiiie et xive siècles" [hereafter cited as "L'enseignement de l'arithmetique elementaire]," in *Homenaje a Millas-Vallicrosa,* 2 vols. [Barcelona, 1954, 1956], 1:93–124), Sacrobosco's was easily the
most popular and retained its primacy until the sixteenth century. Most of Sacrobosco's treatise
has been translated in Grant, *Source Book in Medieval Science,* 94–101. The practical arithmet-
ics and algorisms referred to here were probably studied at medieval universities. But the use of
Arabic numerals also formed part of the curriculum of medieval business schools. In England,
Oxford was the center for business courses that formed no part of the curriculum for university degrees at Oxford University (see Nicholas Orme, *English Schools in the Middle Ages* [Lon-
don, 1973], 75–77). It was in Florence, however, where business schools flourished and played a
significant role in education. From the fourteenth century, and perhaps earlier, private abacus
schools—which, despite the title, made no use of the physical abacus or counters of any kind—
taught young children the use of Arabic numerals, the arithmetic operations, and how to solve
a large variety of problems, including those that we would call algebraic. Most prominent Flo-
rentine Renaissance figures—including Niccolo Machiavelli and Leonardo da Vinci—attended
abacus schools as youngsters. According to Giovanni Villani, writing sometime around 1338,
some one thousand to two thousand children were learning "the abacus and algorism" in six
schools within Florence. My source for the abacus schools of Florence is Warren Van Egmond,
*The Commercial Revolution and the Beginnings of Western Mathematics in Renaissance Florence,
1300–1500* (Ph.D. diss., Indiana University, 1976), 7, 68, 73. The vicissitudes of Arabic numerals in
Europe and the role of arithmetic in medieval European society are brilliantly described by Alexander Murray, *Reason and Society in the Middle Ages, Part 2: Arithmetic* (Oxford: Clarendon
Press, 1978), 141–210.

24. Apparently, treatises on sexagesimal and vulgar fractions were not introduced into the
university curriculum until rather late (Beaujouan, "L'enseignement de l'arithmetique elementaire," 123), probably in the fourteenth century. John of Ligneres (Johannes de Lineriis) (fl. in
France in the first half of the fourteenth century) composed a popular *Algorismus minutiarum,* which treated both sexagesimal (or physical) and vulgar fractions. See Emmanuel Poulle,
"John of Ligneres," *Dictionary of Scientific Biography* (New York, 1973), 7:122–28; for literature
on the *Algorismus minutiarum,* see 127–28.

25. Emmanuel Poulle observes that John of Murs viewed musical problems mathematically and that "his work reveals the pedagogic qualities that assured his musical writings a
wide diffusion until the end of the Middle Ages" ("John of Murs," *Dictionary of Scientific Biography* [New York, 1973], 7:128). The extent to which music was studied in the medieval university is largely unknown. It is not even mentioned in the curriculum lists at Oxford until 1431
(Weisheipl, "Curriculum at Oxford," 171). For a description of a treatise in which traditional themes, techniques, and terms from natural philosophy (motion, intension, and remission
of forms) were applied to the problems of determining the proper subject of "worldly music"
(musica mundana), see John E. Murdoch, "Music and Natural Philosophy: Hitherto Unnoticed
Questiones by Blasius of Parma," *Manuscripta* 20, no. 2 (1976): 119–36.

The significance attached to the exact sciences in the university curriculum does not emerge from curriculum lists, which are at best sporadic and spare of detail. We can best infer their importance from the attitudes of different Scholastic authors who were also university teachers. Geometry was no longer valued merely for its practical use in measurement or even as a vital aid for philosophical understanding. Roger Bacon and Alexander Hales extolled its virtues as a tool for the comprehension of theological truth.[26] Geometry was essential for a proper understanding of the literal sense of numerous passages, descriptions, and allusions in Scripture, as, for example, Noah's ark and the temple of Solomon. Only by interpreting the literal sense with the aid of geometry could the higher spiritual sense be grasped. But it was not spiritual truth alone that was at issue in the study of geometry. Robert Grosseteste, in his treatise *On Lines, Angles, and Figures,* conceived of geometry as essential to natural philosophy.[27] Since the universe was constituted of lines, angles, and figures, it could not be properly understood without geometry. Indeed, geometry was required for comprehending the behavior of light, which was multiplied and disseminated in nature geometrically, as were most physical effects.[28]

Arithmetic was equally valued and was often placed first among the mathematical sciences, although in an imaginary debate between geometry and arithmetic, Nicole Oresme implies that the former ranks higher than the latter.[29] In that interesting and unusual dialogue, arithmetic presents itself as the firstborn of all the mathematical sciences and the source of all rational ratios and therefore the cause of the commensurability of the celestial motions and the harmony of the spheres. Moreover,

26. For references, see Kibre, "The *Quadrivium* in the Thirteenth Century Universities," 184; and David C. Lindberg, *John Pecham and the Science of Optics: "Perspectiva Communis,"* edited with an introduction, English translation, and critical notes (Madison, Wis., 1970), 19.

27. See David C. Lindberg's translation in Grant, *Source Book in Medieval Science,* 385.

28. Roger Bacon explains how geometry is essential to the various sciences, including optics. Only by means of geometry can the multiplication and propagation of the species be explained in optics, astronomy, and other relevant sciences. See R. B. Burke, trans., *The "Opus Majus" of Roger Bacon,* 2 vols. (Philadelphia, 1928), 1:131–36. Indeed Bacon's lengthy section on mathematics in the *Opus Magnus* is intended to show its indispensability for science and theology.

29. *Nicole Oresme and the Kinematics of Circular Motion: "Tractatus de commensurabilitate vel incommensurabilitate motuum celi",* pt. 3, edited with an introduction, English translation, and commentary by Edward Grant (Madison, Wis., 1971), 284–323 for text and translation, 67–77 for analysis (especially 72–73).

prediction of the future depends upon exact astronomical tables, which must be founded on the precise numbers of arithmetic. In a fascinating rebuttal, geometry claims greater dominion than arithmetic since it embraces both rational and irrational ratios. As for the beautiful harmony allegedly brought into the world by the rationality of arithmetic, geometry counters by noting that the rich diversity of the world could only be generated by a combination of rational and irrational ratios, which it alone can produce. Geometry and arithmetic were both valued because they were essential to penetrate the workings of nature and to describe the great variety of motions and actions in the physical world. The medieval emphasis on geometry and arithmetic may come as a surprise to those who are wrongly convinced that medieval Aristotelian natural philosophers and theologians were hostile to mathematics.[30]

The science of astronomy, which included astrology,[31] was also regularly lauded as an essential instrument for the comprehension of the macrocosm. It could predict, though not determine, future events. Bacon judged it essential for church and state, as well as for farmers, alchemists, and physicians;[32] Grosseteste considered it invaluable for many other sciences, including alchemy and botany.[33] The significance of astrology and

30. Mathematics was widely applied to philosophy and theology during the Middle Ages, especially in the fourteenth century. Problems in motion and the intension and remission of forms were frequently mathematized. For an excellent description and assessment of the significant and extensive role of mathematics in philosophy and theology, see John E. Murdoch, "*Mathesis in philosophiam scholasticam introducta:* The Rise and Development of the Application of Mathematics in Fourteenth Century Philosophy and Theology," *Arts libéraux et philosophie au moyen âge,* 215–54. It was not that Galileo and his successors reintroduced mathematics into physics, but rather that they restricted its scope to what was more properly and appropriately mathematizable (see Edward Grant, *Physical Science in the Middle Ages* [New York, 1971; reprint, Cambridge University Press, 1977], 58–59).

31. The terms *astronomia* and *astrologia* were used indifferently in the Middle Ages when referring to the "science of the stars" *(scientia stellarum* or *astrorum).* The latter descriptive phrase actually embraced both astronomy and astrology, which were usually taught together; "Astronomy proper, in our sense, came to be called *scientia motus,* or *motuum,* while astrology in our sense was called *scientia iudiciorum*" (Richard Lemay, "The Teaching of Astronomy in Medieval Universities, Principally at Paris in the Fourteenth Century" [hereafter cited as "Teaching of Astronomy"], *Manuscripta* 20, no. 3 [1976]: 198.

32. See A. G. Little, ed., *Part of the "Opus tertium" of Roger Bacon* (Aberdeen, 1912), 12–14; Kibre, "*Quadrivium* in the Thirteenth Century Universities," 190. In Burke, *Opus Majus,* 1:261–70, Bacon defends "true mathematicians," by whom he means astronomers or astrologers (Burke trans., 1:261–70).

33. Grosseteste, *De artibus liberalibus,* in *Die philosophischen Werke des Robert Grosseteste,*

astronomy for medicine, which Bacon and many others routinely emphasized, was manifested at the University of Paris in the 1360s by the foundation of the College of Maître Gervais,[34] which was endowed with books and instruments by King Charles V and subsequently approved by Pope Urban V. So strong was the interest in astrology that in 1366 candidates for the license in arts were required to read "some books in mathematics," which probably included books on astrology since the latter subject was also implied by the term "mathematics."[35]

Music was also accorded high status. It was significant in medicine since physicians could employ it as part of the overall regimen of health. As a factor in stirring the passions in war and soothing them in peace, the study of the mathematical structure of music was deemed helpful and worthwhile. It was even important for the theologian, as Roger Bacon emphasized. Since musical expressions and instruments are mentioned frequently in Scripture, the wise theologian would do well to learn as much about music as possible.[36]

One as yet unmentioned but significant component of the science curriculum of the medieval university is medicine. As one of the three separate higher faculties, medicine was taught only to those who chose to matriculate for a medical degree. It was not an arts subject as were all of the sciences considered thus far. Prior to its institutionalization in the major medieval universities, especially Bologna and Paris, medicine had been taught during the thirteenth century at specialized centers such as Salerno and Montpellier.[37] With its installation as a higher faculty in the

Bischofs von Lincoln, Beiträge zur Geschichte der Philosophie des Mittelalters, edited by L. Baur, 9:4–7 (Munster, 1912); Kibre, *Arts libéraux et philosophie au moyen âge.*

34. In the official statutes, the college was listed as "Collège Notre Dame de Bayeux" (Lemay, "Teaching of Astronomy," 201, n. 8).

35. For all this, see Lemay, "Teaching of Astronomy," 200–202, 210. Throughout the Middle Ages, a good physician was thought to be one who could determine the present and future positions of the stars and could use that knowledge for the benefit of his patients. That celestial bodies could affect terrestrial matter, including organic entities, was taken as self-evident. Since it was further assumed that the position and relationships of every star and planet affected the nature and intensity of its influence, it is obvious why physicians were thought to require knowledge of astronomy and astrology. For a brief discussion of medical astrology at institutions other than Paris, see Lemay, "Teaching of Astronomy," 206–9.

36. For various references, see Kibre, "*Quadrivium* in the Thirteenth Century Universities," 186–87.

37. For a brief description of the origins and status of these four medical schools, see Vern L. Bullough, *The Development of Medicine as a Profession* (hereafter cited as *Development of*

medieval university, medicine became a profession and was therefore the first science to achieve professional status.[38] Prior to the emergence of universities, medicine had been accorded a modest, and even lowly, place in the hierarchy of the arts and sciences.[39] Its orientation was toward the practical with theories that were rather specific to medicine. With its acceptance into the university, it was soon amalgamated with the newly arrived Aristotelian natural philosophy and developed into a highly theoretical and speculative discipline.[40] Except for Italy, an undesirable consequence of the emphasis on theory was the exclusion of surgeons and surgery from medieval medical schools.[41]

That Italian medical schools generally avoided the divorce of surgery and medicine may perhaps provide a small clue toward the explanation of the reemergence of the practice of human dissection at the University of Bologna after a lapse of approximately one thousand years.[42] Although the first recorded anatomical dissection at Bologna was that of Bartolommeo da Varignana in 1302, the practice probably began in the latter part of the thirteenth century. Human dissection in the medical schools undoubtedly intensified interest in the study of human anatomy. Because of its extraordinary role in medical education, human dissection was occasionally worthy of mention by those who witnessed one or more of them in the lecture hall. The famous surgeon Guy de Chauliac (1298–1368) has described how his master, Bertuccio, proceeded through a dissection in four stages, or cuts, anatomizing first the "nutritive" members, then the "spiritual" members, then the "animal" members, and finally the "extremities."[43] Lacking refrigeration, anatomical dissections were

---

Medicine) (New York, 1966), 46–73; For an interesting and informative summary account of medieval medicine, see Charles H. Talbot, "Medicine," in Science in the Middle Ages, edited by David C. Lindberg, 391–428, esp. 400–405, 408–13;

38. Its development into a profession is the fundamental theme of Bullough's book.

39. Talbot, "Medicine," 400, who cites Hugh of St. Victor's Didascalicon.

40. Talbot, "Medicine," 402.

41. Bullough, Development of Medicine, 81–82.

42. Why the long-standing prejudice against the practice of human dissection should have been overcome first at Bologna is difficult to explain (Bullough, Development of Medicine, 62).

43. La grande chirurgie de Guy de Chauliac, edited by E. Nicaise, 30–31 (Paris, 1890). The passage is translated in Bullough, Development of Medicine, 64. Three of the four members mentioned by Guy de Chauliac are also cited in The Anatomy of Master Nicholas (Anatomia Magistri Nicolai Physici) written around 1200 by a Master Nicholas of the Salernitan school. In this treatise, we learn that the animal members are situated above the epiglottis and include the brain, pia mater, dura mater, and the like; the spiritual members lie between the epiglottis

performed only in winter and, for obvious reasons, were done as quickly as possible. When bodies with internal organs and soft parts were unavailable, anatomies were performed on skeletal remains. Without dissections—and bodies were not easy to come by—Henri de Mondeville (d. ca. 1326) resorted to colored anatomical illustrations, a practice that was probably not widespread.[44]

The anatomies performed in the medical schools of medieval universities were not, however, intended for research but were solely for instructional purposes. Despite the use of so vivid a visual aid, the parts of the body and their relationships were seen through the texts of the great medical authorities such as Galen and Avicenna. Traditional errors were usually perpetuated and new knowledge was minimal. In time, anatomy professors even ceased to teach directly from the cadavers they dissected and instead confined themselves to formal lectures while an assistant actually illustrated the body.[45]

Unfortunately for our knowledge of the quadrivial sciences, no dramatic counterpart to human dissection emerged to prompt an occasional remark on classroom procedure and teaching technique. Although the exact sciences of the *quadrivium* were judged useful for the study of physical nature and Scripture, the texts representing the different sciences appear on required curriculum lists from time to time and we can even occasionally learn the length of time devoted to a particular text, the sources have thus far been silent on the manner in which these subjects were actually taught in the classroom. Did the students memorize some or most texts, which may have been prohibitively expensive?[46] Did they solve problems?

---

and diaphragm and include the heart and lungs; and the nutritive members are between the diaphragm and the kidneys and include liver, spleen, and stomach. By "extremities," perhaps Guy intended the generative members, which, according to Master Nicholas, include the testes and seminal vessels below the kidneys. For Master Nicholas, see *Anatomical Texts of the Earlier Middle Ages*, translated by George W. Corner, 67–70 (Washington, D.C., 1927); the translation is reproduced in Grant, *Source Book in Medieval Science*, 728.

44. Guy de Chauliac mentions this pejoratively in the passage cited in n. 43.

45. See Grant, *Source Book in Medieval Science*, 730, n. 1, by Michael McVaugh. This cooperative procedure of the medical schools is illustrated in numerous woodcuts in early printed texts.

46. The *repetitiones* carried on at most medieval universities seem to have had memorization of lectures as their main purpose. Following the lecture of a master, the students were expected to convene that same afternoon and repeat it as substantially close to the original as possible. According to Weisheipl, Dominican students in the fourteenth century were expected to repeat science and logic lectures on a daily basis and to give a general *repetitio* once a week before the master himself ("Curriculum at Oxford," 152). From this we sense that each student

Were visual aids used in teaching astronomy and geometry? Was the abacus used in practical arithmetic? Were Arabic numerals employed for computations?[47] On these and other vital matters we are largely ignorant.

---

was expected to repeat the lecture each day. At Bologna, the master assigned a *repetitor,* "who," according to Rashdall, "attended the lecture and then repeated it to the students afterwards and catechized them upon it" (*Universities of Europe in the Middle Ages,* 1:249). Whether or not the original lecture was first repeated by an officially assigned *repetitor,* it would appear that the students themselves were expected to repeat the lectures in the hope that they would memorize the whole of it in a form as close to the original as possible.

47. Analysis of university texts used in teaching quadrivial subjects may provide significant information and insight about the possible substantive content of medieval lectures, but such analysis is essentially mute about actual classroom procedures. Analysis may suggest, as it did to Guy Beaujouan ("L'enseignement de l'arithmétique élementaire," 105), that the hexameral verses of the *Carmen de algorismo* were memorized and its obscurities then clarified by appeal to Sacrobosco's *Algorismus vulgaris* (or *prosaicus,* as Beaujouan cites it). If medieval students studied arithmetic merely by memorizing the *Carmen de algorismo,* we would have a reasonable idea of classroom practice, which would consist of the memorization and subsequent verbatim repetition of the text itself (we would, however, still remain ignorant of the precise manner in which the text was repeated and what was actually understood by such rote procedures). But what does it mean to say that the *Algorismus* of Sacrobosco was used in the classroom to clarify the *Carmen?* How were these texts interrelated in actual classroom teaching? Inferences from texts and their possible interrelationships offer little basis for reliable descriptions about actual classroom methods employed to convey the contents of those texts to students. Even the knowledge that problem texts were compiled for the study of arithmetic does not enable us to penetrate the veil that obscures actual classroom practice (Beaujouan, "L'enseignement de l'arithmétique élementaire," 115–23). The mere existence of problem texts does not inform us as to their actual use in the classroom, nor how they may have been used if they were an integral part of classroom instruction. Even Siegmund Günther's four-hundred-page study, which bears the intriguing title "Geschichte des mathematischen Unterrichts im deutschen Mittelalter bis zum Jahre 1525" (*Monumenta Germaniae Paedogogica,* vol. 3 [Berlin: A. Hofmann & Co., 1887]), has virtually nothing of value to say about classroom instruction in mathematics, not even in a brief section (192–97) devoted specifically to "Methods of Academic Instruction" ("Methode des akademischen Unterrichts") where we are told (196), presumably on a priori grounds, that among the seven liberal arts the usual disputational method of teaching would be most risky in mathematics. Since Günther's fine book is actually an analysis of the numerous mathematical texts written and available in the Middle Ages, and therefore is more a history of medieval mathematics than a history of mathematical instruction, the title of his work is obviously misleading. In sum, knowledge of titles and content of science texts used at the universities still leaves unanswered numerous questions about the manner in which the content of those texts was actually conveyed to students. But there is yet much of value that can be said about what was learned in the medieval classroom on the basis of a knowledge of the specific texts involved. For example, although a bias against Arabic numerals is occasionally detected at universities in nonteaching matters (see Murray, *Reason and Society in the Middle Ages,* 171–72), the probable use of Sacrobosco's *Algorismus* and similar treatises as university texts strongly suggests that Arabic numerals were taught and regularly used at the medieval university. Since the Arabic number system was based on place value, it not only supplemented but often supplanted the use of the abacus, which also relied on place value (Murray, *Reason and Society in the Middle Ages,* 163–67; Beaujouan, "L'enseignement de l'arithmétique élemen-

Teaching aids were not unknown, although the specific information available seems confined to the early Middle Ages prior to the universities. Gerbert of Aurillac (946–1003), who became Pope Sylvester II, was reputed to have used visual aids in his teachings. His pupil, Richer, describes globes and spheres designed and constructed by Gerbert solely for instructional purposes. One of these simulated the motions of the constellations, where the latter were shaped and represented by means of wires fixed to the sphere, the axis of which was made from a metal tube through the center of the globe.[48] Thus did Gerbert fix the shapes of the different stellar configurations on the minds of his pupils and also show them how all rotate relative to one another. During the eleventh and twelfth centuries, the game of *rithmomachia,* mentioned by John of Salisbury and Alan of Lille, may have been used as a teaching aid and has been described as "the great medieval number game."[49] It was played upon a table or board divided into a series of squares, and by means of its rules a student could become familiar with arithmetic, geometric, and harmonic proportions, as well as with numerical progressions and the different numerical ratios used in the Middle Ages, such as multiple, superparticular, and superpartient. The educational value of the game lay in its stress on the rules of proportion defined and discussed in the *Arithmetica* of Boethius which, as we saw, was used as an arithmetic text throughout the Middle Ages. Played at first with Roman numerals and later with Arabic, it would undoubtedly have proved useful in the study of music, geometry, and astronomy, since facility with numbers was important in all of the quadrivial subjects. Although rithmomachia texts do not appear in the curriculum lists, they may have been used nonetheless.

Aside from the possible use of teaching aids in the quadrivium, the manner of teaching the exact sciences in the medieval university is virtually unknown. Perhaps it was much the same as the teaching of natural

---

taire," 95). All this does not, however, rule out the possibility that the abacus may have been used to check the accuracy of computations that used Arabic numerals.

48. For the details and references, see Oscar G. Darlington, "Gerbert, the Teacher," *American Historical Review* 52 (1946–47): 467–70; for the title of Richer's work, see 456, n. 2.

49. My remarks on *rithmomachia* are drawn entirely from Gillian R. Evans, "The Rithmomachia: A Mediaeval Mathematical Teaching Aid?," *Janus: Revue international de l'histoire des sciences* 63 (1976): 257–73. For John of Salisbury and Alan of Lille, see 257; for the quotation, see 262.

philosophy from the natural books, or *libri naturales,* of Aristotle about which we know much more and to which we must now turn.

The natural books of Aristotle, which formed the core of the curriculum in natural philosophy at all medieval universities, consisted of the *Physics,* the *De caelo (On the Heavens), On Generation and Corruption, On the Soul, Meteorology, Parva Naturalia (The Small Works on Natural Things),* as well as the biological works such as *The History of Animals, The Parts of Animals,* and the *Generation of Animals.* Here then were the treatises that formed the comprehensive foundation for the medieval conception of the physical world and its operations. Although some students at medieval universities were content to acquire only a bachelor's or master of arts degree and others subsequently entered the higher faculties of law, medicine, and theology, all studied the natural books of Aristotle. More than anything else, it is that shared experience that enables us to characterize medieval education as essentially scientific. That Aristotle's scientific books should have formed the basis of university education for some four centuries comes as a surprise when one contemplates the intense and bitter resistance those books met when initially introduced into the University of Paris in the thirteenth century.[50] For the first time in the history of Latin Christendom, a conceptually rich and methodologically powerful body of secular learning posed a threat to theology and its traditional interpretations. Although many theologians and almost all masters of arts eagerly embraced the new Aristotelian learning, there was a growing uneasiness among certain traditionally minded theologians. With its emphasis on the eternity of the world, the unicity of the intellect, and its naturalistic and deterministic modes of explanation, the Aristotelian world system was not easily reducible to the status of a theological handmaiden, as abortive attempts to ban and then expurgate the texts of Aristotle in the first half of the thirteenth century at Paris bear witness. By the 1260s and 1270s, an intensive effort was made to control the new learning and bring it into conformity with the aims and objectives of traditional theology. This time, however, the weapons

---

50. For a general account of the fate of the Aristotelian corpus at the University of Paris, see Gordon Leff, *Paris and Oxford Universities in the Thirteenth and Fourteenth Centuries* (hereafter cited as *Paris and Oxford Universities*) (New York, 1968), 187–238; Much of what follows in this paragraph is drawn from my article, "The Condemnation of 1277, God's Absolute Power, and Physical Thought in the Late Middle Ages," *Viator* 10 (1979): 211.

employed were not the ban or expurgation, but the outright condemnation or restriction of a whole range of ideas deemed dangerous and reprehensible. The modest Condemnation of 1270 and the massive one of 219 propositions in 1277 by the bishop of Paris and his advisers were an attempt to curb the pretensions of Aristotelian natural philosophy by emphasizing the absolute power of God to do whatever He pleased short of a logical contradiction, even if that meant the invocation of hypothetical and real divine actions that were impossible in the natural world as conceived by Aristotle and his followers.[51]

Despite the effect all this had on the interpretation of Aristotelian natural philosophy, the natural books of Aristotle remained the heart of medieval university education. There was never any serious attempt to dislodge them after 1250. It was because of a worldview derived from Aristotle's natural books that C. S. Lewis could declare that "the human imagination has seldom had before it an object so sublimely ordered as the medieval cosmos."[52] The primary purpose of a medieval university arts education was to enable students to comprehend and interpret the structure and operation of that sublime cosmos.

The manner of achieving this laudable objective was made to depend on lectures and disputations. Lectures were at first largely sequential section-by-section expositions or commentaries on each required text. Here the master read a passage of the text and explained its meaning to the students. When he had finished reading and explaining a number of passages or sections *(textus)* of an Aristotelian work, it became customary to pose a question on those passages and to present the pros and cons of it followed by a proposed solution.[53] These questions frequently formed the basis of the master's *Questiones* on that particular Aristotelian work. In time, however, the questions previously posed toward the end of a lecture came to displace the commentary on the text itself. Thus the mode of teaching came eventually to concentrate on specific questions *(questiones)*

51. For a brief history and background, see John F. Wippel, "The Condemnations of 1270 and 1277 at Paris," *Journal of Medieval and Renaissance Studies* 7 (1977): 169–201; and Roland Hisette, *Enquête sur les 219 articles condamnés à Paris le 7 Mars 1277* (Louvain: Publications Universitaires; Paris, 1977). References to the Latin texts and translations of the Condemnation of 1277 are provided in Grant, "The Condemnation of 1277," *Viator* 10 (1979): 211, n. 1.

52. *The Discarded Image: An Introduction to Medieval and Renaissance Literature* (Cambridge, 1964), 121.

53. Weisheipl, "Curriculum at Oxford," 154.

or problems that followed the order of the required text and developed from it.[54] The written forms of this pedagogical technique that have survived are usually associated with the names of well-known masters who presumably gave some version of the surviving written text in their lectures.

In its public oral version, arts and theology masters were concerned with questions *(questiones disputate)* either in the form of ordinary or magisterial disputations where the master himself posed and answered the questions or in the form of extraordinary or quodlibetal disputations where the questions were raised by the audience and ultimately resolved by the master.[55] In all of these sessions, the undergraduate and/or bachelor was expected to participate either as a respondent *(respondens)* to objections posed during the dispute or as the one who resolves or determines a question under the supervision of a master.[56] Responding to questions and determining them was thus an integral and vital part of the training and education of all who would eventually become masters of arts. For the masters themselves it was a regular feature of intellectual life.

Whatever the roles of masters and students in the disputed questions at the medieval universities, it is clear that the question form of Scholastic literature lay at the heart of the educational system. Science, which constituted the core of the curriculum, was thus taught by the analysis of a series of questions posed by a master and eventually determined by him. Many of these *questiones* on the different works of Aristotle, and other texts as well, were written down and published through university auspices. Each question followed a fairly standard format. The enunciation of the question was always followed by one or more solutions supporting either the affirmative or negative position. If the affirmative position was initially favored, the reader could confidently assume that the author would ultimately adopt the negative position; or conversely, if the negative side appeared first, it could be assumed that the author would

54. This was true for lectures in both arts and theology. For the former, see Weisheipl, "Curriculum at Oxford," and for the latter, Mary Martin McLaughlin, *Intellectual Freedom and Its Limitations in the University of Paris in the Thirteenth and Fourteenth Centuries* (hereafter cited as *Intellectual Freedom and Its Limitations)* (New York, 1977), 208.

55. See Leff, *Paris and Oxford Universities,* 167–68. On the quodlibetal disputes, see P. Glorieux, *La Littérature quodlibétique,* 2 vols. (Belgium, 1925 [vol. 1]; Paris, 1935 [vol. 2]).

56. The ultimate determination of all questions disputed in public at official occasions was the right and privilege of masters alone.

subsequently adopt and defend the affirmative side. The initial opinions, which would subsequently be rejected, were called the "principal arguments" *(rationes principales)*. Following the enunciation of the principal arguments, the author might then describe his procedure and perhaps further clarify and qualify the question or define and explain particular terms in it. He was now ready to present his own opinions, usually by way of one or more detailed conclusions or propositions. Often, in order to anticipate objections he would raise doubts about his own conclusions and subsequently resolve them. At the very end of the question, he would respond to each of the "principal arguments" enunciated at the beginning of the question.

By its very nature, the *questio* form encouraged differences of opinion. It was a vehicle par excellence for dispute and argumentation. Medieval Scholastics were trained to dispute and consequently often disagreed among themselves.[57] Far from a slavish devotion to Aristotle, they were emboldened by the very system within which they were nurtured to arrive at their own opinions. The system would have been very different indeed had it simply provided them with a conclusion and then merely supplied a rationale and defense of that conclusion. But medieval Scholasticism always posed at least two options and often many more. In principle, one was expected to evaluate arguments critically and by a process of elimination arrive at truth. Scholastic ingenuity was displayed by introducing subtle distinctions that, upon further development, might well yield new opinions on a given question. It is thus hardly surprising that centuries of disputation should have produced a variety of opinions on a very large number of questions. Hundreds of questions drawn from Aristotle's natural books formed the basic substance of natural sciences as taught and studied at the medieval university.[58] Not only were they con-

57. Talbot has a low opinion of the medical disputations, or "intellectual wrestling matches," as he calls them ("Medicine," 404–5): "Viewing the subjects of these wrangles with a dispassionate eye and at a distance of some centuries, it is hard to see what all the fuss was about." The medieval disputants, however, took these controversies quite seriously and so must we if we are to understand not only medieval medicine and natural philosophy, but medieval intellectual life in general.

58. For typical questions drawn from *questiones* on Aristotle's *Physics* (by Albert of Saxony), *De caelo* (by John Buridan), *On Generation and Corruption* (by Albert of Saxony), and *Meterology* (by Themon Judaeus), see Grant, *Source Book in Medieval Science*, 199–210. Most of the questions cited below are drawn from this lengthy list of some 266 *questiones*. I have discussed some of them in my article, "Cosmology," in *Science in the Middle Ages*, edited by Da-

cerned with the nature and behavior of the noble and near perfect celestial region and the less perfect generable and corruptible elemental and compound bodies of the sublunar realm, but they also inquired about the eternity of the world and whether other worlds, or an infinite space, might lie beyond ours. The nature of the celestial region, or heaven, was of major concern and elicited such questions as whether it was light or heavy and whether it had absolute directions, such as up and down, front and behind, and right and left. Was celestial matter similar to terrestrial matter and therefore subject to the Aristotelian categories of change in substance, quantity, quality, and place? *Or* was it immutable, in which event the very conception of a celestial "matter" was called into question. Since medieval scholars were almost unanimous in their belief that the planets and stars were carried around on physical spheres, a variety of questions were posed about the nature and motion of those spheres. What is their total number and how are they moved—by angels? forms? souls? or perhaps by some inherent principle? Are the celestial movers integral to the orbs they move, or distinct from them? Do those movers experience fatigue and exhaustion? Does God move the *primum mobile,* or first movable sphere, directly and actively as an efficient cause, or only as a final and ultimate cause? Are all the orbs of the same specific nature? Are they concentric with the earth as center or is it necessary to assume real eccentric and epicyclic orbs? The causative influences of the celestial region on the terrestrial were also of great interest and concern for astrology and natural philosophy, evoking numerous questions about the nature of the forces involved in this unidirectional relationship. Are the celestial and terrestrial regions continuous or discontinuous? How are

---

vid C. Lindberg, 265–302 (University of Chicago Press, 1978). In another article, "Aristotelianism and the Longevity of the Medieval World View," *History of Science* 16 (1978): 93–106, I have attempted to assess the impact of the *questiones* form of literature on medieval concepts of the cosmos and to explain the role of that literature in perpetuating the medieval Aristotelian worldview. The hundreds of written *questiones* mentioned above were almost certainly not in the original form in which they were first discussed and debated in the university classroom. They represent revised and often polished versions of the classroom lectures and debates and therefore do not provide a sense of the actual "give-and-take" that may have occurred at the original classroom presentation or at the public dispute. In the absence of firsthand descriptions of classroom lectures and debates, student annotations of standard texts are helpful as are the lectures of minor or little-known teachers who may not have revised their presentations for "publication." These cautions and insights are provided by John E. Murdoch, "Music and Natural Philosophy: Hitherto Unnoticed *Questiones* by Blasius of Parma(?)," *Manuscripta* 20, no. 2 (1976): 134–35.

the various phenomena of the upper terrestrial region, such as comets, the Milky Way, and the rainbow, formed, and what are they made of? Questions were also posed about the nature of the terrestrial region that was deemed so radically different from the superior and more perfect celestial region. Here the focus was on elements and compounds and their interrelationships and motions. Do elements remain or persist in a compound? Are there only four elements? Is there any pure element? Can one element be generated directly from another? Does a compound or mixed body consist of all four elements? What is the cause of the natural motions of light and heavy bodies? Is there something absolutely heavy and something absolutely light? Finally, questions were also posed about the earth and its relation to the cosmos: Is the earth spherical? Is it always at rest in the center of the world? Is its size as a mere point in comparison to the heavens?

Science at the medieval universities consisted of responses and solutions to questions of the kind just described. Generation after generation of masters of arts taught and wrote on such questions, and generations of students were considered to have been properly educated if they could absorb and master the diverse and often conflicting responses to these seemingly innumerable problems. To understand the nature and content of medieval natural science as taught at the medieval university one must become familiar with the vast *questiones* literature.[59]

The *questiones* on the Aristotelian natural books may have represented the scientific fare of the masters of arts and the hordes of undergraduates and bachelors of arts whom they taught, but it is only one aspect of the natural philosophy and science of the medieval university. Our description would be incomplete and defective without mention of the relevant scientific discussion in the theological faculty. Here masters and bachelors in theology were regularly confronted with problems about the nature of the physical world and its creation. Not only were traditional commentaries produced on the creation and structure of the world as described in Genesis, but even more important were the commentaries and *questiones* on the *Sentences* of Peter Lombard, a theological treatise written around 1150 and divided into four books devoted, respectively, to

---

59. To arrive at a quite reasonable estimate of the number of extant commentaries and *questiones* on the works of Aristotle alone, see Charles H. Lohr, "Medieval Latin Aristotle Commentaries," *Traditio* 23 (1967) and 30 (1974).

God, the Creation, the Incarnation, and the Sacraments.[60] As the standard text on which all theological students had to lecture and comment for some four centuries, the second book on creation afforded ample opportunity to reflect on the origin and operation of the physical world.[61] In considering the six days of creation, medieval theologians, most of whom were also masters of arts thoroughly trained in the natural philosophy of Aristotle and the medieval disputes embedded in the *questiones* literature, injected much contemporary physical theory into their theological deliberations. Problems lurking in the creation account made this almost inevitable, as is evident when they tried to distinguish the heaven, or firmament, created on the first day from the heaven created on the second day; or when they sought to explain the differences, if any, between the light mentioned on the first day and the visible, familiar light associated with the sun and the other celestial luminaries created on the fourth day; or when they were compelled to explain the distinction, if any, between the waters above the firmament and the waters below. But it was not the creation account alone that encouraged theologians to inject science into their explanations, but also problems such as the whereabouts of God and the motions of angels discussed in the first book of the *Sentences*. God's location served as a point of departure for discussions about the possible existence of an infinite extracosmic space; the motion and the positions of angels raised problems about space, place, and the continuum when it was found necessary to distinguish the ways in which angels moved and occupied places from the way bodies did.[62]

60. For the complete text, see *Petri Lombardi Libri IV Sententiarum,* studio et cura PP. Collegii S. Bonaventurae in Iucem editi, 2d ed. (Ad Claras Aquas, 1916). Hereafter cited as *Magistri Petri Lombardi.* A third edition of the first two books (issued as one volume in two parts) appeared in 1971 with the title *Magistri Petri Lombardi Parisiensis Episcopi Sententiae in IV Libris Distinctae,* 3rd ed. (Grottaferrata [Rome]: Editiones Collegii S. Bonaventurae Ad Claras Aquas, 1971).

61. See bk. 2, distinctions 12–15, in vol. 1, pt. 2 of *Magistri Petri Lombardi Parisiensis Episcopi Sententiae in IV Libris Distinctae.*

62. All this is found in bk. 1, distinction 37: "In what ways is God said to be in things" (*Quibus modis dicatur Deus esse in rebus, Magistri Petri Lombardi,* 263–75). Thus Richard of Middleton and Jean de Ripa injected discussions of infinite extracosmic space; Richard and St. Bonaventure considered whether or not angels move with successive motion; and Richard also sought to determine whether an angel is actually in a space. For Richard of Middleton, see *Super quatuor libros Senteniarum Petri Lombardi Quaestiones subtilissimae,* 4 vols. (Brescia, 1591; reprint Frankfurt am Main: Minerva G.m.b.H., 1963), 1:325–34; for de Ripa, see "Jean de Ripa I Sent. Dist. 37: De modo inexistendi divine essentie in omnibus creaturis," edited by

But theologians also eagerly introduced logic and mathematics into their responses to purely theological problems in the *Sentences*.[63] The amounts of grace, merit, sin, and reward that might be dispensed by God were often discussed in a context of the intension and remission of forms and were expressed in the language of proportions and proportionality relations that had been evolved in natural philosophy. Problems of infinity and continuity in a logicomathematical context were frequently introduced into discussions as to whether God's power was capable of producing infinitely intensive qualities and attributes.[64] The widespread acceptance of the doctrine of God's absolute power to do whatever He pleased short of a logical contradiction generated innumerable speculations *secundum imaginationem* in which God was imagined to perform some act according to some given proportional relationship.[65] Many of the acts that God was imagined to perform were couched in logicomathematical terms and concepts imported from natural philosophy or were contrary to traditional Aristotelian conceptions of the physical world.

Theologians played a significant role in developing the character and content of natural philosophy and science in the medieval university. It is no accident that the greatest medieval figures in science were also theologians, as the names of Albertus Magnus, Thomas Aquinas, Duns Scotus, William Ockham, Thomas Bradwardine, Nicole Oresme, Theodoric of Freiberg, and Henry of Hesse, to name only a few, bear witness. Theologians were, of course, not inherently more brilliant in such mat-

André Combes and Francis Ruello, with an introduction by Paul Vignaux, *Traditio* 23 (1967): 231–34; for Bonaventure, see *S. Bonaventurae Opera Omnia*, 1, *Commentaria in primum librum Sententiarum* (Quaracchi, 1882), 657–64. Peter Lombard's consideration of angels in bk. 1, distinction 37, was only incidental to his major concern with God. The extensive treatment of angelic nature and behavior is reserved for bk. 2, distinctions 1–11.

63. As confirmation of this tendency, John Major (1469–1550), in the introduction to his own commentary on the second book of the *Sentences* (1528), could declare that "for some two centuries now, theologians have not feared to work into their writings questions which are purely physical, metaphysical, and sometimes purely mathematical." Although he deplores the practice, Major confesses that he has "not blushed to follow in their footsteps."

64. For a brilliant discussion of these themes and much else, see John E. Murdoch, "From Social into Intellectual Factors: An Aspect of the Unitary Character of Late Medieval Learning," in *The Cultural Context of Medieval Learning*, edited by J. E. Murdoch and E. D. Sylla, 271–348, esp. 298–303 hereafter cited as "From Social into Intellectual Factors."

65. On *secundum imaginationem*, see Murdoch, "From Social into Intellectual Factors," 292, 294, 297, 300, 312, and his *"Mathesis in philosophiam scholasticam introducta . . . ," Arts libéraux et philosophie au moyen âge*, 248; also see Grant, *Physical Science in the Middle Ages*, 34, and "The Condemnation of 1277," *Viator* 10 (1979): 239–40, 241–42.

ters than masters of arts who remained as teachers and scholars in the arts faculty. Theologians were simply better trained than their counterparts in the arts faculty. Not only were they thoroughly versed in Aristotelian science and philosophy, but they were the recipients of some eight or nine years of rigorous training in the subtleties of theology. Since theology and theological considerations played a vital role in many questions of natural philosophy, theologians had a considerable advantage over arts masters. But if that were not enough, masters of arts, who were untrained in theology, were forbidden at the University of Paris to discuss "any question which seems to touch both faith and philosophy" unless they resolved the question in favor of the faith. Required to take an oath to this effect beginning in 1272, masters of arts were intimidated by the theologians and generally omitted theological considerations from their deliberations, even where these might have been relevant.[66] An illustration of the manner in which arts masters might have felt frustrated and intimidated by theologians is available from the works of John Buridan, probably the greatest natural philosopher among the Parisian arts masters in the fourteenth century. Considering the possibility of the existence of vacuum in his *Questions on the Physics* which, according to Buridan himself, touches faith and theology more than any other question, Buridan felt that despite his oath, he had to introduce theological considerations or avoid entirely a range of arguments in opposition to his own position that were yet essential to the whole question.[67] It is clear from the context that Buridan felt constrained to introduce no theological material into the argument, even though this meant that he could not treat the question honestly. Elsewhere, Buridan shows much deference to the theologians, as when he declares, for example, with Aristotle, that no body exists beyond the world, but immediately informs his reader that "you ought to have recourse to the theologians [in order to learn] what must be said about this according to the truth of faith."[68]

---

66. The oath appears in H. Denifle and E. Chatelain, eds., *Chartularium Universitatis Parisiensis*, 4 vols. (Paris, 1889–1897), 1: 499–500 and has been translated by Lynn Thorndike, *University Records and Life in the Middle Ages* (New York, 1944), 85–86. Thorndike's translation has been reprinted in Grant, *Source Book in Medieval Science*, 44–45.

67. The relevant passage appears in Buridan's *Questions on the Eight Books of the Physics of Aristotle*, bk. 4, question 8, and has been translated in Grant, *Source Book in Medieval Science*, 50–51 (the title of the Latin edition appears on 50, n. 1).

68. Translated in Grant, *Source Book in Medieval Science*, 51, n. 4, from Buridan, *Questions*

The oath of 1272 required at the University of Paris of all masters of arts throughout the fourteenth and perhaps most, if not all, of the fifteenth century clearly raises the question of freedom of scientific inquiry at the University of Paris. To the oath of 1272 must, of course, be added the Condemnation of 1277, which was also in effect throughout the fourteenth century and perhaps the fifteenth as well. Many of the articles condemned forbade approval, under penalty of excommunication, of Aristotle's fundamental conviction of the eternity of the world; they compelled arts masters to accept the possibility that God could create other worlds, or that He could move our world with a rectilinear motion despite the vacuum that would be left behind, even though these hypothetical situations were judged impossible within Aristotelian natural philosophy.[69] We must inquire, therefore, whether these and other restrictions contained in the Condemnation of 1277, as well as the denial to masters of arts of the right to discuss purely theological questions, seriously curtailed freedom of inquiry in natural philosophy and restricted investigation of scientific problems.

Despite a degree of intimidation where theological issues might have been relevant to the proper discussion of a scientific question (as with Buridan above), the pursuit of natural philosophy was not really hampered or restricted by theologians, by university authorities, or by church or state. The conflict between philosophy and faith in the thirteenth century produced a situation in the fourteenth in which the arts masters were willing to leave theology to the theologians and hoped, though in vain, that the theologians would leave philosophy and natural science to the arts masters.[70] Although all had to accept the truth of basic Christian doctrine, propositions contrary to those truths could be discussed speculatively under the guise of "speaking philosophically" or "speaking naturally" *(loquendo naturaliter)*.[71] By accepting doctrinal truth on faith

---

on De caelo, bk. 1, question 20 (for the Latin text, see *Iohannis Buridani Quaestiones super libris quattuor De caelo et mundo,* edited by E. A. Moody [Cambridge, Mass., 1942], 93).

69. These articles are discussed at length in Grant, "The Condemnation of 1277," *Viator* (1979): 211–44.

70. See Mary Martin McLaughlin, *Intellectual Freedom and Its Limitations in the University of Paris in the Thirteenth and Fourteenth Centuries* (New York: Arno Press, 1977), 135.

71. Thus Nicole Oresme, after demonstrating that a perfect eclipse of the moon could only occur once through all eternity, explains that "I always understand this 'naturally speaking' [*naturaliter loquendo*] and have even assumed an eternity of motion." Of course, "supernaturally speaking," the world will endure for only a finite time. Hence Oresme qualified his intent

and confining themselves to the domain of natural philosophy or science, arts masters could avoid almost all consideration of the miraculous in nature.[72] As a representative of this approach, John Buridan was probably typical of the arts masters in the fourteenth century. Rather than become preoccupied with supernatural possibilities, which could pose theological difficulties for a master of arts, Buridan devoted himself to the analysis and comprehension of the behavior of natural powers.[73] He sought to defend Aristotelian science as the best means of understanding the physical world, although he disagreed on numerous significant points with Aristotle. Readily conceding that God could interfere at any time and alter the natural course of events, as was demanded by the Condemnation of 1277, Buridan nevertheless assumed that "in natural philosophy, we ought to accept actions and dependencies as if they always proceed in a natural way." Should conflict arise between the Catholic faith and Aristotle's arguments, which, after all, are based only on sensation and experience, it is not necessary to believe Aristotle, as, for example, on the doctrine of the eternity of the world. And yet, if we wish to confine ourselves to a consideration of natural powers only, it is appropriate to accept Aristotle's opinion on the eternity of the world, *as if it were true.* As with most arts masters, Buridan was primarily interested in arriving at truths about the regular operations of the physical world in the "common course of nature" *(communis cursus nature)* and little concerned with all the hypothetical natural impossibilities that God might perform but which He probably hadn't performed and very likely would not perform.

The basis for a "common course of nature" could be established, in Buridan's view, by formulating laws and principles from inductive generalizations aided by reason. Such laws need not be absolute but empirical, "accepted because they have been observed to be true in many instanc-

---

and proceeds as if it were natural to suppose that the world is eternal. The passage cited here appears in Oresme's *De proportionibus proportionum,* ch. 4, in Nicole Oresme, *De Proportionibus proportionum and Ad pauca respicientes,* edited with introduction, English translations, and critical notes by Edward Grant, 305 (Madison, Wis., 1966).

72. McLaughlin, *Intellectual Freedom and Its Limitations,* 312.

73. What follows on Buridan is largely drawn from my article, "Scientific Thought in Fourteenth-Century Paris: Jean Buridan and Nicole Oresme," in *Machaut's World: Science and Art in the Fourteenth Century,* edited by Madeleine Pelner Cosman and Bruce Chandler, *Annals of the New York Academy of Sciences* (New York: New York Academy of Science, 1978), 314: 108–11.

es and to be false in none." Since Buridan's methodology of science was predicated on the "common course of nature," God's intervention in the causal order, which all acknowledged possible, became irrelevant. Thus could Buridan proclaim that "for us the comprehension of truth and certitude is possible." Using reason and experience, Buridan sought to "save the phenomena" in accordance with the principle of Ockham's razor— that is by the simplest explanation that fit the evidence.

Despite the theological condemnation of 1277 and their sworn oath not to dispute theological questions, arts masters were remarkably free to pursue their investigations and to arrive at whatever opinions they pleased. The enormous *questiones* literature with its hundreds of problems demonstrates this beyond any reasonable doubt. The majority of questions taken up in the natural books of Aristotle produced at least two opposing opinions and occasionally more. Some of these alternatives won a consensus among the masters, others did not. Without an atmosphere of intellectual freedom, such diversity could not have been achieved.[74]

In fact, the most famous (or perhaps infamous) of medieval theological condemnations, that of 1277, may have served to stimulate intellectual and scientific curiosity even as it sought to inhibit and curtail ac-

---

74. As for the many theologians who also discussed scientific questions, there were virtually no intellectual restrictions other than acceptance of doctrinal truth. And even doctrinal truth was often uncertain and debatable. On the remarkable degree of intellectual freedom available to medieval theologians of the thirteenth and fourteenth centuries, see McLaughlin, *Intellectual Freedom and Its Limitations*, 170–237. During the thirteenth and fourteenth centuries, individuals and ideas were of course censured. For the most part, however, censures were directed against novice theologians lecturing on the *Sentences* and were usually formulated by the members of the theology faculty itself (McLaughlin, *Intellectual Freedom and Its Limitations*, 209–10). Censures by theological commissions were also frequent enough, as, for example, the one against William Ockham in 1326 when fifty-one articles drawn from his commentary on the *Sentences* were censured, though not condemned (McLaughlin, *Intellectual Freedom and Its Limitations*, 276–77). Generally, it was the university itself—that is, the masters themselves—that exercised control over the intellectual content of lectures and publications; and "if the restrictions imposed were ever effective, it was because they were accepted by the consent of the society, not at the command of an external authority" (McLaughlin, *Intellectual Freedom and Its Limitations*, 310). The concluding sentence of Mary McLaughlin's splendid study admirably conveys the powerful sense of free inquiry that prevailed at the medieval university (317): "Masters of the late thirteenth and fourteenth centuries might indeed exercise, with little or no hindrance, that freedom of the teacher, first explicitly asserted by Siger of Brabant and his colleagues, to discuss and to explore his materials and problems, regardless of the truth of the opinions he considers."

ademic inquiry. By emphasizing God's absolute power to do anything short of a logical contradiction, the Condemnation of 1277 encouraged numerous invocations and applications of God's absolute power to a variety of hypothetical physical situations.[75] The supernatural alternatives that Scholastics at the University of Paris considered in the wake of the condemnation conditioned them to consider possibilities outside the ken of Aristotelian natural philosophy and usually in direct conflict with it, as, for example, the conditions that would obtain if God created a plurality of worlds, or moved the world with a rectilinear motion leaving a vacuum behind, or created an accident without a subject. So widespread was the contemplation of such hypothetical possibilities in the late Middle Ages that it is no exaggeration to view them as an integral feature of late medieval thought. Encouraged to pursue the consequences of hypothetical situations that were naturally impossible in Aristotelian science, Scholastics showed that alternatives to Aristotelian physics and cosmology were not only intelligible but even plausible. Although such speculations did not cause the overthrow of the Aristotelian worldview, they did challenge some of its fundamental principles and made many aware that things could be otherwise than was dreamt of in Aristotle's philosophy. Freedom of inquiry into the physical operations and principles of the world was little hindered and obstructed by theology and theologians during the Middle Ages. To the contrary, theological restrictions may actually have stimulated the contemplation of plausible (and even implausible) physical alternatives and possibilities far beyond those that Aristotelian natural philosophers might otherwise have considered.

Free though it was to pursue almost any lines of inquiry, science at the medieval university remained largely a bookish tradition based primarily on the works of Aristotle and Averroes and the technical treatises associated with the exact sciences of the quadrivium. With a few notable exceptions (e.g., Theodoric of Freiberg's *On the Rainbow* and perhaps Peter Peregrinus's *Letter on the Magnet*),[76] science in the medieval university was neither experimental nor truly empirical. Despite occasional glimmerings of a concept of scientific progress, such an idea was essentially

75. See Grant, "The Condemnation of 1277," 239–40. What follows is based on my article, where evidence is furnished for the claims made here.

76. Translations of both treatises appear in Grant, *Source Book in Medieval Science,* 368–76, 435–41. Whether Peter Peregrinus was university trained is presently unknown.

alien to medieval thought.[77] Scientific knowledge and an understanding of nature's operations and structure were derived primarily from the study of established authors. By careful analysis of such venerable texts, it was possible to gain new insights and to develop further the traditional wisdom. Occasionally, original contributions were even made and the *moderni* were sometimes consciously aware that they had developed a new technique for the treatment of an old or new problem. Moreover, there were always opportunities to conjure up daring and novel imaginative hypothetical physical situations by appeal to God's absolute power. But in its fundamental features, medieval science was essentially a rational inquiry based on the worldview embedded in the natural books of Aristotle. Although Scholastic natural philosophers produced numerous alternative solutions to most of the problems or *questiones* with which they were regularly concerned, they had no mechanisms for choosing among them. As the primary vehicle for the development and expression of scientific ideas and conclusions, the Scholastic *questio* form contained within itself the strengths and weaknesses of medieval science as practiced and taught at medieval universities. By enunciating problems in the form of a question rather than as an already derived conclusion, the Scholastic *questio* encouraged the presentation of the pros and cons of an argument. Each question contained all of the worthy arguments for and against it. Authors not only argued for their own conclusions, but were always expected to refute the contrary positions. In this way careful analysis was encouraged and a reasonably complete picture of all the relevant arguments and conclusions was available to subsequent readers who might then make yet further additions and alterations. At its best, the Scholastic *questio* was a thorough method for the analysis of scientific problems.

But there were serious deficiencies in medieval Scholastic procedure. Although the multiplication of opinions is a sign of free inquiry, there

77. In his article, "Medieval Ideas of Scientific Progress," *Journal of the History of Ideas* 59 (1978): 561–77, George Molland concludes that his account "has done little to disturb the traditional view that saw few conceptions of scientific progress in the Middle Ages" (576). The absence of a sense of scientific progress is perhaps attributable to (576) "the divorce between theory and practice that characterized so much scholastic science." See also Molland's earlier article, "Nichole Oresme and Scientific Progress," *Miscellanea Mediaevalia*, veröffentlichungen des Thomas-Instituts der Universität zu Köln, Band 9: *Antiqui und Moderni* (Berlin/New York, 1974): 206–20.

was no means of deciding most issues other than by consensus, which, often enough, was lacking. For how could one determine the true number of invisible celestial spheres; whether they formed a continuum or were merely contiguous; what was the true cause of the natural motion of elemental bodies; or whether there was an internal resistance in compound bodies, as some believed? The *questio* form of scientific inquiry suffered from another grave deficiency. As the major form of Scholastic literature for the pursuit of science and natural philosophy, the *questiones* produced an atomization of Aristotle's physical treatises into sequences of particular questions and problems that focused attention on the independent question and, as a consequence, tended to sever each question from its connections and associations with other related issues treated in the same work or elsewhere in the Aristotelian corpus.[78] Not only were related topics unintegrated, but even single topics were left in the form of a series of specific questions that were not organized into a larger, coherent, integrated whole. In this way, serious deficiencies and weaknesses of Aristotelian and contemporary science went undetected or overlooked. Primacy of the independent question in medieval Aristotelian physics and cosmology prevented, or at least seriously inhibited, any larger synthesis that might have revealed glaring inconsistencies within the intricate Aristotelian worldview. As long as Aristotelian science dominated the medieval university, the *questio* form of inquiry was its most characteristic feature, with the straightforward commentary also of importance. Even Galileo, while a young professor at the University of Pisa around 1590, found occasion to write *questiones* on Aristotle's *De caelo* and *On Generation and Corruption.*[79] By the late sixteenth century, however, Jesuit scholars developed the *cursus philosophicus,* which largely abandoned the formal procedure of the *questio* in favor of a more developed and integrated narrative account. The subject matter, however, remained much the same. Although the medieval university with its largely Aristotelian curriculum continued into the seventeenth cen-

78. On this point, see my paper, "Aristotelianism and the Longevity of the Medieval World View," *History of Science* 16 (1978): 98–99. My discussion here on the impact of the medieval *questio* has relied heavily on this article.

79. These Latin *Juvenilia,* as they have been called, appear in *Le Opere di Galileo Galilei, Edizione Nazionale,* ed. Antonio Favaro, vol. 1 (1890), and were recently translated by William A. Wallace, *Galileo's Early Notebooks: The Physical Questions, A Translation from the Latin, with Historical and Paleographical Commentary* (Notre Dame, Ind., 1977).

tury, its intellectual dominance was by then at an end. A new science based on a heliocentric astronomy and cosmology and a different physics had come into being. With its emergence, science moved outside its traditional university setting where Aristotelianism continued to reign and control the curriculum. The medieval university was now an anachronism and embarrassment. In time, the new science would reenter the university, but only as one of a number of subjects, where it now had to fight for its place in the curriculum. Never again would science achieve the exalted and almost exclusive status it held in the medieval university.

It is now time to assess the role of medieval science as it was institutionalized in the medieval university. Or to put it another way, what was the significance of the medieval university with its almost exclusive concern with the science of its day? What was its legacy to Western civilization? To understand and appreciate the medieval contribution, we must begin with the massive translations of the twelfth and thirteenth centuries. Near the beginning of this essay were mentioned two consequences of this extraordinary phenomenon. The translations of Greco-Arabic science, with Aristotle's natural books forming the core, laid the foundation for the continuous development of science to the present and also provided a curriculum that made possible the development of the university as we recognize it today.

Without the translations, which furnished a well-articulated body of theoretical science to Western Europe, the great scientists of the sixteenth and seventeenth centuries, such as Copernicus, Galileo, Descartes, and Newton, would have had little to reflect upon and reject, little that could focus their attention on significant physical problems. Many of the burning issues of puzzling scientific problems that were resolved in the Scientific Revolution of the seventeenth century entered Western Europe with the translations or were brought forth by university-trained medieval natural philosophers who systematically commented upon that impressive body of knowledge. The overthrow of one world system by another does not imply a lack of continuity. Medieval science, based on the translations of the twelfth and thirteenth centuries, furnished the physicists and natural philosophers of the seventeenth century with issues, theories, and principles that had to be rejected in order for significant advances to be made. That what emerged was radically different should not blind us to the essential continuity of inquiry between medieval and seventeenth-century science. Although solutions differed, many fundamental problems were common to both. With the introduction of Greco-Arabic science during the twelfth and thirteenth centuries, Western Europe began an unprecedented and uninterrupted concern for the nature and structure of the physical world. To its everlasting glory, the medieval university was the fundamental instrumentality for this epoch-making and still continuing chapter in the history of Western civilization.

3 ❧ The Condemnation of 1277, God's Absolute Power, and Physical Thought in the Late Middle Ages

When Christianity manifested its earliest concern about the physical world, it did so in an atmosphere of fear and hostility toward Greek science and philosophy. Deeply suspicious of these pagan enterprises, the Church Fathers and Christian authors of late antiquity grudgingly came to tolerate them as handmaidens to theology. In time, however, interest in scientific and philosophic thought for their own sake gradually developed. Already evident in the late eleventh century, it is clearly manifested by the prodigious translating activity of the twelfth century, by the close of which the basic intellectual fare for the next four centuries had become available in Latin. In this great storehouse of Greco-Arabic science and learning by far the most significant portion consisted of the works of Aristotle and those of his commentator Averroes. Between them they provided a secular worldview complete with principles and modes of demonstration that was not only suitable for comprehending the physical world but also applicable to problems in theology and Holy Scripture. The Aristotelian system of the world was not easily reducible to the status of a theological handmaiden, as abortive attempts to ban and then expurgate the texts of Aristotle in the first half of the thirteenth century at Paris bear witness. By the 1250s, the effort to control the Aristotelian corpus was abandoned and it became firmly entrenched in the universities, where it formed the basis of the liberal arts curriculum.

For the first time in the history of Latin Christendom, a conceptually rich and methodologically powerful body of secular learning posed a threat to theology and its traditional interpretations. While many theologians and almost all masters of arts eagerly embraced the new Aristotelian learning, there was a growing uneasiness among certain more traditionally minded theologians about the course of events. They feared that Christian theology would be undermined not only by Aristotelian ideas that directly conflicted with the Christian faith, such as the eternity of the world and the unicity of the intellect, but also by modes of thought that were becoming increasingly naturalistic and deterministic. By the 1260s an intensive effort was begun to control the new learning and bring it into conformity with the aims and objectives of traditional theology. This time, however, the weapons employed were not to be the ban or expurgation, but the outright condemnation or restriction of a whole range of ideas deemed dangerous and reprehensible. I am referring here, of course, to the condemnation of 13 articles in 1270 and of 219 in 1277, the latter condemnation being the basic concern of the discussion to follow.[1]

1. For the Latin text of the 219 articles presented in their original order, see H. Denifle and E. Chatelain, *Chartularium universitatis Parisiensis* 1 (Paris 1889), 543–55; for a methodical regrouping of the articles aimed at facilitating their use, see Pierre F. Mandonnet, *Siger de Brabant et l'Averroisme latin au XIIIme siècle*, 2nd ed. rev., pt. 2: *Textes inédits* (Louvain, 1908), 175–91. Using Mandonnet's reorganized version, Ernest L. Fortin and Peter D. O'Neill translated the condemned articles into English in *Medieval Political Philosophy: A Sourcebook*, edited by Ralph Lerner and Muhsin Mahdi, 337–54 (New York, 1963). Their translation has been reprinted in Arthur Hyman and James J. Walsh, eds., *Philosophy in the Middle Ages: The Christian, Islamic, and Jewish Traditions* (Indianapolis, 1973), 540–49. Also following Mandonnet's order is the translation by J. Wellmuth in *Philosophy in the West: Readings in Ancient and Medieval Philosophy*, edited by J. Katz and R. Weingartner, 532–42 (New York, 1965). For a translation of selected articles deemed relevant to medieval science, see Edward Grant, ed., *A Source Book in Medieval Science* (Cambridge, Mass., 1974), 45–50. Descriptions of the events leading up to, and including, the Condemnation of 1277 appear in Mandonnet, *Siger de Brabant et l'Averroisme latin au XIIIme siècle*, pt. 1: *Étude critique* (Louvain, 1911), 214–51, chap. 9, "Condemnation du Péripatétisme 1277"; Fernand Van Steenberghen, *Siger de Brabant d'après ses oeuvres inédites* 2, Les Philosophes belges 13 (Louvain, 1942,) 357–497, chap. 2, "La Philosophie à l'Université de Paris avant Siger de Brabant"; Pierre Duhem, *Le Système du monde: Histoire des doctrines cosmologiques de Platon à Copernic*, 10 vols. (Paris, 1913–1959), pt. 4: "Le Reflux de l'Aristotélisme: Les condemnations de 1277," 6.3–69, and 8.7–120; Gordon Leff, *Paris and Oxford Universities in the Thirteenth and Fourteenth Centuries* (New York, 1968), 187–238; and, finally, John F. Wippel, "The Condemnations of 1270 and 1277 at Paris," *Journal of Medieval and Renaissance Studies* 7 (1977): 169–201.

In this seven hundredth anniversary of the Condemnation of 1277, the only major study of its impact on the course of medieval natural philosophy appears in Duhem's *Le Système du monde*. Generally, Duhem subordinated the articles of the condemnation to the larger topics with which he dealt. Moreover, since he tended to treat authors in successive sections, his treatment is fragmented. While Duhem's monumental effort will probably remain the standard for the foreseeable future, my purpose here is to concentrate on the influence of certain articles and to consider only those Scholastics who appear to have had, directly or indirectly, one or more of them in mind.

Promulgated seven hundred years ago, the Condemnation of 1277 was the outcome of doctrinal, philosophical, and personal animosities that rocked Paris in the 1260s and 1270s. It pitted theologians against masters of arts, or philosophers, and theologian against theologian. At issue was the manner in which God's relationship to the world and its physical operations was to be understood. Was the physical world to be interpreted in rigorous conformity with the principles and laws of Aristotelian natural philosophy even where this conflicted with traditional Christian views of the world? The 219 articles condemned by the bishop of Paris, Étienne Tempier, were intended to resolve this dilemma following a request by Pope John XXI to investigate the intellectual unrest that had beset the University of Paris. A diverse collection of 219 propositions that were neither to be held nor defended under pain of excommunication, the Condemnation of 1277 was intended to subvert the philosophical necessitarianism and determinism that had become characteristic of philosophical thought in the thirteenth century and that had been derived from Greco-Arabic sources, especially from the works of Aristotle and his ardent admirer and commentator Averroes.

Inadvertently or not, Aristotelian natural philosophers were thought to have severely restricted God's power by a seeming overreliance on a naturalistic determinism rooted in Aristotle's physical and metaphysical principles. If the condemned articles are an accurate reflection of contemporary opinion, some Scholastic natural philosophers were prepared to deny the divine creation of the world, that God could create more than one world, that he could move the world in a straight line leaving behind a void space, that he could create an accident without a subject, and so on. God's power to perform these and other feats that were impossible in the natural world as conceived by Aristotle and his followers was thus denied and severely restricted. It was with all this in mind that the theologians who drew up the condemnation sought to curb the pretensions of Aristotelian natural philosophy by emphasizing the absolute power of God *(potentia Dei absoluta)* to do whatever He pleased short of a logical contradiction.[2]

2. For Thomas Aquinas's remark that not even God could produce a logical contradiction, see the translation of his *De aeternitate mundi* in *St. Thomas Aquinas, Siger of Brabant, St. Bonaventure, On the Eternity of the World (De aeternitate mundi),* translated and introduced by Cyril Vollert et al., 22 (Milwaukee, 1964). It was an opinion that was widely accepted.

Since it touched upon major issues in philosophy and theology, the condemnation was bound to have significant influence and impact. Viewed by some as a necessary antidote to the poison of deterministic, Aristotelian philosophy[3] and by others as a dangerous restriction to philosophical and theological inquiry,[4] the condemnation was generally effective at Paris throughout the fourteenth century,[5] despite a declaration by the bishop of Paris in 1325 rendering null and void all articles directed specifically against St. Thomas Aquinas.[6] Although the legal force of the condemnation was technically confined to the region under control of the bishop of Paris, its influence occasionally spread to England where eminent English Scholastics found occasion to cite one or more of the articles as if relevant to, and authoritative in, England.[7]

Scholars in the twentieth century are generally agreed that the Con-

3. A staunch supporter of the Condemnation, Ramon Lull, in 1298, defended each of the 219 articles in his *Declaratio Raymundi per modum dialogi edita*. For the Latin edition, see Otto Keicher, ed., *Raymundus Lullus und seine Stellung zur arabischen Philosophie, mit einem Anhang, enthaltend die zum ersten Male veröffentlichte "Declaratio Raymundi per modum dialogi edita,"* Beiträge zur Geschichte der Philosophic des Mittelalters 7.4–5 (Minister, 1909), 95–221; see also J. N. Hillgarth, *Ramon Lull and Lullism in Fourteenth-Century France* (Oxford, 1971), 230–31.

4. A bitter opponent was Godfrey of Fontaines, who, sometime around 1296, in a question "Whether the bishop of Paris sins by failing to correct certain articles condemned by his predecessor" ("Utrum episcopus parisiensis peccet in hoc quod omittit corrigere quosdam articulos a praedecessore suo condemnatos"), noted the questionable and contradictory nature of the condemned articles and the grave burdens it laid upon students and masters who fell victim to its uncertainties and confusions. See *Les Quodlibets onze-quatorze de Godefroid de Fontaines (Texte inèdit)*, edited by J. Hoffmans, Les Philosophes belges, Textes et études 5.1–2, 100–105, quodlib. 12, q. 5 (Louvain, 1932). For a discussion of this, and Godfrey's seventh quodlibet, question 18, where he considered "Whether a master in theology ought to speak against an article of the bishop if he believes the opposite to be true" ("Utrum magister in theologia debet dicere contra articulum episcopi si credat oppositum esse verum"), see Duhem (n. 1 above), 6.70–76.

5. Duhem 6.80. Hillgarth (n. 3 above), 251, claims that "[i]n the decade after Lull's *Declaratio . . .* in the Faculty of Arts the tide was setting more and more against the condemnation of 1277." As evidence, he observes that in the works Lull wrote against the "Averroists" between 1309 and 1311, he "does not appeal to the condemnation and does not refer directly to his *Declaratio,* though he repeats all its arguments." However, since Lull did not abandon the arguments, but indeed repeated them, surely we ought to conclude that the condemnation still exerted a powerful influence on him. Although it is highly probable that the arts masters generally disapproved of the condemnation, it is not likely that they willfully repudiated its separate articles. As we shall see below, Buridan, in the 1340s and 1350s, not only upheld them, albeit reluctantly, but occasionally used them to advantage.

6. See Denifle and Chatelain (n. 1 above), 2 (Paris, 1891), 280–81.

7. E.g., John Duns Scotus and William Ockham (see below).

demnation of 1277 severely reduced the scope and pretensions of philosophy as an independent discipline and that, as the powers of natural reason and experience were circumscribed, reliance on God's omnipotence was increased. In place of the waning Greek naturalism, John Duns Scotus and William Ockham, and their numerous followers, stressed the contingency of God's operations and his omnipotence to do as He pleased short of a logical contradiction.[8] Thus whatever Aristotle may have demonstrated for the natural world could easily be negated or altered by God's absolute power.

It appears, then, that the major consequence of the Condemnation of 1277 was to manifest and emphasize the absolute power of God, a doctrine that was hardly novel in 1277. After all, God's absolute power to effect whatever He pleased had already been proclaimed by St. Peter Damian in the eleventh century,[9] enunciated more effectively in the *Sentences* of Peter Lombard in the twelfth century,[10] and declared unequivocally by Thomas Aquinas in the thirteenth century.[11] But why did the doctrine of God's absolute power acquire special significance in 1277 and thereafter? Largely, it would seem, because prior to the thirteenth century there had been no serious internal intellectual threat to Christian theology. With the introduction of Greco-Arabic natural philosophy and metaphysics in the thirteenth century, all this changed, as we have seen. The doctrine of God's absolute power was now invoked in fear and anger by theologians who viewed it as an ultimate defense against the dangerous inroads of pagan thought. Many of the condemned articles forced all to concede

8. Gordon Leff, *The Dissolution of the Medieval Outlook: An Essay on Intellectual and Spiritual Change in the Fourteenth Century* (New York, 1976), 28–29, and Étienne Gilson, *History of Christian Philosophy in the Middle Ages* (London, 1955), 410.

9. See the translation of part of Peter's *De divina omnipotentia* in *Medieval Philosophy: From St. Augustine to Nicholas of Cusa*, edited by John F. Wippel and Allan Wolter, O.F.M., 143–52, esp. 148–49 (New York, 1969).

10. Peter declared that "God truly and properly is called an omnipotent trinity because by Himself, that is by His natural power, He can accomplish whatever He wishes to do [*quidquid vult fieri*] and whatever He wishes to be able to do [*quidquid vult se posse*]. . . . For if, indeed, He wishes something, it happens because nothing can resist His will." My translation (unless otherwise specified, all translations are mine) from *Magistri Petri Lombardi Parisiensis episcopi Sententiae in IV libris distinctae*, 3rd ed., 1.2 (Grottaferrata, 1971), 297–98, bk. 1, dist. 42, chap. 3, par. 6. In the next distinction (bk. 1, dist. 43), par. 1 (p. 298), Peter emphasizes that no restrictions can be placed on God's infinite power. See also Leo Sweeney, S.J., "Divine Infinity: 1150–1250," *The Modern Schoolman* 35 (1957): 42.

11. For example, in his *De aeternitate mundi* (n. 2 above), 20.

that God could do something that had previously been denied Him. And if that failed to convey the new message, Article 147 made it as explicit as possible by condemning the opinion "[t]hat the absolutely impossible cannot be done by God or another agent," which is judged "[a]n error, if impossible is understood according to nature."[12]

But it was one thing to concede God's absolute power to perform the naturally impossible any time whatever, and quite another to suppose that He actually performed such impossibilities. For if God chose to exercise His absolute power and execute naturally impossible acts, would He not have violated those very laws of nature which He Himself had ordained for a world of His own making? A world subject to such divine alterations would prove unknowable and uncertain. In fact, the medieval world was not conceived as a stage on which an inscrutable, and even capricious, God performed seemingly random acts that made a mockery of lawful regularity. The lawfulness of the cosmos was made compatible with God's absolute power by a distinction originating in the eleventh century between God's *potentia absoluta* and His *potentia ordinata*, or ordained power. The former "referred to the total possibilities *initially* open to God, some of which were realized by creating the established order; the unrealized possibilities are now only hypothetically possible." By contrast, the *potentia ordinata* is restricted "to the complete plan of God for his creation."[13] From this crucial distinction, it followed that once God had decided the natural order of our world from among the innumerable, initial possibilities, He would not tamper with the plan by substituting from the store of unused possibilities.[14] Thus John Buri-

12. "147. Quod impossibile simpliciter non potest fieri a Deo, vel ab agente alio.—Error, si de impossibili secundum naturam intelligatur"; Denifle and Chatelain (n. 1 above), 1.552.

13. See William J. Courtenay, "Nominalism and Late Medieval Religion," in *The Pursuit of Holiness in Late Medieval and Renaissance Religion*, edited by Charles Trinkaus and Heiko A. Oberman, 39 (Leiden, 1974).

14. Ibid., 43. A presumably typical interpretation of the distinction was that of Gabriel Biel (d. 1495). According to Heiko Oberman, *The Harvest of Medieval Theology: Gabriel Biel and Late Medieval Nominalism* (Grand Rapids, Mich. 1967 [1963]), 37, Biel understood the distinction to mean "that God can—and, in fact, has chosen to—do certain things according to the laws which he freely established, that is, *de potentia ordinata*. On the other hand, God can do everything that does not imply contradiction, whether God has decided to do these things [*de potentia ordinata*] or not, as there are many things God can do which he does not want to do. The latter is called God's power *de potentia absoluta*." The brackets are Oberman's. The divine acts discussed in this article are among those that God was thought to have excluded from his ordained plan of creation, but which He was nonetheless capable of enacting by His absolute power.

dan might well have had this significant distinction in mind when he declared that although God could indeed create corporeal spaces and substances beyond the world, and to any degree He pleased, it did not follow that He had actually done so. Therefore, we ought not to assume, for example, that God had created an infinite space beyond the world unless we have independent evidence for so believing, evidence drawn from one or all of the ordinary sources, that is, from the senses, experience, natural reason and the authority of Sacred Scripture.[15]

But if the distinction between God's *potentia absoluta* and *potentia ordinata,* as described above, was generally, and perhaps even universally, accepted in the late Latin Middle Ages, so that few, if any, would have believed that God did, or ever would, actually use His absolute power to alter the natural laws and structure of the world He had created—that is, to perform natural impossibilities[16]—then it follows that the Condemnation of 1277, with its glorification of God's absolute power, could have had virtually no effect on medieval conceptions of the actual operation and structure of the physical world. It merely compelled all to concede that, contrary to the principles of Aristotelian natural philosophy, God could, if He wished, create worlds other than ours, move our world rectilinearly, create an accident without a subject, and do anything else contrary to those accepted principles. But once that concession was made, whether voluntarily or under the duress of possible excommunication, all were free to retain the traditional opinions, as indeed they usually did. Thus it

15. John Buridan, *Questions on De caelo,* bk. 1, q. 17, in *Quaestiones super libris quattuor De caelo et mundo,* edited by Ernest A. Moody, 79, lines 1–9 (Cambridge, Mass., 1942).

16. If at some future time, a natural impossibility should actually occur, medieval theologians would have interpreted it as part of God's ordained plan *(potentia ordinata).* It is this ordained power that is at issue in an interesting discussion by the seventeenth-century Jesuit Bartholomaeus Amicus, who allowed that, although vacuum was contrary to universal order, God could act against that order and introduce a vacuum. Amicus believed that, because of divine goodness, God could not diverge from all order and introduce total disorder into the world; but there was nothing to prevent Him from substituting for one system of natural order that prevented the formation and existence of vacua, another that allowed the formation and existence of vacua. These were not incompatible acts, since God did not exhaust His creative possibilities in this present world. Thus to introduce vacua into our world would be compatible with His creative power and plan, although it would be contrary to the present customary course of nature. See Bartholomaeus Amicus, *In Aristotelis libros De physico auditu dilucida textus explicatio et disputationes . . . ,* 2 vols. (Naples, 1626–1629), 2.746E and 747B–C, bk. 4, q. 2, dub. 3. If, however, God had not ordained the introduction of vacua into our world, the possibility of so doing would, according to medieval conceptions, belong to His absolute power *(potentia absoluta).*

would appear that Alexandre Koyré was right when, in rebuttal of Pierre Duhem's extravagant claims made for the ultimate impact of the Condemnation of 1277,[17] he insisted that the condemned articles relevant to cosmology were of no genuine significance to scientific development because Scholastics would have preferred to study the world as it really was rather than be compelled by theological fiat "to study the conditions of possibility of the universes which God could have created had he wished to do so, but which he did not create because he did not wish to do so."[18]

Although it may be true that Scholastics would have preferred to study the world as it really is—a claim not easily substantiated—Koyré is mistaken when he argues that undue concern with unrealizable possibilities was unproductive and sterile for the history of medieval and early modern science. It will be one of the objectives of this essay to counter Koyré's claim as one aspect of a broader purpose that seeks to demonstrate the powerful impact of, and preoccupation with, God's absolute power not only as it was manifested in specific articles of the condemnation, but also, more generally, in its capacity as a powerful analytic tool in natural philosophy. The latter role, as we shall see, was as much a part of the history of the overall influence of the Condemnation of 1277 as are the specific articles with which we shall be concerned.

Despite the exaggerated and indefensible character of Pierre Duhem's claim that the Condemnation of 1277 was "the birth of modern science," he was right to emphasize the special significance of two articles, Article 34, which made it mandatory to concede that God could make more than one world, and Article 49, which compelled assent to the claim that God could move the heavens, or world, with a rectilinear motion even though such motion might leave behind a vacuum. Since these two articles struck at fundamental ideas in Aristotelian natural philosophy, it is appropriate to commence our study with them, turning first to Article 34 and the possibility of a plurality of worlds.[19]

17. In Pierre Duhem, *Études sur Leonard de Vinci, ceux qu'il a lus et ceux qui l'ont lu*, 3 vols. (Paris, 1906–1913; reprint 1955), 2.412, Duhem attributed the birth of modern science to the influence of Articles 34 and 49, which will be discussed below. For a translation of the passage, see Edward Grant, "Late Medieval Thought, Copernicus, and the Scientific Revolution," *Journal of the History of Ideas* 23 (1962): 200, n. 8. Duhem moderated his opinion (n. 1 above), 8.7–8.

18. Alexandre Koyré, "Le vide et l'espace infini au XIVe siècle," *Archives d'histoire doctrinale et littéraire du moyen âge* 24 (1949): 51. See also Grant (n. 17 above), 200, n. 9.

19. In dealing with the influence of the Condemnation of 1277, we must ask whether it is

Sometime between 1230 and 1235, and long before 1277, the most significant arguments against the existence of a plurality of worlds had already been clearly formulated in a commentary on the *Sphere* of Sacrobosco ascribed to Michael Scot.[20] One of the most widely used arguments involved void space.[21] For if several worlds existed, they would exist either in one place or in different places. Since it was axiomatic that two or

reasonable and plausible to assume that most, if not all, instances where God's absolute power is made the basis of a physical argument are also instances of the influence of the Condemnation of 1277. At the risk of being accused of arguing "post hoc, ergo propter hoc," I have assumed an affirmative response, since it was only after 1277, and because of the condemnation, that the principle of God's absolute power came to be used widely in the analysis and discussion of numerous physical problems involving both corporeal and spiritual entities.

In determining the influence of specific articles on the subsequent course of natural philosophy, it is obviously important to ascertain whether a particular article is actually intended. Where the phrase "article condemned at Paris" appears, we can be reasonably certain that it was among those condemned in 1277. More often, however, the substance of an article is given without any mention of its condemnation, as when, for example, William Ockham declared that the divine power could locate several bodies in the same place ("Ad tertium dico quod sic quia plura corpora possunt esse in eodem loco per potentiam divinam . . ."; William Ockham, *Quotlibeta septem; Tractatus de sacramento altaris* [Strasbourg, 1491; reprint in facsimile Louvain, 1962], quotlib. 1, q. 4, sig. a4v, col. 1 [the folios are unnumbered]). Since, elsewhere in the same work, Ockham mentions specific articles "condemned at Paris," and therefore indicates an awareness of the Condemnation of 1277, it would appear reasonable to suppose that he had in mind Article 141, which declared it an error to claim that God could not make several bodies exist simultaneously in the same place (Denifle and Chatelain [n. 1 above], 1. 551). Judgments of this kind must frequently be made and will play a role in what follows.

Ideally, any assessment of the impact of the Condemnation of 1277, and its intimately associated notion of the absolute power of God, on the course of medieval natural philosophy should rest on evidence gathered by a thorough study of the whole range of relevant Scholastic literature. Only on such empirical foundations can we arrive at a definitive and conclusive evaluation of the impact of the condemnation as a whole and of each of its numerous articles. Since inspection of the whole mass of medieval Scholastic literature is obviously impractical and unfeasible, the evidence gathered here is the by-product of my own research interests on the concepts of place, space, and vacuum, and cosmology in general. Although it is very likely that many other specific citations and unspecified allusions to relevant articles of the condemnation lie as yet undetected in the largely unstudied mass of Scholastic literature, those that have thus far come to light will be used to serve the objectives of this essay.

20. Edited by Lynn Thorndike, *The "Sphere" of Sacrobosco and Its Commentators* (Chicago, 1949), 247–342. For the dates, see 48. The most extensive discussion on the problem of a plurality of worlds is by Duhem (n. 17 above), 2.57–96, 408–423, and (n. 1 above) 9.363–430. For a recent summary and evaluation of medieval views, see Steven J. Dick, "Plurality of Worlds and Natural Philosophy: An Historical Study of the Origins of Belief in Other Worlds and Extraterrestrial Life," Ph.D. diss., Indiana University, 1977, 71–108. Of the two basic types of plurality frequently discussed—i.e., successive or simultaneous worlds (see Buridan's distinction in his *Questions on De caelo*, bk. 1, q. 19 [n. 15 above], 88)—we shall focus on the latter.

21. All the arguments cited here from Michael Scot's commentary on the *Sphere* of Sacrobosco occur on 252–54 of Thorndike's edition.

more bodies could not exist naturally in the same place simultaneously, it followed that they must exist in different places, which implied the existence of intervening space, a condition that would obtain even if two of the worlds were in contact at a single point. Now either this intervening space is filled with body or it is void. But no body can exist there since it would belong to none of the worlds around it. Nor indeed could the space be void, since Aristotle had demonstrated the impossibility of void space in the fourth book of the *Physics*.[22]

Michael Scot drew his second fundamental argument against a plurality of worlds from the first book of Aristotle's *De caelo*,[23] where Aristotle argues that if other worlds existed, their elements, and the motions of those elements, would be identical with those in our world. With identical natures, Aristotle insisted that all these elements in the different worlds could have only one center and circumference. But if there is only one center and circumference for all elements and things, there can only be one, unique world, for if many centers and circumferences existed, many worlds would exist, a world for each center and circumference. Indeed, if many worlds existed each with identical elements possessed of the same natural motions and, furthermore, if each world had its own center and circumference, particles of earth from our world would tend to move toward the center of another world and thus rise with a violent, or unnatural, motion toward the circumference of our world; particles of earth in other worlds would behave similarly. The same reasoning would, of course, apply to the element fire. Both earth and fire would thus be capable of rising and falling "naturally," even though an element could only have one natural motion, either up or down. Since a plurality of worlds would obviously play havoc with Aristotelian physics and cosmology, the possibility was rejected by Michael Scot and others.[24]

Despite these powerful physical arguments, however, Michael Scot ac-

---

22. Sometime between 1231 and 1236, when Michael Scot wrote his commentary on the *Sphere* of Sacrobosco, William of Auvergne repeated much the same argument against a plurality of worlds, as did Roger Bacon decades later. See William of Auvergne, . . . *Opera omnia*, 2 vols. (Paris, 1674; reprint in facsimile Frankfurt a.M., 1963), 1.607–608, *De universo*, first part of pt. 1, chaps. 13–14, and Roger Bacon, *Opus majus*, translated by Robert B. Burke, 2 vols. (Philadelphia, 1928), 1.186.

23. For Aristotle's arguments against a plurality of worlds, see *De caelo* bk. 1, chaps. 8, 9.

24. Aquinas also found this a compelling argument. See Thomas Aquinas, *In Aristotelis libros De caelo et mundo; De generatione et corruptione; Meteorologicorum expositio*, edited by Raymond M. Spiazzi, O.P., 80 (Turin, 1952), *Expositio in Aristotelis libros De caelo et mundo* bk. 1, lect. 16.

knowledges that some believed it possible for an omnipotent God to make other worlds, even an infinite number of them, and to create them from identical or different elements indifferently.[25]

In response, Michael Scot readily assents that by his absolute power God could, if He wished, create a plurality of worlds. But although God has the power to create many worlds, nature itself, as a caused entity, is incapable of receiving them, since it has not been endowed with a capacity to receive many worlds simultaneously.[26] Like Michael Scot before him, Thomas Aquinas also acknowledged that, by His absolute power, God could create other worlds.[27] But Aquinas insisted that the best and most noble ends would not be served by many worlds. For if these other worlds were similar to ours, they would be superfluous; and if dissimilar, none would be perfect, since none could incorporate within itself the totality of natures of sensible bodies. Under these conditions, it would require a combination of all the separate worlds to make a perfect world, a state of affairs that could be achieved by a single world. It is better, therefore, to make a single, perfect world than many that are imperfect and better also to assign goodness to a single world than to diminish that goodness by division. For Aquinas, "[T]hose can posit many worlds who do not assume any guiding wisdom as the cause of the world, but [rather] chance, as Democritus, who said that this world, and the infinite number of other worlds, was made from a [chance] concourse of atoms."[28]

---

25. "Et dicendum, et quidam dicunt, quod deus potuit et potest ita cum isto mundo alium et alios facere vel etiam infinitos cum sit omnipotens, et hoc ex elementis eiusdem speciei et nature vel etiam diverse"; Thorndike (n. 20 above), 253. Since the authors embraced by the term "quidam" are not identified, we cannot assess the importance of Michael Scot's dramatic remark. Prior to 1277, however, their importance could only have been minimal and the content of their discussions meager.

26. Ibid., 253–54. In his *De universo,* William of Auvergne similarly denied a plurality of worlds and attributed this to a deficiency on the part of the other potential worlds rather than to God (n. 22 above), 1.611, *De universo,* first part of pt. 1, chap. 16. Duhem argues that in opposing to God's creative power an already determined nature, Michael Scot reveals acceptance of the God of Averroes rather than the God of the Christians (n. 1 above), 9.365. Prior to 1277, Parisian masters, according to Duhem (380), denied God the power to create a plurality of worlds and then argued that this was not a limitation on his creative powers. Michael Scot and William of Auvergne seem to fit this pattern, since they allow that God has the power to create other worlds, but somehow these other potential worlds cannot be realized because of nature's inherent deficiencies.

27. Aquinas (n. 24 above), 94–95, *De caelo et mundo* bk. 1, lect. 19.

28. For the Latin text, see Thomas Aquinas, *Summa theologiae* (Ottawa, 1941), vol. 1, pt. 1, q. 47, art. 3 ("Utrum sit unus mundus tantum"); also see Dick (n. 20 above), 80–81.

With the promulgation of the Condemnation of 1277 and Article 34, which condemned those who believed that God could not create more than one world,[29] the content and character of the arguments changed drastically. Although all at Paris were compelled to acknowledge that God could create other worlds if He wished, their reactions fall into two broad categories. There were those who, while forced to acknowledge God's absolute power to create as many worlds as He wished, found even the possibility of a plurality of worlds physically untenable and invoked arguments that were intended to show that it was naturally impossible. For this group, among whom John of Jandun is perhaps the most significant representative,[30] Aristotle's arguments remained in force and all attempts to accommodate physics and cosmology to the possible existence of other worlds were avoided or refuted.

But there were those who took seriously the possibility that God could create other worlds than our own and, on the assumption that He did create them, sought to counter those of Aristotle's arguments that had previously been accepted more or less routinely. It is from among this group of Scholastics[31] that the impact and significance of Article 34 is revealed. For despite an almost unanimous conviction that God had not actually created other worlds,[32] they formulated arguments that sought to make the possible existence of other worlds intelligible.

Let us consider first the kinds of responses that were made to the Aristotelian claim that formally identical elements in the separate worlds would move to a single center and circumference, thus producing, as we

29. "34. Quod prima causa non posset plures mundos facere"; Denifle and Chatelain (n. 1 above), 1.545.

30. For Jandum, see Duhem (n. 1 above), 9.387–89.

31. The group includes Godfrey of Fontaines, Richard of Middleton, Ramon Lull, Johannes Bassolis, William Ockham, Walter Burley, Robert Holkot, William of Ware, Gaietanus de Thienis, Nicole Oresme, and Thomas of Strasbourg. As we see below, while John Buridan and Albert of Saxony accepted as plausible one of the major departures from Aristotle and both conceded the supernatural possibility of a plurality of worlds, they rejected most, if not all, of the other alternatives to Aristotle's defense of a single world.

32. John Major seems an extraordinary exception. In his *Propositum de infinito*, published in 1506, he first brought the question of a plurality of worlds to a stalemate, arguing that, "naturally speaking" *(naturaliter loquendo)* one could no more prove the existence of a unique world than the existence of an infinite number, as Democritus would have it. But he went on to proclaim that "I believe, as the master Democritus thought, that there are an infinite number of excentric and perhaps concentric worlds." See Dick (n. 20 above), 100; the translation is Dick's. Major's discussion appears in *Le Traité "De l'infini" de Jean Mair*, edited and translated by Hubert Elie, 60–62; see also 56–58, and 114 (Paris, 1938).

saw above, the absurd consequence that a particle of earth would move naturally up in one world and naturally down in another and thereby violate Aristotle's fundamental principle that an element can possess only one natural motion. The basic response, on which numerous elaborations would be made, was already formulated by Richard of Middleton sometime around 1300. In commenting on book 1, distinction 44, of Peter Lombard's *Sentences,* which was concerned with the problem of whether God could make something better than He had made it, a problem that frequently prompted discussion of a plurality of worlds, Richard maintained that even if all the worlds that God might create were identical, as Aristotle had assumed, it would not follow that the earth and any of its parts would have a natural inclination to move upward in one world toward the center of another. Indeed, each earth would remain at rest in the center of its own world and any parts of it that might be removed would, if unimpeded, always tend to return to the whole of it at the center. Moreover, if it were possible to remove the earth of another world and place it at the center of ours, that earth would remain at rest in the center of our world; and, conversely, if our earth were removed to the center of another world, it would remain at rest there with no inclination to move toward its former place. All this, Richard concludes, is also "the opinion of Lord Stephen, bishop of Paris and doctor of sacred theology, who excommunicated those who dogmatized that God cannot make more worlds."[33] Contrary to Aristotle, Richard of Middleton, and many others subsequently, conceived each world as a self-contained, closed system having its own proper center and circumference. It followed that if God should create more than one world, no unique and privileged center could exist.

During the fourteenth century, further elaborations and refinements were made. William Ockham, for example, in commenting on the same distinction of the *Sentences,* derived the possibility of other worlds from

33. Richard of Middleton, *Super quatuor libros sententiarum Petri Lombardi Quaestiones subtilissimae,* 4 vols. (Brescia, 1591; reprint Frankfurt a.M., 1963), 1.392, col. 2 (bk. 1, dist. 44, art 1, q. 4: "Utrum Deus posset facere aliud universum"). Although he makes no mention of Article 34, Thomas of Strasbourg repeated and approved much the same argument, adding only that each earth would rest in its own world naturally, not violently. God's absolute power must not, he insisted, be limited by sophisms and Aristotelian arguments. See Thomas of Strasbourg, . . . *Commentaria in IIII libros sententiarum* (Venice, 1564; reprint 1965 by the Gregg Press, Ridgewood, N.J.), fols. 117v–118r, bk. 1, dist. 44, art. 4; and also Duhem (n. 1 above), 9.385–87.

God's ability to produce "an infinite number of individuals of the same species as those which now exist." Since "He is not limited to producing them in this world, therefore He could produce them outside this world and make a world of them just as He made this world from those things which He produced here now."[34] In each possible world that God could create to contain some, or all, of these additional individuals, elements such as fire and earth would move only within their own worlds. Their behavior would be analogous to that of two fires, one moving toward the circumference of the heaven over Oxford, the other moving toward heaven over Paris. If these masses of fire were switched, the fire now over Paris, but formerly over Oxford, would move directly upward toward the part of the celestial circumference over Paris, with no inclination to move back over Oxford.[35]

In his *Questions on De caelo,* John Buridan characterized as "nondemonstrative" Aristotle's defense of the claim that heavy bodies in other worlds would move toward the center of ours. This would be true, countered Buridan, only if the inclinations of heavy bodies depended solely on their common tendency to move toward a single center. But motions are also dependent on celestial bodies and God. Since every world would have its own celestial bodies and God's presence and control would be equal in all, the heavy earthy bodies of a particular world would fall only to the center of their own world.[36]

---

34. Translation by Dick (n. 20 above), 91, from William Ockham, *Opera plurima,* 4 vols. (Lyons, 1494–1496; reprint in facsimile London, 1962), vol. 3: *Super 4 libros sententiarum,* bk. 1, dist. 44, sig. bbv, verso (E) (unfoliated).

35. Dick, 92, and Ockham (F).

36. See Buridan (n. 15 above), 86–87, bk. 1, q. 18. In qq. 18 and 19, where he considered a plurality of worlds, Buridan emphasized that God could, if he wished, create other worlds, both similar and dissimilar. Indeed, in an obvious reference to Article 34 of the condemnation, Buridan asserts that "we hold on faith that just as God made this world, so could he also make another, or others" (84; see also 89 for the same sentiment). Although compelled to concede God's power to create other worlds, Buridan preferred to believe that if God wished to create additional creatures such as appear in our world, he would simply enlarge our world to double, or one hundred times, its present size. See Buridan, *Questions on the Physics,* bk. 3, q. 15, in John Buridan, *Questiones super octo Phisicorum libros Aristotelis diligenter recognite et revise magistro Johanne Dullaert de Gandavo* (Paris, 1509; reprint in facsimile under the title *Johannes Buridanus, Kommentar zur Aristotelischen Physik,* Frankfurt a.M., 1964), fol. 57v, col. 2.

In a similar manner, Albert of Saxony, like Buridan only a master of arts and not a theologian, conceded that supernaturally many worlds were possible, whether simultaneous or successive, eccentric or concentric. He also supported the opinion that if other worlds like ours existed, the earth of each would remain naturally at rest. But he too refused to believe that

A significant departure from Aristotelian cosmology that emerged ˙ from the assumption of a plurality of worlds was formulated by Nicole Oresme in a French commentary on Aristotle's *De caelo* written in 1377. Here, in denying the movement of heavy bodies to a unique center,[37] Oresme redefined the meanings of "up" and "down" in terms of their relationships to "light" and "heavy," respectively. Abandoning Aristotle's absolute sense of up, down, light, and heavy, Oresme argued that a body is to be judged "heavy," and may be said to be "down," when it is surrounded by light bodies, where the surrounding "light" bodies are said to be "up." Thus heavy and light, with the associated, and interrelated, concepts of down and up, could be conceived independently of the natural places of bodies,[38] as Oresme illustrates by reference to an earlier example in which this independence of natural place was graphically demonstrated. After imagining that a tile, or copper, pipe extends from the center of the earth to the heavens, presumably to the concave surface of the lunar sphere, Oresme argues that "if this tile were filled with fire except for a small amount of air at the very top, this air would drop down to the center of the earth for the reason that the less light body always descends beneath the lighter body. And if this tile were full of water save for a small quantity of air near the center of the earth, this air would mount up to the heavens, because by nature air always moves upward in water."[39] In this context, earth is heavy and down because it comes to rest naturally in the center of the lighter bodies that surround it. From this Oresme infers

---

any such worlds could exist naturally beyond ours. See his *Questions on De caelo,* bk. 1, q. 11 ("Utrum sint vel possint esse plures mundi"), in *Questiones et decisiones physicales . . . Alberti de Saxonia in octo libros Physicorum; tres libros De celo et mundo . . . recognitae . . . Georgii Lokert Scotia* (Paris, 1518), fol. 95, col. 2.

37. *Nicole Oresme: Le Livre du ciel et du monde,* edited by Albert D. Menut and Alexander J. Denomy, translated by Albert D. Menut, 174, 175 (Madison, 1968).

38. After distinguishing two senses of "up" and "down" (one with regard to us, as when we say that half the heavens lie up, above us, and the other half down, below us; and the other used with respect to heavy and light bodies, the latter sense being Oresme's sole concern here), Oresme declares that "up and down in this second usage indicate nothing more than the natural law concerning heavy and light bodies, which is that all the heavy bodies so far as possible are located in the middle of the light bodies without setting up for them any other motionless or natural place"; ibid., 173.

39. Ibid., 71. Although Oresme makes certain qualifications where he first formulates this illustration, they seem not to apply in the context of our discussion where the example is intended only to show "how a portion of air could rise up naturally from the center of the earth to the heavens and could descend naturally from the heavens to the center of the earth" (ibid., 173).

further that if an earthy, or heavy, body were not surrounded by lighter bodies, that heavy body could not be said to be up or down, from which he concludes that if a vacuum existed between our world and any other, a particle of earth from that world could not possibly move to the center of ours. For even if it could rise up and depart beyond the circumference of its own world, it would enter the void between worlds and come to rest, since lighter bodies would no longer surround it and all directionality would have vanished. Thus did Oresme conclude that "if God in his infinite power created a portion of earth and set it in the heavens where the stars are or beyond the heavens, this earth would have no tendency whatsoever to be moved toward the center of our world."[40]

Thus did Nicole Oresme propose a new explanation of directionality which was utterly opposed to Aristotle's. Despite a conviction that "there never has been nor will there be more than one corporeal world,"[41] Oresme was ultimately motivated by Article 34 of the Condemnation of 1277, which he reiterated when he declared that "God can and could in his omnipotence make another world besides this one or several like or unlike it."[42]

In turning to the second major argument against a plurality of worlds, namely, the presumed impossibility of void space between distinct and simultaneously existing worlds, we find that in his discussion Oresme simply assumed the existence of intercosmic void space. Prior to 1277, as we have already seen, discussants of the plurality of worlds question, such as Michael Scot, William of Auvergne, and Roger Bacon, had rejected the existence of other worlds by virtue of their conviction that intervening void space was a necessary consequence of plurality. After all, Aristotle had shown in the fourth book of his *Physics* that void space was impossible; therefore, they concluded that it could no more exist beyond the world than within.[43]

---

40. Ibid., 173. Since the ethereal substance of stars was conceived as neither light nor heavy and offered no resistance to bodies, Oresme could equate stars and void. It is of interest that Oresme cites Aristotle's conclusion that in a vacuum there is no up or down so that a heavy body could not move itself in a vacuum. For Aristotle, however, the lack of directionality in a vacuum was a direct consequence of its homogeneity, whereas for Oresme it resulted from the absence of surrounding light bodies.

41. Ibid., 179.

42. Ibid., 177–79.

43. Although not cited in the pre-1277 arguments we have described here, Aristotle had also denied that body or void could exist beyond the world (*De caelo* 1.9.279a.12–18).

With the proclamation of Article 34, however, not only was it necessary to admit the possibility that God could create other worlds, but that very admission, coupled with Aristotle's definition of vacuum as a place deprived of body but capable of receiving body,[44] implied the existence of something beyond the world, either body or void. If body existed where God might create a world, it would follow that upon creation of a world there, two bodies would coexist simultaneously in the same place, a natural, but not a supernatural, impossibility. In this alternative, one had only to invoke Article 141 of the Condemnation of 1277, which made it necessary to concede that God could create several dimensions, or bodies, in the same place simultaneously.[45] Should the region beyond our world be devoid of body, then the mere possibility that a world, or body, could be created there implied, by Aristotle's very definition, that a void space existed there.

Robert Holkot appears to have seen these implications clearly, although he makes no reference to condemned articles. In a significant argument, he declared that if God could make another world, He could create it anywhere. Holkot then inquires whether there is anything now in the place where God could create that world. If there is, it must be something, presumably a body, or nothing. If something exists there, then a body exists beyond the world. If nothing, then Holkot argues as follows: "Beyond the world nothing exists; but beyond the world a body can exist [since God can create a world there]; therefore a vacuum exists beyond the world because a vacuum exists where a body can exist, but does not. Therefore a vacuum is there now."[46] For Holkot, the existence

---

44. For this definition, see Aristotle, *De caelo* 1.9.279a.14–15. In *Physics* 4.1.208b.25, 4.7.213b.32, and 4.8.214b.18–19, Aristotle defines vacuum simply as a place without a body. As Dick observes (n. 20 above), 76–77, Aristotle did not use the possible existence of intercosmic void space as an argument against a plurality of worlds. Such an argument would have been circular, since "one of the proofs of the nonexistence of void was that there could be no body outside the world."

45. For the text and translation of Article 141, see below.

46. Because of its importance, I cite the whole of Holkot's brief argument: "Praeterea, si Deus posset modo facere alium mundum ab isto, posset facere ilium esse alicubi, sicut iste est modo, ita quod partes illius mundi distarent abinvicem extra mundum istum. Quero ergo quid est ibi modo: an aliquid an nihil. Si aliquid, ergo extra mundum de facto est aliquid. Si nihil, tunc arguitur sic: extra mundum nihil est, et extra mundum potest esse corpus; ergo extra mundum est vacuum, quia ubi potest esse corpus, et nullum est, ibi vacuum est. Ergo vacuum modo est"; Robert Holkot, *In quatuor libros sententiarum Quaestiones* (Lyon, 1518; reprint in facsimile Frankfurt a.M., 1967), bk. 2, q. 2, sig.bii, recto, col. 2 (no foliation).

of a body or void beyond the world was not merely hypothetical, but real. The mere possibility that God could create another world, as required by Article 34, enabled him to infer these significant alternatives, one of which must be true, with extracosmic void as the more likely candidate. Perhaps it is a measure of the degree to which opinions and attitudes had changed since the thirteenth century that Walter Burley could declare that Christian theologians, and those generally who believed in the creation of the world, could hardly avoid the conclusion that a vacuum existed beyond our world since they also conceded that God could create another world.[47]

In the context of a discussion on a supernaturally created plurality of worlds, Nicole Oresme would also proclaim the reality of extracosmic void space, when he insisted that "the human mind consents naturally, as it were, to the idea that beyond the heavens and outside the world, which is not infinite, there exists some space whatever it may be, and we cannot easily conceive the contrary."[48]

When it was realized that an extracosmic void space was a plausible consequence of a plurality of supernaturally created worlds, inquiries were made as to its nature. Was it possible, for example, to measure distances in such a space?[49] If a body were located beyond our world, could

47. "Difficile tamen ut mini videtur est vitare quin loquentes nostre legis et generantes mundum habeant ponere vacuum extra mundum quia ipsi dicunt quod sicut Deus creavit hunc mundum ita posset creare alium mundum"; Walter Burley, *Super octo libros Phisicorum* (Venice, 1501; reprint in facsimile with the title *In Physicam Aristotelis Expositio et quaestiones* [Hildesheim, 1972]), fol. 89r, col. 1. Burley indicates that he will consider the problem of extracosmic void further in his commentary on Aristotle's first book of *De caelo*, which I have not seen. It is perhaps worth reporting that, in the fifteenth century, Gaietanus (or Caietanus) de Thienis, who summarized Burley's opinion and mentioned the latter by name, denied that Christians must concede a vacuum beyond the world ("nec oportet Christianos concedere vacuum extra celum").

See Gaietanus, *Recollecte . . . super octo libros Physicorum cum annotationibus textuum* (Venice, 1496), bk. 4, fol. 28v, col. 2. Even before Gaietanus, Albert of Saxony refused to concede that any space or vacuum existed beyond the world even though he admitted that God could create other worlds (see his *Questions on De caelo*, bk. 1, q. 9 [n. 36 above], fol. 93v, col. 2).

48. Oresme (n. 37 above), 177. Although, as Oresme explains, "we cannot comprehend nor conceive this incorporeal space which exists beyond the heavens," its existence is confirmed by reason and truth ("rayson et verite nous fait congnoistre que elle est").

49. For a fuller discussion and references, see Edward Grant, "Place and Space in Medieval Physical Thought," in *Motion and Time, Space and Matter: Interrelations in the History of Philosophy and Science,* edited by Peter K. Machamer and Robert G. Turnbull, 147–48, 151–52 (Columbus, 1976); also see n. 128 below.

its distance to the convex surface of the outermost sphere of our world be measured? Reared in an intellectual tradition in which distances were always measured over intervening material dimensions, which characterized the Aristotelian plenum of our cosmos, most Scholastics denied the possibility of measurements in empty space. For them, a stone located anywhere beyond the world in a void space would immediately come into contact with the outermost convex surface of our world. Only if God created intervening bodies, insisted Marsilius of Inghen, could the stone lie at a measurable distance from that surface. Indeed even those other worlds would come into contact at a single point with our world and with others.

But efforts were made to confer intelligibility on a sense of measurement in void space. Henry of Ghent, for example, assumed extracosmic vacuum to be three-dimensional and therefore able to function as if it were an intervening body in a plenum. Jean de Ripa[50] utilized a series of paradoxes to demonstrate that if the matter between two bodies was destroyed, the distance previously separating them would remain the same even in the imaginary space, or vacuum, that now separated them.[51] And, finally, William of Ware (Guilelmus Varonis) insisted that God could create two distinct, spherical worlds with an intervening distance, just as one part of a heaven in our world is separated from part of another heaven also in our world. Indeed, William assumes that prior to the creation of our world absolutely nothing existed, a situation that could be imagined as an infinite space with nothing in it. Within that infinite space, God created our world and could add as many more as He pleases, even to infinity.[52]

Thus it was that the absolute power of God to make as many worlds as He pleased raised physical problems that evoked interesting solutions

50. On Henry of Ghent and Jean de Ripa, see ibid.

51. Gaietanus de Thienis denied that two worlds could be materially distant from each other—that is, have a body between them (no body exists there)—or that they could be separated by a vacuum, the existence of which he denied. But they could be separated formally (formaliter). His opinion was in reaction to that of Walter Burley, who, according to Gaietanus, believed that two distinct worlds would indeed be separated by a divisible void space receptive of body, but in which there was no body. See Gaietanus (n. 47 above), fol. 28v, col. 2.

52. Duhem (n. 1 above), 9.381–82. Duhem's discussion and French translations of William of Ware's ideas are based on the latter's commentary on the *Sentences*, bk. 2, q. 8 ("Quaeritur utrum Deus posset facere alium mundum simul cum isto").

most of which conflicted with, or were alien to, the principles of Aristotelian physics and cosmology, or involved the adaptation of an Aristotelian principle, or principles, to situations and conditions that Aristotle had never seriously contemplated and that he would probably have considered absurd.

Article 49 produced an even greater range of interesting responses relevant to the history of medieval physical thought. The contexts of the subsequent discussions of Article 49—that is, discussions in which God is assumed to move the outermost sphere of the heaven *(ultimum celum)*, or the world itself *(mundus)*[53]—indicate that as many as four major reasons may have induced the bishop of Paris to condemn it. Two of these are mentioned by Richard of Middleton in the second book of his commentary on the *Sentences,* written only eight or nine years after issuance of the condemnation.[54]

In considering the question "whether God could move the last heaven rectilinearly," Richard presents two arguments as to why God could not do this, arguments that were perhaps standard responses prior to 1277. The first of these concerns the Aristotelian principle that every rectilinear motion is necessarily from place to place. Since Aristotle had shown that the last heaven of the world is not in a place and had further claimed that no body, and therefore no place, could exist beyond the world, it followed that without a place to depart from and a place to arrive at, not even God could move the last heaven, or world, rectilinearly.[55] The second argu-

---

53. The text of Article 49 states: "Quod Deus non possit movere celum motu recto. Et ratio est, quia tunc relinqueret vacuum"; Denifle and Chatelain (n. 1 above), 1.546. Although, as will be seen below, Richard of Middleton interpreted "celum" as "heaven," and assumed that the outermost sphere was intended, the term "celum" was also taken to represent the world, or "mundus." Indeed, the latter term was often substituted for "celum," as when Buridan, in a specific reference to the error condemned by the bishop of Paris, declared: "Sed de potentia divina determinatum fuit per episcopum Parisiensem et per studium Parisiense, quod error esset dicere quod Deus non posset movere totum mundum simul motu recto"; (n. 15 above), 75, bk. 1, q. 16. Nicole Oresme (see below), and many others, would make the same substitution.

54. A. B. Emden, *A Biographical Register of the University of Oxford to A.D. 1500,* 3 vols. (Oxford, 1957–1959), 2.1254, dates the composition of the second book of Richard's *Sentences* commentary between 1285 and 1286. Richard studied at the University of Paris, but whether he was French or English is uncertain (pp. 1253–54).

55. Richard explains that God cannot move the world "quia omnis motus corporis rectus est de loco ad locum. Sed secundum Philosophum 4. *Physicorum* ultimum celum non est in loco; secundum etiam eundem primo *Caeli et Mundi,* extra ultimum celum non est locus, neque plenitudo, neque vacuitas. Ergo impossibile est Deum movere ultimum caelum motu

ment, since it conforms to the literal sense of the text, was perhaps the most important reason for the condemnation of Article 49. It denied to God the ability to move the world with a rectilinear motion because a vacuum would be left behind, which clearly implied that although God might possess the power to move the world with a translatory motion, He would be prevented from doing so by nature's abhorrence of a vacuum, which Aristotle had demonstrated in a variety of ways. Therefore, the world must of necessity remain where it is to prevent formation of the dreaded vacuum, which not even God could create.[56]

A third relevant reason for the condemnation of Article 49 may have derived from Averroes, who argued that the existence of an absolutely motionless body was a necessary precondition for motion. As an immobile body incapable of translatory motion, the world itself served this essential function, a situation which, presumably, not even God could alter. To counter this restriction of God's absolute power, and deprive the Averroists of their absolutely immobile body, the bishop of Paris and his theologians forced the concession that God could, if He wished, move the last heaven, or world, rectilinearly.[57]

Thomas Bradwardine, in a work of 1344, suggests yet a fourth reason when he explained that those who follow Aristotle's arguments in the first book of De caelo "assume that every local motion is necessarily upward, downward, or circular—that is, away from the center [of the world], toward the center, or around the center." Since the rectilinear motion of the world qualifies as none of these, "they say that it is impossible for the world to be moved."[58]

Compelled after 1277 to concede that God could move the last heav-

---

recto"; Richard of Middleton (n. 33 above), 2.186, col. 1, bk. 2, dist. 14, art. 3, q. 3 ("Utrum Deus posset movere ultimum caelum motu recto").

56. "Item Deus non posset facere vacuum . . . . Sed si Deus posset movere celum ultimum motu recto posset facere vacuum, quia si moveretur motu recto, aliqua pars eius recederet a loco in quo est; nec succederet in ilium locum aliud corpus. Ergo Deus non posset ultimum celum movere motu recto"; ibid.

57. This plausible reason was presented, without supporting evidence, by Max Jammer, Concepts of Space: The History of Theories of Space in Physics, 2nd ed. (Cambridge, Mass. 1969), 60. As we shall see below, the necessity of an immobile reference body for the occurrence of motion was challenged within the context of discussions on Article 49.

58. Thomas Bradwardine, De causa Dei contra Pelagium et De virtute causarum . . . (London, 1618), 177 (C–D). Cited from my own translation of Bradwardine's discussion of infinite void space in Grant (n. 1 above), 557.

en, or whole world, rectilinearly, Scholastics had to reformulate their responses to the arguments described above and devise new interpretations to meet the altered circumstances. How this was done, we must now describe.

In the aftermath of the condemnation, Scholastics had to determine how it was possible for a material entity like the last heaven, or the world itself, to be moved rectilinearly without occupying successive places. Aristotle had defined the place of a body as "the [innermost] boundary [or surface] of the containing body at which it is in contact with the contained body."[59] Since Aristotle's cosmos is a material plenum without vacua, every body within that world was assumed to have a place because it was surrounded by a body or bodies that differed from it. In this plenum, the rectilinear motion of a body from place to place always involved departure from the innermost surface of one surrounding body to the innermost surface of another surrounding body. Local motion was thus a series of successive abandonments and acquisitions of different material containing surfaces, or places, until the body came to rest in its natural place.[60] Since every part of the world was said to be in a place, it was common to say that the world was in a place accidentally by virtue of all its parts. But most admitted, with Aristotle, that the last, or outermost, sphere was not itself in a place because no body existed beyond the world to serve as its container, or place. Indeed, no places could exist beyond the world because no bodies were there to constitute those places. Since motion was from place to place, it seemed to follow that motion of the last sphere, or the world, was impossible. And yet, Article 49 demanded the concession that God could, if he wished, move that last sphere, or the world itself, with a rectilinear motion. But how could the world be moved if there were no places to receive it and from which it could depart?

It was undoubtedly with all this in mind that Walter Burley distinguished two kinds of worlds that God could have created. The first is our kind, namely, a heterogeneous, or discontinuous, plenum with different bodies each forming a distinct part with its own place. In this type, those

59. Aristotle, *Physics* 4.4.212a.5–6. The translation is that of R. P. Hardie and R. K. Gaye in the Oxford English translations of Aristotle. I have added the bracketed qualifications. Aristotle assumed that the place, or container, is distinct and separate from the thing in place.

60. Because they are irrelevant to our purposes, I shall ignore the numerous paradoxes and difficulties associated with Aristotle's concept of place.

who believe, as some Arabs did, that the world is in a place *per se,* and not accidentally by reason of its parts, are committed to a precreation vacuum into which the world was placed at creation, since there is no body beyond the world that can function as a true place.[61] But Christian theologians *(loquentes nostre legis)* who assume that the world is not in a place *per se,* but only accidentally by virtue of its parts, do not assume a precreation vacuum for the world, but rather assume that God created the place of the world simultaneously with the world.[62]

These responses depend on the creation of a heterogeneous, or discontinuous, world. But what if God had created a wholly continuous, spherical world without distinct parts—that is, what if God had created a completely homogeneous universe? Such a world could have no distinct and different parts and therefore could not be in place by its parts; and since it has no external material container, the whole of it cannot be in a place. But every body must be in some place. Therefore, since this world is a single, homogeneous body, it too must be in a place. The only place it could have, Burley concludes, is a vacuum.[63]

But what if God should now move this homogeneous and continuous world in a straight line? Without external places, Burley concludes that God would first have to create a new place to which the body is moved. Thus Burley allows that God could move the world with a rectilinear motion, but not before creating an external place, or places.[64]

61. "Sic igitur apparet quod ponentes totum mundum esse per se in loco et etiam de novo generari habent ponere vacuum"; Burley (n. 47 above), fol 89r, col. 1.

62. "Sed recte philosophantes(?) et loquentes nostre legis qui ponunt totum mundum non esse per se in loco, sed solum per partes, vel per accidens, non habent ponere vacuum quia non ponunt locum precedere generationem mundi, sed dicunt locum simul generari cum mundo"; ibid.

63. The passage now quoted follows immediately after the text in the preceding note: "Sed dubitatur quia ponentes mundum de novo generari habent dicere quod sicut Deus creavit mundum discontinuum in partibus propter quarum discontinuationem partes mundi sunt per se in loco. Ita Deus potuit creasse unum corpus continuum omnino in omnibus partibus ita quod nihil aliud creasset quam illud rotundum continuum. Ponamus igitur quod Deus, quando creavit istum mundum, creasset loco istius mundi unum corpus rotundum omnino continuum; et cum omne corpus sit in loco, illud corpus rotundum etiam fuisset in loco et non per partes quia nulla pars esset in loco, cum locus sit continens divisum et illud corpus est omnino continuum. Ergo relinquitur quod illud corpus sit in vacuo"; ibid.

64. "Dicendum quod ponendo tale corpus continuum et nihil extra illud continuum, Deus non posset illud corpus movere motu recto nisi crearet locum novum ad quem moveretur"; ibid. Making no distinction between types of worlds, Richard of Middleton (n. 33 above), 2.186, who specifically mentions the condemnation of Article 49 by the bishop of Paris, insisted

Although none would have dared deny that, if needed, God could create those places beyond the world, a number of Scholastics were unconvinced of their necessity. Indeed, they were of the opinion that place was in no way essential for motion. Article 49 and God's absolute power provided the natural impossibilities that enabled them to imagine the conditions that would make motion without place possible. In a major discussion, John Buridan[65] assumed, as Burley had, that God created a homogeneous universe in which the world is one continuous body, the potential parts of which are identical. Under these circumstances, as we saw, places as defined by Aristotle could not exist, so that no places would exist inside or outside the world. And yet, if He wished, God could move the world circularly and produce a motion without the existence of places.[66] That God could move such a world circularly is inferred from Article 49, which Buridan cites as "an article condemned at Paris." For if God can move the world rectilinearly, He surely can move it circularly,[67] and do this whether it be continuous or discontinuous. But if it is continuous, the circular motion of the last sphere would not depend on any changing relationship to the earth at its center, or to any other body. For on the assumption that all things in the world share the circular motion of the last sphere and are moved simultaneously with it, all relationships between any parts of the world would remain constant and no relative motions would be detectable. And yet a motion takes place since God is assumed to move the whole world circularly.[68]

As with circular motion, Buridan argues that rectilinear motions are also conceivable without relationships to other bodies. Here again it is

---

that God could not move the whole last heaven, which he identifies with the Empyrean sphere, unless he also created an external space for it. Nor could God move an angel, who is assumed to be the sole existing creature in the world, unless he also created for that angel a surrounding external space.

65. Buridan (n. 36 above), fols. 50r–51r, considers the problem in bk. 3, q. 7 ("Utrum motus localis est res distincta a loco et ab eo quod localiter movetur"). Occasionally emending the 1509 edition from manuscript sources, Anneliese Maier has analyzed this question in *Zwischen Philosophie und Mechanik: Studien zur Naturphilosophie der Spätscholastik* (Rome, 1958), 121–31.

66. Nicholas Bonetus, a fourteenth-century Franciscan, agreed with this. See Pierre Duhem, "Le temps et le mouvement selon les scholastiques," *Revue de philosophie* 23 (1913): 459–60.

67. Buridan (n. 36 above), fol. 50v, col. 1; for the Latin text, see Maier, *Zwischen Philosophie und Mechanik*, 122.

68. Buridan (note 36, above); Maier, *Zwischen Philosophie und Mechanik*, 124–25, for the Latin text.

Article 49 that provides the basic illustration. For if God moved this continuous world rectilinearly, the outermost sphere would not change its
relative positions to the earth resting at the center. And since it is assumed that nothing lies outside the world, it follows that the world's rectilinear motion is absolute and independent of any relationships to an
immobile body. The independence of the world's motion is emphasized
in yet another way. If, in our continuous and homogeneous world, we
now assume the earth's rotation with respect to the heavens, the latter
would bear different relationships to the earth. But from this one cannot
properly infer a rectilinear motion of the world. Thus, in the one case, assumption of the world's rectilinear motion does not affect the relative position between the last heaven, or sphere, and the earth; and, conversely,
different positional relationships between the last heaven and the earth
tell us nothing as to whether the world is actually in rectilinear motion.
The world's supernatural motion was thus conceivable without reference
to any other body.

Some years later, Nicole Oresme agreed that the local motion of a
body from place to place was not determined by its relationship to external bodies, whether immobile or mobile. The supernatural motion of
the world provided an indubitable illustration of an absolute motion, one
which, for Oresme, occurred in an imaginary, infinite space beyond our
world, a space in which our unique world is moved as if it were a single body. For Oresme it was no contradiction to declare "that the whole
world moves in this space with a rectilinear motion. To say the contrary
is to maintain an article condemned in Paris. With this assumption,
no other body exists with which the world could vary with respect to
place."[69] After further demonstrating the independence of local motion
from change of position relative to another body, Oresme declares that
for a body "to be moved with respect to place is for it to bear different
relationships with respect to the imagined immobile space, for it is with
regard to this space that the speed of the motion and of its parts are measured."[70] Thus did Article 49 lead Nicole Oresme to view the change of

---

69. Oresme (n. 37 above), 367, 369. I have altered Menut's translation.

70. Ibid., 373. I have again changed the first part of Menut's translation. Perhaps it was with
Oresme in mind that Marsilius of Inghen reported the opinion of those who "posit that there
is place outside of the heavens, or an infinite space. Therefore, if God were to move the whole
world rectilinearly or circularly, the world would be differently disposed with respect to the

position of a single body in the universe against the backdrop of an absolute, infinite empty space, the existence of which for him was not hypothetical but real.

But how was an absolute rectilinear motion of the whole world to be classified? As we saw earlier, such a motion could not be identified with any of the natural motions within the world, for which reason Aristotelians had denied the possibility of it and, according to Thomas Bradwardine, evoked the wrath of the bishop of Paris sufficiently to bring on the condemnation of Article 49.[71] It was undoubtedly with this problem in mind that Gaietanus de Thienis, in the fifteenth century, inferred that if God did move the world rectilinearly, as indeed he could, its motion would not be classifiable as either up or down. Rather it would belong to another species of motion, which is left unspecified and undiscussed.[72]

It was almost inevitable that an idea such as the one that God could move the world rectilinearly should have posed difficult, and even unanswerable, questions about motion and place in the context of Aristotelian natural philosophy. Its most significant impact, however, was as further reinforcement of the concept of extracosmic void which, as we have seen, was for some an obvious consequence of Article 34 on the plurality of worlds. The very text of Article 49 speaks of a void being left behind if God moved the world. Since, after 1277, this consequence of the world's motion could not be invoked to deter God from moving the world if He pleased, it followed that if God moved the world, a vacuum would remain. Thus the world could be conceived as located in a vacuum.

But as the world moves from its old place, what did it move into? The obvious response was that it moved into other void places, which had existed outside the world prior to its motion. Since no good reason could be offered for a finite termination of extracosmic void, its infinitude was easily inferred, though hardly necessary, since God could just as easily

---

place or separate space in which it would be. Therefore, they concede that the world moved locally [in the fashion posited in the case] is disposed differently than before with respect to that which is nothing [i.e., separate space]"; translated by Marshall Clagett from Marsilius's *Questions on the Physics*, in Clagett, *The Science of Mechanics in the Middle Ages* (Madison, 1959), 623. The additions in brackets were added by Clagett. Although Marsilius did not reject this interpretation, he preferred another, described earlier in his treatise.

71. Bradwardine (n. 58 above), 177 (C–D), and Grant (n. 1 above), 557.

72. "Possumus concedere Deum posse movere mundum motu recto nullum locum creando et quod talis motus rectus non esset sursum, nec esset deorsum, sed alterius speciei"; Gaietanus (n. 47 above), fol. 28v, col. 2.

create a finite space beyond the world.[73] Around 1344, Jean de Ripa used Article 49 to establish the plausibility of extracosmic void. He achieved this by extending the notion of place from our material cosmic plenum to void space. He argued that since all positive places or spaces existed only within the material plenum of our world, it was necessary to assume the existence of imaginary, or void, places or spaces beyond our world in which bodies or angels could be received. For if only positive places existed, none of which could exist outside the world, God could not move the world with a rectilinear motion, since there would be no place or places into which the world could be moved. But this would place a restriction on God's power to move the world and was indeed condemned.[74] It was therefore necessary to concede that imaginary void places could also exist which were capable of receiving the whole world, or any part thereof. Although they did not formulate specific arguments of this kind, Bradwardine and Oresme had implicitly assumed much the same position.[75]

In Articles 139, 140, and 141, the bishop of Paris struck at some of the most basic ideas in Aristotelian natural philosophy. Not only did these articles condemn the seemingly self-evident principle that an accident could not exist without a subject in which to inhere and also condemn

73. Although Buridan did not himself accept the existence of extracosmic void space, he reports that those who held this opinion believe that if such a space did exist it would, by the principle of sufficient reason, be infinite. For why should it be of one size rather than another greater size? Whatever finite size is assumed for this space, we may always properly inquire whether another space lies beyond. Thus, "if it could be shown that there is a space there, it ought to be conceded that it is infinite." For the Latin text, see Buridan (n. 36 above), fol. 57r col. 2, bk. 3, q. 15. In reply, Buridan emphasizes that it is not necessary that such a space be infinite, since by his supernatural power, God could, if it pleased him, create only a finite space beyond the world. It was Buridan's opinion, however, that God probably did not create any space beyond the world (fol. 58r, col. 2; see also fol. 57v, col. 2).

74. "Jean de Ripa I Sent. dist. XXXVII: De modo inexistendi divine essentie in omnibus creaturis," edited by André Combes and Francis Ruello, introduction by Paul Vignaux, *Traditio* 23 (1967): 232, lines 66–68; the argument is repeated on 234, lines 6–9. De Ripa also argued for the existence of imaginary space by appeal to angels. If an angel existed alone with everything else corrupted, including all positive places, that angel would be nowhere and unable to change positions. Since this is patently absurd, de Ripa infers the existence of imaginary places and spaces wholly independent of material bodies and media. See 232, lines 60–66, and 234, lines 6–8.

75. That Bradwardine and Oresme believed that extracosmic space was necessarily infinite and real derives from their belief that infinite space is God's infinite immensity and therefore in no way created. For de Ripa, God's creation of an actual infinite space would be an instantiation of His general power to create actual infinites. To deny God the power to create an actual infinite would be a restriction of His absolute power. See Grant (n. 49 above), 149–50.

the widely held belief that not even God could create an accident without a subject, but they also censured the Aristotelian axiom that no quantity or dimension could exist independently of a material body—for this would make it a substance—and the equally basic principle that two or more dimensions could not exist simultaneously in the same place.[76]

It was indeed with these articles and their implications that Jean Buridan concerned himself in a question devoted to the possibility that a certain power might create a vacuum.[77] Painfully aware of the theological ramifications of the question, Buridan provides a rare insight into the strained relations between theology and philosophy. Admitting that masters from the arts faculty were sworn not to discuss any purely theological questions,[78] and readily conceding that the question of the existence of a vacuum touched faith and theology, Buridan insists that he be allowed to pursue the theological aspects of the question or be guilty of perjury and evasion.[79] Adopting a position by then conventional, he admits that God can create an accident without a subject, or that He can separate, and independently conserve, accidents from their subjects. As a special case, Buridan concedes that God could create a separate dimension independently of any substance or other accidents. Moreover, God could also create several bodies in the same precise place and cause the interpenetration of dimensions by creating a separate, three-dimensional void space that could receive natural bodies.[80]

---

76. "139. That an accident existing without a subject is not an accident, except equivocally; and that it is impossible that a quantity or dimension exist *per se,* since this would make it a substance." ("Quod accidens existens sine subjecto non est accidens, nisi equivoce; et quod impossibile est quantitatem sive dimensionem esse per se; hoc enim esset ipsam esse substantiam.")

"140. That to make an accident exist without a subject is an impossible argument implying a contradiction." ("Quod facere accidens esse sine subjecto, habet rationem impossibilis, implicantis contradictionem.")

"141. That God cannot make an accident exist without a subject, nor make several dimensions exist simultaneously [in the same place]." ("Quod Deus non potest facere accidens esse sine subjecto, nec plures dimensiones simul esse.") See Denifle and Chatelain (n. 1 above), 1.551.

77. Buridan (n. 36 above), fols. 73v–74r, bk. 4, q. 8 ("Queritur octavo utrum possibile est vacuum esse per aliquam potentiam").

78. The reference is probably to the statute of 1272. For the Latin text, see Denifle and Chatelain (n. 1 above), 1.499–500; the English translation by Lynn Thorndike appears in his *University Records and Life in the Middle Ages* (New York, 1944), 85–86.

79. For my translation of Buridan's remarks, see Grant (n. 1 above), 50–51.

80. After declaring his belief that the divine power can create the two different types of vacuum imagined in the preceding question, Buridan says that this is not proved by natural

But there is a second kind of vacuum, one that results from the annihilation of all matter between the surfaces of the containing body that serves as the place of the annihilated body. In order that a vacuum result, the bounding surfaces of the containing body, or place, must retain their figure and not collapse together in abhorrence of a vacuum.[81] God could not only create such a vacuum by annihilating all the matter within the confines of the concavity of the lunar orb, but He could also preserve the capacity to see and hear in this vacuum by retaining and preserving the spatial configurations of air and water while destroying the air and water. God could then place two men in the region where air formerly existed and they would be able to see and speak to each other without the material air serving as a medium of visual and audio transmission.[82]

Buried amid all these natural impossibilities, and, to many in the Middle Ages, undoubted absurdities, was the idea of a separate space existing independently of material body. What Articles 34 and 49 did for the existence of extracosmic space, Articles 139, 140, and 141 did, somewhat less intelligibly perhaps, for possible dimensional spaces within the world. Thus Walter Burley was prepared to argue that if such a separate quantity, or dimensional vacuum, existed in the same manner as Catholics assume the existence (in the Host) of a quantity separated from every substance and from all qualities, light and heavy bodies would be able to move through it successively as if they were moving through a medium.[83] In this manner did Burley link the condemned articles on the su-

---

reason. In turning to the first mode, which is a three-dimensional separate vacuum, Buridan says: "Primo ergo quantum ad primum modum imaginandi vacuum esse ego pono quod Deus potest facere accidens sine subiecto et potest accidentia separare a subiectis suis et separatim conservare. Ideo potest simplicem dimensionem creare absque hoc quod cum ea sit aliqua substantia vel etiam aliquid accidens distinctum ab ea. Secundo, videtur michi quod non est apud Deum impossibilis penetratio dimensionum. Immo ipse potest plura corpora facere esse simul in eodem subiecto vel in eodem loco absque hoc quod differant ab invicem secundum situm, scilicet absque hoc quod unum sit extra alterum secundum situm. Ergo Deus potest facere simplicem dimensionem sive spacium ab omni substantia naturali separatum in quo vel cum quo absque hoc cedat recipi possunt corpora naturalia. Et hoc vocabatur vacuum secundum primam imaginationem"; Buridan (n. 36 above), fol. 74r, bk. 4, q. 8. For the two kinds of vacuum, see q. 7 on fol. 73r, col. 2.

81. "Deinde de secundo modo imaginandi credo sicut prius arguebatur quod Deus posset annichilare istum mundum inferiorem conservando celum, magnitudines et figuras quales et quantas nunc habet, et concavum orbis lune esset vacuum et de hoc et de dubitationibus circa hoc accidentibus dictum fuit satis in decimaquinta questione tertii libri"; ibid., fol. 74r, col. 1.

82. Ibid.

83. "Ponendo tamen quantitatem separatam eo modo quo Catholici posuerunt quanti-

pernatural separation of accidents from their subjects, or attributes from their substances, with the much discussed medieval problem of motion in a separate space.[84]

Although Jean Buridan probably opposed the Condemnation of 1277 and endorsed, privately at least, those articles that were supportive of Aristotelian physics and cosmology, he was not above using certain of the natural impossibilities to his own advantage, as we already saw with Article 49. He would do the same with Article 141, which he found a convenient support for his conception of the true nature of motion. Convinced that the motion of the last celestial sphere could not be identified with either the sphere itself or the place of that sphere, if it had any, Buridan[85] rejected William Ockham's definition of motion, which equated motion with the mobile and the successive places it occupied.[86] For Buridan, motion was a nonpermanent, pure flow—a *res pure successiva*—in which each part passes away when its successor comes into being. The changing relationships exhibited by the last sphere, or heaven, arise from the impossibility of prior and posterior parts of motion existing simultaneously, for if successive parts of motion could exist simultaneously, motion would be a permanent quality, or accident, similar to the magnitude or shape of a body, which can remain constant over successive moments.[87]

---

tatem posse separari a substantia et ab omni qualitate, sic esset dicendum, ut opinor, quod grave vel leve posset moveri motu successivo in tali quantitate tanquam in medio"; ibid., fol. 116v, col. 2. Earlier, Burley had designated this type of vacuum as "less impossible" *(minus impossibilis)* than a vacuum that is neither a natural body or a separate quantity. Of this "less impossible" kind of vacuum, he says: "quod sit aliqua quantitas longa, lata, et profunda separata ab omni qualitate [corrected from "quantitate"] sensibili quoniam secundum theologos illud est possibile apud Deum, sicut in Sacramento altaris est quantitas sine omni substantia corporea in qua sit sicut in subjecto. Ita Deus posset facere quantitatem esse sine omni qualitate et talem quantitatem separatam receptivam corporis; et hanc dixerunt aliqui antiquorum esse vacuum"; ibid., fol. 116v, col. 1. It would appear that Articles 139, 140, and 141 were condemned because they would have denied the Eucharist, which required that qualities exist without subjects in which to inhere.

84. See the section "Motion in a Hypothetical Void," in Grant (n. 1 above), 334–50.

85. Buridan considers the problem in the same question discussed above, where, in a fourth conclusion, he declares "quod motus ultime spere non est illa spera nec locus eius" (n. 36 above), fol. 50v, bk. 3, q. 7; see also Maier (n. 65 above), 126.

86. For Ockham's definition, see *Tractatus de successivis Attributed to William Ockham,* edited by Philotheus Boehner, O.F.M., Franciscan Institute Publications 1, 46 (St. Bonaventure, N.Y., 1944); for a translation of the passage, see Herman Shapiro, "Motion, Time, and Place According to William Ockham," *Franciscan Studies* 16, no. 3 (1956): 251–52.

87. Buridan (n. 36 above), fol. 50v, col. 2; Maier (n. 65 above), 126–27.

Motion is thus a disposition, or an accidental form, inhering in a body. But if motion is an accidental form or quality, would it not, then, be possible for God to destroy a body and the places it might occupy and yet conserve its motion as an independent entity? Whoever posed this question, whether Buridan himself or opponents of his concept of the nature of motion, was undoubtedly aware of Article 141 of the Condemnation of 1277, which had denounced those who would argue "[t]hat God cannot make an accident exist without a subject." The bishop of Paris, Étienne Tempier, could have found no fault with Buridan's response some sixty years later, when the latter declared that "I do not consider it more absurd that there could be a motion and nothing would be moved than that there could be whiteness and nothing that would be white. Neither is possible naturally but each is possible supernaturally."[88]

Thus not only did Buridan use Article 49 to help establish the absolute nature of motion, but he found Article 141 useful in a negative sense in winning acceptance for the idea that motion is a quality and therefore subject to the same fate as any other quality. If it was required at Paris to concede that God could make a quality exist separately from its subject, then so also could He make a motion exist separately from its subject. While Buridan may have invoked Article 141 negatively, and perhaps even defensively, the idea that a quality could exist supernaturally without its subject, and the further idea, embodied in Article 140, that the separate existence of a quality did not imply a contradiction, may have played a role in the development of the medieval doctrine of the configuration of qualities in which the addition of qualities without subjects was a basic concept.[89] It was within the context of this doctrine that the famous Merton College "mean speed theorem" was derived and which eventually served as the foundation of Galileo's new mechanics.[90]

Buridan's concept of the nature of motion, in which Article 49 played a significant role, and Article 141 a minor role, has been interpreted by Anneliese Maier as containing the germ of an inertial theory.[91] For if

---

88. Buridan (n. 36 above), fol. 51r, col. 1; see also Maier (n. 65 above), 129.

89. William of Ware may have been the first to propose the idea. See Edith Sylla, "Medieval Quantification of Qualities: The 'Merton School,'" *Archive for History of Exact Sciences* 8, nos. 1–2 (1971): 11–12, n. 9.

90. See Clagett (n. 70 above), 255–418, chaps. 5–6.

91. Maier (n. 65 above), 133.

motion inheres in a body just as a quality does, the former would be independent of external or internal movers and ought, therefore, to remain in existence until destroyed by external resistances. Although Buridan drew no such inference and failed to exploit the inertial possibilities, it should be obvious by now that the articles of the Condemnation of 1277 mentioned thus far were influential primarily in discussions about space and motion.

But there were yet other articles that played a role in discussions about space and motion, as well as a wide range of other topics. Certain articles concerned the location, movement, and activities of angels and intelligences.[92] Of these, some were especially controversial because Thomas Aquinas had held them. Thus Article 204[93] condemned the opinion that a separate substance could be everywhere by means of the action it exercised. This meant that an angel could not move from one extremity to another, or move through the middle of anything unless it wished to act in any of these places. If it wished to act only in the middle it could do so directly without having arrived there from either extremity. With the condemnation of Article 204, it also seemed plausible to condemn Article 219, which held that separate substances are nowhere, that is, not in a place according to their substance.[94] Although he had died some three years prior to the Condemnation of 1277, Aquinas's works contained opinions that were clearly in violation of Articles 204 and 219. With respect to location and movement, he had treated spiritual substances in a manner radically different from bodies. For Aquinas, a body is in place by the contact of its volume with the innermost surface of the containing body that surrounded and touched it at every point. However, since an angel is not a corporeal volume, Aquinas concluded that it could not be in a place, and inferred from this that it acted in places by its will, or

92. See Duhem (n. 1 above), 6.22, 29–59; on condemned articles concerned with celestial motions, see 59–66.

93. "204. Quod substantie separate sunt alicubi per operationem; et quod non possunt moveri ab extremo in extremum, nec in medium, nisi quia possunt velle operari aut in medio, aut in extremis.—Error, si intelligatur, sine operatione substantiam non esse in loco, nec transire de loco ad locum"; see Denifle and Chatelain (n. 1 above), 1.554.

94. "219. Quod substantie separate nusquam sunt secundum substantiam.—Error, si intelligatur ita, quod substantia non sit in loco. Si autem intelligatur, quod substantia sit ratio essendi in loco, verum est, quod nusquam sunt secundum substantiam." See Denifle and Chatelain (n. 1 above), 1.555.

desire, and not by the presence of its substance.[95] Consequently, an angel could, if it wished, also apply itself to a place by its power without passing through all the intervening points with a continuous motion.[96]

In upholding the condemnation of Article 204, Duns Scotus[97] was critical of those, including Aquinas, who would assign to an angel, which possessed only finite power, unlimited power to act wherever it pleased, thus conferring on it a power that was only appropriate to God. It also appeared absurd to Scotus that if an angel were to pass from heaven to earth, it could do so without passing through, or acting on, the intermediate places.[98] Rejecting "action at a distance" for separate substances other than God, Scotus, and those who supported the condemnation, required that an angel act in a place only by "occupying" that place. To reach that place, however, it must pass through all the intervening points between its *terminus a quo* and *terminus ad quem*—that is, its motion must be successive and continuous. Although there was disagreement, the condemnation of Articles 204 and 219 was upheld in a variety of interpretations, by numerous other fourteenth-century Scholastics, including Peter Aureoli,[99] William Ockham,[100] Thomas of Strasbourg,[101] and John Baconthorpe.[102]

95. Thomas Aquinas (n. 28 above), q. 52, art. 1 ("Utrum angelus sit in loco"), p. 326a.

96. Ibid., q. 53, art. 2 ("Utrum angelus transeat per medium"), p. 330b.

97. John Duns Scotus, *Quaestiones in lib. II sententiarum*, dist. 2, q. 6 ("An locus angeli sit determinatus, punctualis, maximus, et minimus?") in *Opera omnia* 6.1 (Lyons, 1639; reprint Hildesheim, 1968), 189. Scotus's discussion is especially significant because he says that this article was condemned by the bishop of Paris, and that although it might be said that the penalty of excommunication does not "cross the sea"—i.e., does not apply to England—an article that is judged heretical is to be condemned everywhere, not only by authority of the diocese, but also by authority of the pope; at the very least, he adds, an opinion condemned at a university ought to be suspect.

98. Ibid., 189–90. See also Edward Grant, "Medieval and Seventeenth-Century Conceptions of an Infinite Void Space beyond the Cosmos," *Isis* 60 (1969): 50, n. 50, where Scotus allows that God could act at a distance in places remote from His presence. Such power was not, however, possessed by angels.

99. Petrus Aureoli, *Commentariorum in secundum librum sententiarum tomus secundus* (Rome, 1605), 52–53, bk. 2, dist. 2, art. 2 ("Utrum angeli sint creati in caelo Empireo sicut in loco"). Art. 204 is explicitly alluded to as "articulus excommunicatus Parisijs" (52, col. 2).

100. Ockham (n. 19 above), quotlib. 1, q. 4: "Utrum angelus sit in loco per suam substantiam," sig.a4r–a4v.

101. Thomas of Strasbourg (n. 33 above), fols. 107r–108r, bk. 1, dist. 37, q. 1 ("An substantia quaelibet spiritualis in loco existat?"), where Articles 204 and 219 are specifically cited.

102. John Baconthorp not only cites Articles 204 and 219, but actually refers to them by

In his discussion of the question "whether an angel is in place by its substance," Ockham cites Article 219 and resolves the problem by opting for one of two traditional ways in which a thing could be conceived to be in a place, namely, circumscriptively or definitively.[103] The former is excluded because it assumes that every part of a thing is in a part of the place, and that the whole thing is in the whole place, a description applicable only to bodies. But an angel can be in a place definitively, since a thing is said to be in a place definitively "when the whole is in the whole place and not outside it, and the whole is in any part of the place, as the body of Christ is in place definitively in the Eucharist because his whole body exists [*coassistit*] in the whole place of the consecrated species and his whole body exists [*coassistit*] in any part of the place."[104] Ockham thus interpreted Article 219 to mean that an angel is only in a place definitively and not circumscriptively.[105] With angels assumed to exist in a place by their substances, albeit in a special way, Ockham and many others would then consider whether angels could be moved locally[106]— that is, from place to place—and whether they could move successively through a vacuum.[107]

Many other articles of the Condemnation of 1277 played a role in the physical and cosmological discussions of the fourteenth century with varying degrees of impact. There were articles relevant to the generation and creation of things,[108] celestial movers (angels and intelligences),[109] and the eternity of the world and things in the world.[110] Indeed, Bradwardine had invoked Article 52,[111] that "many things are eternal," in order not only to argue that this limited God's absolute power to destroy

---

number, although erroneously (he lists them as 22 and 218, respectively). See Baconthorp, *Super quatuor sententiarum libros* (Venice, 1526), fol. 144v, bk. 2, dist. 3, q. 2, art. 3 ("An ex parte angeli sit possibilitas ut locetur").

103. Ockham (n. 19 above), quotlib. 1, q. 4, sig.a4r, col. 1.

104. Ibid.

105. "Ad principale dico quod angelus non est circumscriptive in loco per substantiam; et sic intelligitur articulus Parisiensis; sic est verus et non aliter"; ibid., sig.a4v, col. 1.

106. Ibid., q. 5 ("Utrum angelus possit moveri localiter"), sig.a4v.

107. Ibid., q. 6 ("Utrum angelus possit moveri per vacuum"), sig.a5r.

108. See Mandonnet (n. 1 above), 181.

109. Ibid., 179–80, for a lengthy list of errors on intelligences; see also Duhem (n. 1 above), 6.29–59, where some of them are discussed.

110. Mandonnet (note 1 above), 182–83.

111. "52. Quod id, quod de se determinatur ut Deus, vel semper agit, vel numquam; et, quod multa sunt eterna"; see Denifle and Chatelain (n. 1 above), 1.546.

them, and was therefore rightly condemned, but also to apply it against a famous Aristotelian dilemma, made fully explicit by Averroes, that either the world is eternal, which Aristotle believed, or that an independent, uncreated, and presumably eternal, precreation void space must have existed in which the world was created.[112] Since neither of these alternatives was acceptable to a Christian, for whom only God could be eternal, the condemnation of Article 52 made both alternatives untenable.[113]

Since the concept of a regular, lawful, and deterministic world had great appeal for many astrologers and for those who followed Greek tradition as described in Stoic and Aristotelian thought, quite a few articles were directed against deterministic astrology[114] and the idea that not even God could intervene in the natural order by creating new effects.[115] Denial of new effects and the assumption of deterministic astrology were embodied in the Stoic concept of a Great Year which assumed the complete recurrence of all events and individuals over fixed periods of time, usually 36,000 years based on the Ptolemaic value for precession of the equinoxes of 1° in 100 years. It was clearly the Great Year which the bishop of Paris had in mind, when, in Article 6, he condemned the belief "[t]hat when all celestial bodies have returned to the same point—which will happen in 36,000 years—the same effects now in operation will be repeated."[116]

---

112. See Aristotle, *De caelo* 3.2.301b.31–302a.9. For Averroes's version, see his *Commentary on Aristotle's Physics*, bk. 4, comm. 6, in *Aristotelis omnia quae extant Opera*, 9 vols. and 3 supplements (Venice, 1562–1574; reprint Frankfurt a.M., 1962, as *Aristotelis Opera cum Averrois commentariis*), vol. 4, fol. 123v, col. 2–124r col. 1, a commentary on Aristotle's *Physics* 4.1.208b.25–33.

113. For Bradwardine's argument, see the translation from the *De causa Dei* in Grant (n. 1 above), 558, col. 1. Surprisingly, Bradwardine did not cite Article 201, which condemned the opinion "[t]hat one who generates the whole world [i.e., one who believes the world is divinely created or has come into being naturally] assumes a vacuum because place necessarily precedes what is generated in a place; and so before the generation of the world, there was a place without a thing located in it, which is a vacuum." ("201. Quod, qui generat mundum secundum totum, ponit vacuum, quia locus necessario precedit generatum in loco; et tunc ante mundi generationem fuisset locus sine locato, quod est vacuum"; see Denifle and Chatelain [n. 1 above], 1.554.) Article 201 seems clearly directed against the arguments of Aristotle and Averroes cited in the preceding note.

114. For a list, see Duhem (n. 1 above), 8.419–23, and, for a discussion, 423–501.

115. E.g., Article 48 declared "[t]hat God cannot be the cause of a new act [or thing], nor can he produce something anew." ("Quod Deus non potest esse causa novi facti, nec potest aliquid de novo producere"; see Denifle and Chatelain [n. 1 above], 1.546.) The translation is from Grant (n. 1 above), 48. For other relevant articles, see numbers 21, 87, and 88.

116. "6. Quod redeuntibus corporibus celestibus omnibus in idem punctum, quod fit in xxx sex milibus annorum, redibunt idem effectus, qui sunt modo"; see Denifle and Chatelain (note 1, above), 1.544.

Well aware of the existence of Article 6, Nicole Oresme formulated a series of mathematical propositions by means of which "many errors about philosophy and faith could be attacked ... as [for example] that [error] about the Great Year which some assert to be 36,000 years, saying that celestial bodies were in an original state and then return [to it in 36,000 years] and that past aspects are arranged again as of old."[117] The mathematical propositions, which Oresme had devised and discussed in a number of treatises,[118] demonstrated the probability that any two or more celestial motions were probably incommensurable. From this probable celestial incommensurability, Oresme argued that if the celestial spheres commenced their motions from some particular configuration, it was highly unlikely that they would again arrive at that same arrangement in any fixed interval of time. He further inferred that, if terrestrial events are caused by celestial events, as was commonly believed, unique celestial dispositions, such as conjunctions, that would occur as a consequence of probable celestial incommensurability, could cause unique effects or events, as, for example, a new species. Moreover, precise astrological prediction would be impossible as would the determination of the length of the solar year and the construction of an exact calendar.[119] Although the condemnation of Article 6 may not have been the sole inspiration for Oresme's fascinating venture into the realm of mathematical incommensurability and its possible physical consequences, it probably played a significant role[120] and was therefore ultimately responsible for subsequent discussions by authors who drew their knowledge of this subject directly or indirectly from the works of Oresme.[121]

117. See chap. 4 of Oresme's *De proportionibus proportionum*, in *Nicole Oresme "De proportionibus proportionum" and "Ad pauca respicientes,"* edited and translated by Edward Grant, 307 (Madison, 1966). For a discussion of the Great Year, see *Nicole Oresme and the Kinematics of Circular Motion: Tractatus de commensurabilitate vel incommensurabilitate motuum celi,* edited, translated, and with an introduction by Edward Grant, 103–24 (Madison, 1971).

118. Primarily in his *De proportionibus proportionum, Ad pauca respicientes,* and *Tractatus de commensurabilitate vel incommensurabilitate motuum celi,* but also in his *Questiones super de celo* and *Le Livre du ciel et du monde.* The contributions of all these treatises are discussed by Grant, *Kinematics* (n. 117 above).

119. For Oresme's arguments in these instances, see ibid., 55–67. For Marsilius of Inghen's conclusion that celestial incommensurability could generate new effects, see ibid. 127–28.

120. Duns Scotus may also have been influenced by Article 6, when he denounced, as contrary to faith, the concept of an exact cyclical return and argued further that if the celestial motions are incommensurable, a Great Year could not occur; ibid., 118–20.

121. On Henry of Hesse, Marsilius of Inghen, Pierre d'Ailly, and Jerome Cardano, see ibid., 124–61; see also 122–24, n. 99.

Frequent citation of, and implicit allusions to, numerous articles of the Condemnation of 1277 should convince us that it was taken seriously throughout the fourteenth century and that it encouraged innumerable invocations of God's absolute power in a variety of hypothetical physical situations. The supernatural alternatives that medieval Scholastics considered in the wake of the condemnation conditioned them to consider possibilities outside the ken of Aristotelian natural philosophy, and usually in direct conflict with it. So widespread was the contemplation of such hypothetical possibilities in the late Middle Ages that it is no exaggeration to view them as an integral feature of late medieval thought. The infinite power of God to perform certain acts as specified in the various articles of the condemnation was soon extended to almost any impossible act, which is not surprising in view of the license to formulate impossibilities granted by Article 147. Scholastics were thus encouraged to explore the consequences of such acts, and, as we have already seen, frequently did so.

Of all the areas of physical thought that were affected by the Condemnation of 1277 and its concomitant idea of God's absolute power, none was influenced more than the concept of vacuum. Not only was the possible existence of vacuum a theological problem because, as Gregory of Rimini put it, every Catholic had to concede the possibility that God could create one,[122] but also because it raised the question as to whether God required an empty space in which to create the world, a view condemned by Article 201, which we cited earlier. We have already seen that by postulating other worlds and the motion of our world, Articles 34 and 49 generated serious discussion of the possible existence of extracosmic void space. Although no articles of the Condemnation of 1277 concerned vacua within our cosmos, it seemed obvious that if God could create a vacuum beyond the world, he surely could do so within the world. And so it was that God was frequently imagined to annihi-

---

122. "Similiter si vacuum foret, sicut possibile esse per potentiam Dei, saltem quilibet Catholicus habet concedere, quodlibet grave per quantamcumque finitam distantiam descenderet in instanti"; Gregory of Rimini, *Super primum et secundum sententiarum*, edited by E. M. Buytaert, Franciscan Institute Publications, Text Series 7 (1522; repr. St. Bonaventure, N.Y., 1955), fol. 50v, bk. 2, dist. 6, q. 2. Gregory completed his lectures on the *Sentences* in 1344 and added to them in 1351. Earlier (at n. 79 above), we saw that Buridan called the question of the existence of vacuum a theological problem. See also Henry of Ghent's statement that a Catholic ought not to deny that God could make a vacuum if he wished; Koyré (n. 18 above), 58 n. 1.

late all or part of the matter within the material plenum of our world.[123] Within this wholly or partially empty space all sorts of situations were imagined after 1277, and the questions raised came to be commonly discussed in the large literature on the nature of vacuum and the behavior of bodies placed within it. Would the surrounding celestial spheres collapse inward instantaneously as nature sought to prevent formation of a vacuum?[124] Indeed, would the empty interval, or nothingness, be a vacuum or space?[125] With all the matter destroyed within the concave surface of the last celestial sphere, would it be meaningful to consider that concave surface a place?[126] Could a stone placed within such a void be ca-

123. Because of their essentially theological context, medieval discussions of the annihilation of matter to create a vacuum were probably not derived from Aristotle, who is not even mentioned in this connection, despite his brief summary of earlier opinions that place must precede the things in it and therefore be "a marvellous thing, and take precedence of all other things. For that without which nothing else can exist, while it can exist without the others, must needs be first; for place does not pass out of existence when the things in it are annihilated"; *Physics* 4.1.208b.33–209a.1, in the translation by Hardie and Gaye. In Averroes's commentary on the *Physics*, the text appears in bk. 4, text 7 (see *Aristotelis omnia quae extant Opera* [n. 112 above], vol. 4, fol. 124 cols. 1–2).

124. Ockham, for example, argued that if God preserved the heaven and destroyed all else within it, the sides of the heaven would not collapse inward and come into contact to prevent formation of a vacuum. For such a collapse would be either instantaneous or successive; if the former, it would not be called a local motion because a body in instantaneous motion would not pass through the midpoint of the distance to be traversed, but would arrive at its terminus in a durationless moment, which is impossible. But if the motion were successive, it would occur in a measurable time; therefore, in the first part of that time, a vacuum would occur before the sides of the heaven came into contact. See Ockham (n. 19 above), *Quotlibeta septem,* quotlib. 1, q. 6, sig. a5v, col. 1. Buridan, along with other Scholastics, assumed that when God annihilated everything within the lunar orb, he also preserved the shape and configuration of the lunar orb and the heavens. Hence the celestial spheres would not collapse inward, since God preserved them as before. See Buridan (n. 36 above), fol. 57v, bk. 3, q. 15 ("Utrum est aliqua magnitudo infinita"). For other interpretations, see Edward Grant, "Medieval Explanations and Interpretations of the Dictum that 'Nature Abhors a Vacuum,'" *Traditio* 29 (1973): 331 and n. 7.

125. Buridan argues that if God destroyed all matter within the lunar orb, there would be "nothing" there, not even a vacuum or space, since the latter are not positive things. Thus it implies a contradiction to say that "nothing" *(nihil)* is left after God destroys everything within the lunar orb, and then to infer that a vacuum or space is there. It is not the intervening nothing that is to be described as a vacuum, but only the surrounding concave lunar surface, "for the concave surface of the heaven is now a place filled with body or bodies, and then [after God destroyed everything] it would be a place not filled with body; thus this surface is a vacuum"; Buridan (n. 15 above), 95, bk. 1, q. 20. Buridan arrived at the same conclusion in his *Questiones* (n. 36 above), fol. 57v, col. 2, bk. 3, q. 15.

126. Johannes Canonicus concluded that if God destroyed everything within the last sphere, a possibility which all Catholics had to concede, and the inner surface of the last sphere remained, it would no longer constitute a place since the nature of a place is to be in contact

pable of rectilinear motion?[127] Would it be possible to measure distances within such a vacuum?[128] If people were placed in it, would they be able to see and hear each other?[129] And, on the commonly accepted assumption that the behavior of bodies in the sublunar region is ordained and governed by celestial and superior causes, what would happen to a spherically shaped piece of earth located in the air enclosed within a house, if God destroyed everything, including the celestial spheres, outside that house?[130] Analysis of these, and similar, "thought-experiments" in the late Middle Ages was often made in terms of Aristotelian principles even though the conditions imagined were "contrary to fact" and impossible within Aristotelian natural philosophy.

God's absolute power had thus become a convenient vehicle for the introduction of subtle and imaginative questions, which generated novel replies; and though the speculative responses did not replace, or cause the overthrow of, the Aristotelian worldview, they did challenge some of its fundamental principles and force their attention on the medieval mind. They made many aware that things might be quite otherwise than were dreamt of in Aristotle's philosophy.

But if the oft-discussed natural impossibilities and their consequences failed to overthrow Aristotelian physics and cosmology during the late Middle Ages, their influence and utility outlived the age that spawned them. Although knowledge of the Condemnation of 1277 had long disappeared,[131] some of the problems and solutions that had emerged as a

---

with a body; Johannes Canonicus, *Questiones super VIII lib. Physicorum Aristotelis perutiles . . .* (Venice, 1520), fol. 40r, q. 1.

127. See Buridan (n. 36 above), fol. 58r, col. 1, bk. 3, q. 15.

128. Although Richard of Middleton denied that the heaven could be said to be distant from the earth if everything between them was destroyed (n. 33 above), 2.186 (bk. 2, dist. 14, q. 3), Henry of Ghent, Jean de Ripa, and William of Ware thought such measurements were possible (see above, notes 49, 50, and 52). While Buridan denied that the distance between the poles of the lunar orb would be measurable rectilinearly if all things were annihilated within it, he did allow that under such circumstances curvilinear distances could still be measured along the surface of the lunar orb (see Buridan [n. 36 above], fol. 58r, bk. 3, q. 15; for a similar opinion by Marsilius of Inghen, see Grant [n. 49 above], 151).

129. A problem discussed by Buridan (see above at n. 82).

130. In considering this problem, Buridan concluded that the piece of earth would not move at all "because there is no more reason why it should be moved toward one part than to another, since one part of air would be no more up or down than another; nor would there be another power in one [part] than another because the governance of the heaven would have been removed"; Buridan (n. 15 above), 86–87, bk. 1, q. 18.

131. That an occasional reference might still appear in the seventeenth century is evident

direct consequence of it continued to exercise influence in the late sixteenth and seventeenth centuries, not only on Scholastic authors, who
were preserving and extending traditional arguments, but also on non-
Scholastic authors, who were not unaware of the topics debated by their
Scholastic contemporaries.

God's absolute power to create matter beyond our world and to annihilate that matter, as well as some or all of the matter within our world,
was widely discussed in the seventeenth century. Two Jesuits, Francisco
Suarez and Bartholomaeus Amicus, allowed that God could create other
noncontiguous worlds beyond ours between which a vacuum would exist.[132] Suarez also invoked God's power to annihilate matter within the
world and made it the basis for demonstrating the existence of a void
space in which measurements could be made just as in a plenum.[133] John
Locke took a similar approach in arguing for the existence of a three-
dimensional void space. Since no one could deny that God was capable of
annihilating any part of matter, it followed that a vacuum would remain
if God did indeed destroy a body, "for it is evident that the space that was
filled by the parts of the annihilated body will still remain, and be a space
without a body."[134] Somewhat more intricately, but following the same
pattern, Pierre Gassendi arrived at the actual existence of an infinite, dimensional void space by first imagining the supernatural annihilation
of all matter within the sublunar region, then in the celestial region beyond, and then in a world imagined successively larger and larger. For "if
there were a larger world, and a larger one yet, on to infinity, God successively reducing each of them equally to nothingness, we understand
that the spatial dimensions would always be greater and greater, on to

---

in Thomas Campanella's *Defense of Galileo*, where Thomas Aquinas's attempt to unite Aristotle
and Christian theology is said to have been "rebuked in the Articles of Paris." See "The Defense
of Galileo of Thomas Campanella . . . ," edited and translated by Grant McColley, *Smith College
Studies in History* 22, nos. 3–4 (April–July 1937): 40.

132. See Francisco Suarez, *Disputationes metaphysicae*, 2 vols. (Paris 1866; reprint in facsimile Hildesheim, 1965; first printed 1597), 2.106, col. 1; and Bartholomaeus Amicus (n. 16
above), 2.746 col. 2(C), tract. 21 *(De vacuo)*. In his *Elements of Philosophy*, Thomas Hobbes also
allowed that God could create other worlds, but, rather than discuss the void between them,
Hobbes emphasized that these other worlds would have to be created in empty spaces beyond
our world. See *The English Works of Thomas Hobbes of Malmesbury*, edited by William Molesworth, 1.93 (London, 1839).

133. Suarez (n. 132 above), 2.106, cols. 1–2.

134. John Locke, *An Essay Concerning Human Understanding*, bk. 2, chap. 13, par. 22, in *The
Philosophical Works of John Locke*, edited by J. A. St John, 2 vols., 1.295 (London, 1903–1905).

infinity," so that we can "likewise conceive that the space with its dimensions would be extended in all directions into infinity."[135] With Thomas
Hobbes, the annihilation of matter became a principle of analysis. Despite omission of God as annihilator, Hobbes paid unwitting tribute to
his Scholastic predecessors when he declared that "[i]n the teaching of
natural philosophy, I cannot begin better (as I have already shewn) than
from *privation;* that is, from feigning the world to be annihilated,"[136] a
process, which, among other things, enabled him to formulate his concepts of space and time.

And there is more than an echo in the seventeenth century of that
condemned article which made it mandatory after 1277 to concede that
God could move the world rectilinearly despite the vacuum that might
be left behind. Thus Bartholomaeus Amicus (1562–1649) speaks of God
moving the world rectilinearly in connection with problems and paradoxes about the Aristotelian concept of place. If God moved the world
toward the antipodes both the earth and the concave surface of the lunar
orb would move simultaneously. However, since the earth is conceived as
immobile and the place we call "down," and the positionally immobile
concave surface of the lunar orb is the place we call "up," it follows that
the rectilinear motion of the world would cause the "immobile" places
"up" and "down" to move simultaneously.[137] Indeed, if God preserved
the positions of all the sublunar elements while He moved the world rectilinearly, then, if air is assumed to be the containing place of the earth,
the motion of the world would cause a simultaneous motion of earth and
air even though the earth did not change its relative position with respect
to the air. Thus we would have a motion without a change of place![138]

Pierre Gassendi found the supernatural motion of the world a convenient support for the absolute immobility of infinite space when he declared that "it is not the case that if God were to move the World from
its present location, that space would follow accordingly and move along
with it."[139] In his dispute with Leibniz, Samuel Clarke, speaking for New-

---

135. From Gassendi's *Syntagma philosophicum,* Physica, sect. 1, lib. 2, "De loco et dura-
tione rerum," as translated by Walter Emge and Milie Čapek in *The Concepts of Space and Time:
Their Structure and Their Development,* edited by Milie Čapek, 91–92 (Dordrecht, 1976).

136. Hobbes (n. 132 above), 1.91.

137. Bartholomaeus Amicus (n. 16 above), 2.659 col. 1(B), bk. 4, tract. 20 *(De loco),* dub. 9.

138. Ibid., 660, col. 1(D).

139. Čapek (n. 135 above), 93.

ton in that famous controversy, also defended the existence of absolute space by arguing that "if space was nothing but the order of things coexisting [as Leibniz maintained]; it would follow that if God should remove the whole material world entire, with any swiftness whatsoever; yet it would still always continue in the same place."[140] Rejecting this consequence, Clarke insisted that such a motion would be absolute even though unrelated to any other external body, a judgment strikingly similar to that which Buridan and Oresme had made in the fourteenth century.

But even as the seventeenth century provides additional instances of imaginary and hypothetical situations derived from the concept of God's absolute power, the mechanical universe that was fashioned in that century heralded the end of divine intervention. The divine possibilities, or natural impossibilities, which played a significant and interesting role for some four centuries of Western thought terminated with a changed conception of God's power. The God of the Middle Ages, who could do anything He pleased short of a logical contradiction, was replaced by a God of constraint, who, having created a perfect clocklike universe, rested content merely to contemplate His handiwork ever thereafter. The era of possible divine intervention and action, and the imaginative speculations it provoked, had come to an end.

140. *The Leibniz-Clarke Correspondence,* edited by H. G. Alexander, 32 (New York, 1956). The bracketed qualification is mine.

4 ∾ God, Science, and Natural Philosophy in the Late Middle Ages

Andrew Cunningham and Roger French have made important and provocative claims about natural philosophy and science to which I should like to reply.[1] The major claim asserts that the object of natural philosophy as a discipline was the study of God's creation and God's attributes. So powerful was this objective, that Cunningham proclaims that natural philosophy was not just "about God" and His creation at those moments when natural philosophers were explicitly talking or writing about God in their natural philosophical works or activities. It was, by contrast, "about God" and His creation the whole time.[2]

Cunningham acknowledges that he has here taken "something of a blank cheque . . . to make any claim I like without having to produce any evidence."[3] It is indeed a "blank cheque." We cannot know what was in the minds of medieval or early modern natural philosophers as they wrote their treatises. In view of the celibate status of medieval natural philosophers, their thoughts, as they wrote their treatises, may just as plausibly have been filled with sexual fantasies, along with, or in lieu of, God and His creation. Or, perhaps, they simply filled their minds with the problems of natural philosophy. It is best to leave such matters to psychohistorians. In this essay, I shall evaluate only the exant writings of medieval natural philosophers. When they write about God and faith, then that segment of their writings is about God and faith. But when there is no mention of God and faith, or allusions to them, then it is not about God and faith.

1. See A. Cunningham, "How the *Principia* Got Its Name; Or, Taking Natural Philosophy Seriously," *History of Science* 29 (1991): 377–92; and also R. French and A. Cunningham, *Before Science: The Invention of the Friar's Natural Philosophy* (Aldershot, 1996).

2. Cunningham, "How the *Principia* Got its Name," 388.

3. Ibid., 382.

Because they are convinced that medieval natural philosophy was primarily about God and the creation, French and Cunningham conclude that what has been interpreted as science is really natural philosophy, from which it follows that nothing we could properly call "medieval science" existed that is in any way comparable to modern science. Indeed, the very title of their book, *Before Science,* makes this first claim quite apparent.[4] The claim is based on the assumption that, unlike natural philosophy, "modern science does *not* deal with God or with the universe as God's creation," an assumption that "is one of the most basic things that the members of the modern scientific community hold in common."[5] In what follows, my main concern will be to argue that natural philosophy is *not* primarily about God and His creation. Before turning to that topic, however, I want to deny the claim that there was no science in the Middle Ages, and also to reject the sharp dichotomy that Cunningham and French draw between medieval natural philosophers, who allegedly always thought about God and His creation in all their works, and modern scientists, who supposedly eliminated God and His creation from their works.

To show this, I offer a comparison of two treatises, *The Book of Jordanus de Nemore On the Theory of Weight (Jordani de Nemore Liber de ratione ponderis),*[6] a thirteenth-century treatise by Jordanus of Nemore, and "On the Electrodynamics of Moving Bodies," an article composed by Albert Einstein in 1905.[7] Inspection of both works shows at a glance that they are highly mathematical and spare in their exposition (Jordanus's, ironically, being more spare than Einstein's). Jordanus uses geometry, and

---

4. Cunningham and French, *Before Science,* 273, where they declare that "there was no scientific tradition (in the modern sense of the term 'scientific') of looking at nature in the thirteenth century, only a religio-political way of doing so. Natural philosophy was not the same as modern science."

5. Cunningham, "How the *Principia* Got its Name," 382–83.

6. For the Latin text and English translation, see *The Medieval Science of Weights ("Scientia de ponderibus"), Treatises Ascribed to Euclid, Archimedes, Thabit ibn Qurra, Jordanus de Nemore, and Blasius of Parma,* edited and translated by E. A. Moody and M. Clagett, 174–227 (Madison, 1952). E. A. Moody edited and translated Jordanus's *On the Theory of Weight.*

7. A. Einstein, "On the Electrodynamics of Moving Bodies," translated from "Zur Elektrodynamik bewegter Körper," *Annalen der Physik* 17 (1905), in H. A. Lorentz, A. Einstein, H. Minkowski, and H. Weyl, *The Principle of Relativity: A Collection of Original Memoirs on the Special and General Theory of Relativity,* translated by W. Perrett and G. B. Jeffrey, with notes by A. Sommerfeld, 37–65 (New York, 1952 [first published 1923]).

Einstein the calculus, but that is irrelevant. Both treatises would meet reasonable and appropriate criteria for being scientific. In the Middle Ages, Jordanus's treatise would have been regarded as a "middle science," that is, a science that is neither natural philosophy nor pure mathematics, but one that lies between them because it involves the application of mathematics to natural philosophy.[8] Jordanus's work also posseses the other attributes of a modern scientific treatise: it makes no mention of God or His creation, and indeed has nothing in it about the faith or the supernatural. And yet we may plausibly assume that Jordanus, like Einstein, believed in a supreme being. It is evident, however, that the religious beliefs of Jordanus and Einstein, exercised *no detectable influence* on their respective treatises, which are wholly devoid of religious or theological content or sentiment. Einstein's article is "modern science" by definition. Jordanus's treatise cannot be "modern science" by definition. It is "medieval science," just as Ptolemy's *Almagest* is "ancient science." But the treatises of Jordanus de Nemore and Ptolemy deserve the title "science" just as much as Einstein's article on electrodynamics. One can multiply similar examples.[9]

Finally, Cunningham's claim that "modern science does not deal with God or with the universe as God's creation," and that this assumption "is one of the most basic things that the members of the modern scientific community hold in common," is quite misleading. The fact that modern scientists do not mention God or His creation in their publications and in their professional lives does not mean that God and His creation

---

8. The basis for distinguishing middle sciences lies in Aristotle's *Posterior Analytics,* chs. 13 and 14. For discussions of the relationships between natural philosophy, mathematics, and middle sciences, see S. J. Livesey, *Theology and Science in the Fourteenth Century: Three Questions on the Unity and Subalternation of the Sciences from John of Reading's "Commentary on the Sentences"* (Leiden, 1989), 22–29; W. R. Laird, "The Scientiae mediae in Medieval Commentaries on Aristotle's *Posterior Analytics"* (Ph.D diss., University of Toronto, 1983), ch. 2; E. Sylla, "Autonomous and Haidmaiden Science: St. Thomas Aquinas and William of Oakham on the Physics of the Eucharist," in *The Cultural Context of Medieval Learning,* Boston Studies in the Philosophy of Science 26, edited by J. E. Murdoch and E. D. Sylla, 355 (Dordrecht, 1975); and C. Day, "Jean Buridan and the Classification of the Sciences" (Ph.D. diss., Indiana University, 1986), 137 and n. 51, 164–65.

9. For example, the *Perspectiva communis* of John Pecham (ca. 1230–1292), a Franciscan theologian, who wrote the treatise around 1277–1279. Pecham's work includes nothing whatever about God, theology, the faith, or anything supernatural, which stands in contrast to his earlier *Tractatus de perspectiva.* For a comparison of the two treatises, see *John Pecham Tractatus de perspectiva,* edited by D. C. Lindberg, 13 (St. Bonaventure, N.Y., 1972).

may not play a significant role in their lives and influence their thoughts about the universe. In a recent survey of one thousand scientists (one-half were biologists, one-quarter mathematicians, and one-quarter physicists and astronomers) whose names were drawn at random from *American Men and Women of Science* (1995), six hundred responded to a series of queries about their religious beliefs.[10] The questions posed to the scientists were exactly the same as those put to one thousand scientists by James Leuba in 1916, with virtually the same results: "about 40 percent of scientists still believe in a personal God and an afterlife."[11] Although he never participated in any of these surveys and may not have believed in an afterlife, Albert Einstein, who claimed that religious beliefs motivated his scientific research, clearly belongs among those scientists who admit to belief in a supreme being. Einstein insisted "that the cosmic religious feeling is the strongest and noblest motive for scientific research"[12] and was convinced that religious feeling for the scientist "takes the form of a rapturous amazement at the harmony of natural law, which reveals an intelligence of such superiority that, compared with it, all the systematic thinking and acting of human beings is utterly insignificant. This feeling is the guiding principle of his life and work, in so far as he succeeds in keeping himself from the shackles of selfish desire. It is beyond question closely akin to that which has possessed the religious geniuses of all ages."[13]

Despite these sentiments, Einstein was not moved to mention God in his scientific treatises. On Cunningham's approach, we may infer from the absence of God and the creation from Einstein's published works that God played no role in his thinking. But we saw that would be patent-

10. E. J. Larson and L. Witham, "Scientists Are Still Keeping the Faith," *Nature* 386 (3 April 1997): 435–36. "Our survey," the authors explain, "polled only about 3 percent of the biological and physical scientists and mathematicians listed in the 1998 American Men and Women of Science." For a summary of the report, see the New York Times of April 3, 1997.

11. Ibid., 435. The authors of the survey report (436) that their "findings do corroborate a large survey done in 1969 by the Carnegie Commission, asking 60,000 professors in the United States questions such as 'how religious do you consider yourself?'" The commission found that 34 percent of physical scientists were 'religiously conservative' and about 43 percent of all physical and life scientists attended church two or three times a month—on a par with the general population."

12. Albert Einstein, *Ideas and Opinions. Based on "Mein Weltbild,"* edited by C. Seelig and others, with new translations and revisions by S. Bargmann, 3rd ed, 39 (New York,1982).

13. Ibid., 40.

ly false. It is equally objectionable to infer that medieval natural philosophers and theologians are thinking about God and the creation even when they do not mention God or the creation. We would be well advised to make no inferences about the role of God in a treatise where God and the creation are not mentioned, whether that treatise was composed by medieval natural philosophers or by modern scientists. In what follows, I shall rightly assume that all medieval natural philosophers and theologians believed on faith that God created the world and was the ultimate cause of all effects. From this assumption, however, I shall not infer, in the absence of explicit citations and discussion, that an author had God and the creation in mind while writing on this or that topic. Without some evidence, we may not argue from silence.

God and Natural Philosophy

When investigating connections between God and natural philosophy in the Middle Ages, it is essential to distinguish two quite different aspects of this relationship: (1) the intrusion of God, His creation, and theology into the commentaries and questions on Aristotle's natural books, and therefore into natural philosophy; and (2) the intrusion of natural philosophy into theology, that is, the importation of natural philosophy into theological treatises by theologians (especially in *Commentaries on the Sentences of Peter Lombard*), where natural philosophy was treated in traditional terms as a "handmaiden to theology," or to combat heretical opinions.[14] Natural philosophers readily admitted that God had created our world—indeed, they could hardly have done otherwise—and governed it. But they never regarded it as their primary aim to focus on God and His supernatural creation. That was a task better left to professional theologians in their theological treatises. The objective of natural philosophers was to explicate the workings of the physical world within the framework of Aristotle's natural books and to do so in the manner exemplified by Aristotle and his great Islamic commentator Averroes, that

---

14. The authors of *Before Science* base their arguments for a natural philosophy that is about God and His creation largely on the second way. By design, their study ignores the massive body of commentary literature on Aristotle's natural books (see Epilogue, 269–72; for the literature on natural philosophy, see below, note 16), which represents natural philosophy done for its own sake rather than as an aid to understanding theology.

is, to do so by natural reason and not by invocation of the supernatural. Natural philosophy was an independent discipline taught in the medieval universities to all who were interested, but especially to those who wished to pursue a career as teachers of natural philosophy, a relatively small group; or to a far larger group who required it as a prerequisite for matriculation toward a higher degree in law, medicine, or theology. Theologians were expected, if not required, to attain a master's degree in natural philosophy. As a consequence, they were quite familiar with natural philosophy, sufficiently familiar so that we may appropriately describe them as "theologian-natural philosophers."

Those who commented on the natural books of Aristotle were usually teaching masters in an arts faculty, although some would eventually matriculate in a theology faculty and become professional theologians. When they wrote their Aristotelian commentaries, they had every incentive to keep their natural philosophy "natural." Under pressure from theologians and the theological faculty of the University of Paris, who were alarmed at the manner in which some arts masters taught natural philosophy, the arts faculty itself, in 1272, instituted an oath that made it mandatory for arts masters to avoid theological discussions in their questions. Where this was unavoidable, they were sworn to resolve the issue in favor of the faith.[15]

In the course of teaching and studying Aristotle's natural books in the arts faculties of medieval universities for more than three centuries, a vast body of commentary literature was produced.[16] Those who wrote these treatises firmly believed that God had created the world from nothing, and that He was the ultimate cause of all events or effects, the First Cause (prima causa), as He was frequently called. How did these Christian beliefs affect the way medieval scholars wrote natural philosophy? Did it mean that their objective in doing natural philosophy was essentially theological or religious? That their aim was to transform natural philosophy into an instrument for the defense of the faith and therefore

15. For the relevant document, see L. Thorndike, *University Records and Life in the Middle Ages* (New York, 1944), 85–86.

16. To obtain a good sense of the extant number and range of these commentaries from 1200 to 1650, see C. H. Lohr, "Medieval Latin Aristotle Commentaries," *Traditio* 23 (1967), 33–413; 24 (1968): 149–245; 26 (1970): 135–216; 27 (1971): 251–351; 28 (1972): 281–396; 29 (1973): 93–197; "Supplementary Authors," 30 (1974): 119–44; and Lohr, *Latin Aristotle Commentaries, Vol. 2: Renaissance Authors* (Florence, 1988).

to intrude as much religious material as possible into their investigations into natural questions?

The most effective way to respond to these questions and judge the impact of God and religion on this body of natural philosophy is to examine the extant texts of natural philosophers. We must determine what they actually said and did. Our judgments and interpretations must be based on the texts of natural philosophy as written by those who were consciously doing natural philosophy, not theology. That is, we must carefully inspect treatises on natural philosophy per se, not treatises on theology that used natural philosophy in the service of theology. As we shall see, a remarkable feature of medieval natural philosophy is that most theologians, who did not hesitate to import natural philosophy into their theological works to resolve scriptural dilemmas and problems of the faith, refrained from needlessly introducing God and the supernatural when they themselves wrote treatises on natural philosophy, that is, when they wrote questions and commentaries on the natural books of Aristotle. Albertus Magnus and Thomas Aquinas are prime examples of this tendency.

The core of medieval natural philosophy lay in the five major treatises of Aristotle's natural books, namely, *Physics, On the Heavens (De caelo), On Generation and Corruption (De generatione et corruptione), On the Soul (De anima),* and the *Meteorology.* Although comments about God and the faith might be inserted almost anywhere if an author chose to do so, occasions for so doing were almost unavoidable in parts of the *Physics, On the Heavens,* and *On the Soul.* To assess the role of God and the faith in natural philosophy, it is essential to examine commentaries and questions on all five books, but especially the three just cited, that were composed during the thirteenth and fourteenth centuries.

Scholastic Attitudes toward Natural Philosophy

For this study, I have examined a number of commentaries and questions on Aristotle's natural books. Included are commentaries by Albertus Magnus and Thomas Aquinas in the thirteenth century, and by John Buridan, Nicole Oresme, Themon Judaeus, and Albert of Saxony in the fourteenth century. The most important point about medieval natural philosophy that emerges from these commentaries and questions on Aristotle's natural books is that natural philosophy was about Aristotle's principles, ideas, and concepts, and was therefore about natural phenomena and not about God, faith, and the supernatural. Although it was inevitable within the context of medieval Christendom, and by virtue of the relationship between the Catholic Church and the universities, that theological concepts would intrude into natural philosophy, the overwhelmingly rational character of Aristotle's logic and natural books restricted and discouraged such intrusions, making them occasional and limited, rather than customary and extensive. Inspection of the numerous works on natural philosophy by the authors just mentioned makes it apparent that in most instances where God and matters of faith are intruded into commentaries and questions on Aristotle's natural books, they occur in one or more of five categories or contexts:

Category 1. When medieval Aristotelian commentators report opinions of Greek and Roman pagan philosophers on some issue that bears on Christian doctrine and faith; or where Aristotle himself mentions God, the gods, or something about divinity.

Category 2. Where Aristotle's arguments are contrary to Church doctrine, as, for example, on the eternity of the world, resurrection of the body, immortality of the soul, and so on. In such instances, by the statute of 1272, which all arts masters were sworn to uphold, natural philosophers were expected to indicate that Aristotle and the faith were in opposition, or to show that Aristotle was somehow in accord with the faith. Whatever the decision, the author was required to support the faith. Also in this category, I include statements in which something is stated to be held or supported according to the faith, but where Aristotle may not be mentioned, and may or may not have been in the author's mind.

Category 3. God and articles of faith were sometimes useful in an analogical, or exemplary, sense to serve as a basis of comparison with natural phenomena, or simply to illustrate something about the natural world.

Category 4. A fourth and major source for the introduction of God and matters of faith was the Condemnation of 1277 and its aftermath. These instances were largely concerned with God's absolute power. Natural philosophers frequently found it necessary to acknowledge that God could do things that Aristotle had said were naturally impossible. Moreover, they often distinguished between what God could do by His absolute power and what could be done by natural powers. When opting for the latter, they frequently used the expression "speaking naturally" (using some form of *loqui naturaliter*). This fourth condition was operative only for the fourteenth-century authors discussed here, and was not a factor for Albertus Magnus and Thomas Aquinas in the thirteenth century.

Because not all citations of God and faith fit appropriately into the four categories just described, it is necessary to add a fifth.

Category 5: Mentions of God and faith that do not fit any of the four categories just described have been placed in this fifth category, along with references to God as a cause of natural events.

In what follows, I shall refer to one or more of these five conditions as *Category 1, 2, 3, 4,* or *5,* respectively.

To determine the extent to which concerns about God, faith, and theology played a role in medieval natural philosophy, I have investigated the use of some key terms in commentaries and questions on Aristotle's works that were produced by the authors mentioned above. Of these key terms, the most significant are those for God, such as *deus,* the most important of them, and some of its synonymous versions such as First Cause *(causa prima),* Prime Mover *(primus motor),* and Immobile Mover *(immobilis motor).* Although there are other interesting terms,[17] I shall be concerned with only one more: "faith" *(fides).*

The Thirteenth Century: Roger Bacon, Albertus Magnus, and Thomas Aquinas

Roger Bacon (ca. 1220–ca. 1292) was the first, or one of the first, to lecture on Aristotle's natural books at the University of Paris after they had been banned for most of the first half of the thirteenth century. Bacon not only contributed to natural philosophy, but also encouraged the application of natural philosophy to theology and of theology to natural philosophy.[18] But Bacon did not take his own advice, as is obvious by inspection of his great work on perspective. In that treatise, Bacon provides a lengthy sci-

---

17. The term "intelligence" *(intelligentia)* occurs fairly frequently in commentaries on the *Physics.* But such occurrences are largely related to the Aristotelian association of intelligences (or angels, as they were sometimes called) as movers of celestial orbs. Only occasionally do they have theological significance.

18. See *The "Opus Majus" of Roger Bacon,* 2 vols., translated by R. B. Burke (New York, 1962; originally published 1928), 73 and 65.

entific account of many aspects of the nature of light. In none of this does he mention God or the faith. His work is strictly naturalistic and rationalistic. When Bacon completed all that he would say about the science of perspective as a natural phenomenon, he turns to the faith in the final four chapters, for which he provides the general title: "Concerning the relationship of *perspectiva* to sacred wisdom and mundane utility, in four chapters."[19] Here Bacon distinguishes between natural philosophy and science, on the one hand, and divine wisdom and faith, on the other.

Like John Pecham, Bacon wrote perspective treatises that did not include appeals to God, faith, or theology. But because perspective certified "natural things" *(res naturales),* and Bacon was convinced that knowledge of natural things was essential for understanding divine things within and outside of Sacred Scripture, perspective was regarded as an invaluable discipline for a Christian. It was not only legitimate to use perspective in the service of divine wisdom, but essential to do so. In this sense, perspective (and all of natural philosophy) was regarded as a handmaid to theology. The investigation of perspective did not require the aid of theology and faith. It was to be studied for its own sake, after which one could certify what within it is useful for understanding divine wisdom.

In his *Questions on the Eight Books of Aristotle's Physics,*[20] Bacon includes a relatively small number of references to the deity and to spiritual entities. Of the 461 brief questions on the eight books, Bacon mentions something about God and the supernatural in only 23, or in 4.9 percent of the questions. The religious import and content in most of these 23 questions is minimal. The term God *(deus)* is mentioned in five different questions.[21] In one of these, Bacon also mentions the Father, Son, and Holy Spirit,[22] and in another he speaks of the First Cause.[23] The First Cause *(prima causa)* appears by itself in four questions,[24] and with other expres-

---

19. See *Roger Bacon and the Origins of "Perspectiva" in the Middle Ages: A Critical Edition and English Translation of Bacon's "Perspectiva" with Introduction and Notes,* edited and translated by D. C. Lindberg, 321 (Oxford, 1996).

20. See *Opera hactenus inedita Rogeri Baconi,* 16 fascicules, edited by R. Steele and F. M. Delorme (Oxford, 1905–1940), Facsicule 13: *Questiones supra libros octo Physicorum Aristotelis,* edited by F. M Delorme, with the assistance of R. Steele (Oxford, 1935). The questions are unnumbered.

21. See Bacon, *Questions on the Physics,* 43, 101, 125, 373, 375, and 390. Pages 373 and 375 are part of the same question.

22. Ibid., 375.   23. Ibid., 124.

24. See ibid., 126, 129, 345, and 415.

sions, such as God and intelligences, in three more.[25] Not surprisingly, the term intelligence(s) (*intelligentia[e]*)—a term used frequently by Aristotle himself—appears in at least eleven questions in a role that is largely that of a motive cause for the celestial orbs and has therefore no theological significance.[26] Terms and concepts such as intelligence, first cause, and prime mover were so common in Aristotelian commentaries that they had lost theological significance. Indeed, the only genuine references to anything that would remind us that Roger Bacon was a Christian is a mention of Father, Son, and Holy Spirit[27] and, in a question as to whether Aristotle and other philosophers believe that motion has an end, his invocation of the Resurrection and the faith.[28] Even where it became customary to introduce something about God and the faith in discussions about the infinite in the third book of the *Physics,* Bacon is silent. When we further take into account the fact that in the questions in which Bacon does mention something that seems relevant to theology, the terms, phrases, and discussions occupy only a small fraction of the questions in which they occur. Thus the percentage of theologically relevant material is miniscule.

In the 147 pages of the printed edition of Bacon's *De celestibus,*[29] which is based on Aristotle's *On the Heavens,* he offers thoughts about the faith on only two of those pages. A treatise on cosmology was an ideal place to intrude thoughts about God and the faith. Yet Bacon chose not to avail himself of this opportunity. What is most remarkable about the *De celestibus* is what Bacon does not say. He makes no mention of the empyrean heaven, which was a purely theological invention. One would fully expect someone as interested in the interrelations of theology and natural philosophy as was Bacon to mention this commonly accepted dwelling place of God and the elect. In his discussion of the possibility of other worlds,[30] Bacon, writing before the Condemnation of 1277, also

25. See ibid., 124, 125 (one question) and 54 and 145.

26. See Ibid., 54, 125, 145 (twice, in two different questions), 146 (the second occurrence on p. 145 is in the same question as the occurrence on p. 146), 204–5, 207, 271–72, 331, 412, 416–18.

27. See ibid., 375.

28. See ibid., 389.

29. The *De celestibus* is the second book of Bacon's *Communia naturalium.* See *Opera hactenus inedita Rogeri Baconi,* edited by R. Steele, fascicule 4 (1913). The *De celestibus* extends over pages 309 to 456. The *Communia naturalium* was probably written in the late 1260s or early 1270s. See *Roger Bacon's Philosophy of Nature: A Critical Edition, with English Translation, Introduction, and Notes, of "De multiplicatione specierum" and "De speculis comburentibus,"* edited and translated by D. C. Lindberg, xxv (Oxford, 1983).

30. See Bacon, *De celestibus,* 374.

makes no mention of God and therefore does not suggest that by His omnipotence God could, contrary to Aristotle's denial of the possibility of other worlds, create other worlds if He so wished.[31]

Unlike Roger Bacon, who never became a theologian, Albertus Magnus and Thomas Aquinas were already masters of theology when they wrote their commentaries on the natural books of Aristotle. As professional theologians, both were free to insert thoughts about God and the faith in their treatises on natural philosophy, wherever such thoughts might be deemed appropriate. It is of importance, therefore, to see how they viewed the relations between natural philosophy and theology, and to determine the extent to which they were prepared to theologize natural philosophy. The evidence shows unequivocally that both chose to keep the theologization of natural philosophy to a minimum.

In the opening words of his commentary on Aristotle's *Physics*, Albertus declares that his Dominican brothers had implored him to "compose a book on physics for them of such a sort that in it they would have a complete science of nature and that from it they might be able to understand in a competent way the books of Aristotle."[32] Perhaps thinking that his fellow friars would expect him to intermingle theological ideas with natural philosophy, Albertus declares that he will not speak about divine inspirations, as do some "extremely profound theologians," because such matters "can in no way be known by means of arguments derived from nature." And he then explains that

[p]ursuing what we have in mind, we take what must be termed "physics" more as what accords with the opinion of Peripatetics than as anything we might wish to introduce from our own knowledge . . . for if, perchance, we should have any opinion of our own, this would be proffered by us (God willing) in theological works rather than in those on physics.[33]

Albertus thus believed that Aristotle's natural philosophy was to be treated naturally, in the customary manner of Peripatetics. Where theo-

---

31. On the plurality of worlds, see E. Grant, "The Condemnation of 1277, God's Absolute Power, and Physical Thought in the Late Middle Ages," *Viator* 10 (1979): 217–226; reprinted in E. Grant, *Studies in Medieval Science and Natural Philosophy*, ch. 12 (London, 1981).

32. Translated in E. Synan, "Introduction: Albertus Magnus and the Sciences," in *Albertus Magnus and the Sciences*, edited by J. A. Weisheipl, O.P., 9 (Toronto, 1980).

33. See ibid., 10. Synan presents the section of this passage that follows the ellipsis before the lines that precede it. But the order of the passages in Albertus's *Physics* is as they appear here.

logical issues might be involved, they were to be treated in theological treatises. In his *Commentary on De caelo* Albertus makes it evident that he wished to uphold his basic conviction that, unless unavoidable, theology should not be intruded into natural philosophy. In discussing whether the heaven is ungenerable and incorruptible, Albertus explains that

[a]nother opinion was that of Plato who says that the heaven was derived from the first cause by creation from nothing, and this opinion is also the opinion of the three laws, namely of the Jews, Christians, and Saracens. And thus they say that the heaven is generated, but not from something. But with regard to this opinion, it is not relevant for us to treat it here.[34]

Because he sought to avoid theology, Albertus says that he will therefore only inquire about a third opinion,

which says that the heaven is generated from something preexisting and is corrupted into something that remains after it, just as natural things are generated and corrupted by the actions of qualities acting and being acted on mutually. And because these things alone proceed naturally and from principles of nature, we inquire about this mode, [namely,] whether the heaven is generated.[35]

Thus Albertus will speak not about the generation of the heaven from nothing, which is only possible supernaturally, but about its generation from something preexisting, which is naturally possible, even though it conflicts with a fundamental doctrine of his faith.

It is undoubtedly because of his conviction that a theologian doing natural philosophy should avoid theological discussions to the greatest extent possible that we find relatively little about God and the faith in Albertus's *Commentary on De caelo*.[36] The subject of the third tractate of the first book is "whether there is one world or more" *(Utrum mundus sit unus vel plures)*,[37] a theme that often produced mentions of God. Albertus, however, explains that

---

34. Albertus Magnus, *Alberti Magni Ordinis Fratrum Praedicatorum Opera Omnia,* edited by B. Geyer; *Monasterii Westfalorum in aedibus Aschendorff,* vol. 5, pt. 1, *De caelo et mundo,* edited by P. Hossfeld, 1971, bk. 1, tract. 1, ch. 8, 19, col. 2–20, col. 1; hereafter *Commentary on De caelo.)* Plato did not hold that the world was created from nothing. The translation is mine. Unless indicated otherwise, the translations below are mine.

35. Albertus Magnus, *Commentary on De caelo,* 20, col. 1.

36. As inspection of the index under "deus" (300, col. 1) and "fides" (304, col. 3) reveals.

37. Ibid., bk. 1, tract. 3, chs. 1–10, 55–77.

[i]f ... someone should say that there can be more worlds but there are not, because God could have made more worlds if He wished and even now could make more worlds, if He wishes. Against this, I do not dispute, since here I conclude that it is impossible that there be several worlds and that it is necessary that there be one [world] only. Here we understand about [i.e., we are concerned about] the impossible and necessary—that is, [we are concerned about] the world with regard to its essential and proximate causes. And there is a great difference between what God can do by means of His absolute power and what can be done in nature [or by nature].[38]

With respect "to the nature of the world," Albertus says that "there cannot be more worlds, although God could make more, if He wishes."[39] It is not, however, what God can do that interests Albertus in his commentary on De caelo, but what nature can do. He concludes that nature cannot produce other worlds by its own powers. At the conclusion of the first book, Albertus emphasizes that investigators into nature do not inquire about how God uses the things He has created to make a miracle in order to proclaim His power; but, rather, they investigate "what could be done in natural things according to the inherent causes of nature."[40]

Albertus kept theological references in his natural philosophy to a minimum, as is evident in his Aristotelian commentaries. In the 261 chapters that comprise the eight books of his Commentary on the Physics, Albertus mentions God (deus and its variants) in 24, or in approximately 9 percent of his chapters; and in the 111 chapters that make up the four books of his Commentary on De caelo, he mentions God in 9, or in approximately 8 percent of the total. Most of Albertus's uses of the term God in his Commentary on the Physics are in direct response to Aristotle's text, especially in the eighth book. Thus of the 64 occurrences of primus motor, 55 occur in book eight; of the 69 occurrences of causa prima, 37 occur in book eight; and of the 78 occurrences of deus, 40 occur in book eight.

38. Ibid., bk. 1, tract 3, ch. 6, 68, col. 2.
39. "Et ideo quantum est de natura mundi, dico non posse fieri plures mundos, licet deus, si vellet, posset facere plures." Albertus Magnus, Commentary on De caelo, bk. 1, tract. 3, ch. 6, 69, col. 1.
40. "Et ideo supra diximus, quod naturalia non sunt a casu nec a voluntate, sed a causa agente et terminante ea, nec nos in naturalibus habemus inquirere, qualiter deus opifex secundum suam liberrimam voluntatem creatis ab ipso utatur ad miraculum, quo declaret potentiam suam, sed potius quid in rebus naturalibus secundum causas naturae insitas naturaliter fieri possit." Albertus Magnus, Commentary on De caelo, bk. 1, tract. 4, ch. 10, 103.

Most of these occurrences of key terms fall into *Category 1* because they are not a defense of the faith, or about the faith as such. But Albertus unhesitatingly defended the faith against those who offered conflicting interpretations. One of the most serious claims that required a defense was Aristotle's arguments for the eternity of the world, which, if ignored, would have denied the creation. A major locus for these arguments was the eighth book of Aristotle's *Physics,* where Aristotle argued more specifically for the eternity of motion. To these kinds of arguments, Albertus replies in a chapter in which he demonstrates that the world began by a creation.[41]

Many mentions of the Christian God are minimal, little more than passing references, as when Albertus, in presenting eight ways in which something can be in another, says that "sometimes it is internal, namely when form is a mover with respect to place, just as the soul in a body and God [*deus*] in the world"[42]; or, in a discussion of time, when Albertus says that "they say that, when it is said that God is 'now' [*nunc*], and an intelligence is 'now,' and a motion is 'now,' the same 'now' is denoted."[43] In the two instances just cited, Albertus's usage conforms to *Category 3* (see above), where theological terms and concepts are used analogically, or to exemplify and illustrate things and processes in the natural world. The parts of their respective chapters which these two instances comprise are very small indeed.

As a theologian-natural philosopher, Albertus could easily have inserted passages about God almost anywhere in his physical commentaries. For example, in his lengthy commentary on the infinite, extending over thirty-two double-columned pages,[44] it might have been tempting to elaborate on God's infinite powers. But Albertus mentions God only twice: once in a context describing the way in which pre-Socratic philosophers used the term infinite[45] and again, by way of example, in the first of five ways in which the infinite is described, a privative one, "God [*deus*] is

41. Albertus Magnus, *Alberti Magni Ordinis Fratrum Praedicatorum Opera Omnia,* edited by B. Geyer, *Monasterii Westfalorum in aedibus Aschendorff,* vol. 4, *Physica,* edited by P. Hossfeld, part 1 (bks 1–4), 1987; part 2 (bks. 5–8), 1993, *Commentary on the Physics,* bk. 8, tract. 1, ch. 13, 574–77.

42. Ibid., bk. 4, tract 1, ch. 6, 211. This is the only mention of God in a lengthy chapter that extends over pages 210 to 214.

43. Ibid., bk. 4, tract. 4, ch. 5, 299, lines 16–18.

44. Ibid., bk. 3, tract 2 *(De infinito),* 168–200.

45. Ibid., bk. 3, tract. 2, ch. 2, 172, lines 58–62.

said to be infinite [*infinitus*] and incorporeal [*incorporeus*] and immense [*immensus*],"[46] that is, God is "not finite"; God is "not a body"; and God is "not measurable." Indeed, Albert ignores a good opportunity to invoke God when, within the context of the infinite, he launches into a discussion of extracosmic space, place, and vacuum.[47] In theological treatises, it was common to involve God with space, place, and vacuum. Albertus could easily have done so had he wished. Also surprising is the fact that in his discussion of the celestial orbs in his *Commentary on De caelo*, where he speaks of ten orbs, Albertus makes no mention of the crystalline orb and the empyrean heaven, the traditional theological spheres.[48]

Thomas Aquinas (ca. 1224–1274) preserved the approach that Albertus Magnus developed toward Aristotelian natural philosophy. Like Albertus, Thomas sought to minimize theological intrusions into his commentaries on the natural books of Aristotle. The relatively few occurrences of key terms such as "God," "faith," "creation," "first mover," and "first cause" in Thomas's commentaries on the *Physics* and *On the Heavens*, and their near total absence from his commentaries on *On Generation and Corruption (De generatione et corruptione)* and the *Meteorology* strongly support this interpretation. In Thomas's *Commentary on Aristotle's Physics*, we find almost all mentions of God and its medieval Scholastic synonyms, as well as all appeals to faith, in the eighth book, a feature that is also true of Albertus Magnus's *Commentary on the Physics* (see above). Only a few isolated citations occur in the rest of his lengthy commentary. This is striking, but not startling, since Aristotle's major demonstration of a first mover in the eighth book caused Thomas, and all who commented on that book, to speak frequently of the first mover and, consequently, to find occasions to mention God. In view of long-held attitudes and opinions about the role of theology and faith in natural philosophy, the relatively few citations that Thomas made involving theology and the faith come as a surprise when we realize that Thomas found occasion to mention God in only 21 paragraphs out of 2,550[49]; that the 54 occurrences of

46. Ibid., bk. 3, tract 2, ch. 4, 175, lines 63–65.

47. Ibid., bk. 3, tract. 2, ch. 3, 174–175.

48. See Albertus Magnus, *Commentary on De caelo*, bk. 2, tract 3, ch. 11, 166–167. Also surprising is the absence of anything of a religious nature in a chapter titled "On the Perpetuity of Life That Exists in the External Convexity of the Heaven" (bk. 1, tract 3, ch. 10, 75–77).

49. The data is drawn from *S. Thomae Aquinatis In octo libros De physico auditu sive Physicorum Aristotelis commentaria*, edited by P. Fr. Angeli-M. Pirotta, O.P. (Naples, 1953).

"Prime Mover" and its variants occur in 43 paragraphs; that the 10 usages of "First Cause" occur in 10 paragraphs; and that matters of faith are mentioned in only 8 paragraphs. If we sum 21, 43, 10, and 8, we arrive at a total of 82 differently numbered paragraphs. Allowing for overlap in two paragraphs, the total number of paragraphs in which some version of God's name or mention of the faith appears is 80, of which 69 are in the eighth book, leaving 11 for the other seven books. The 80 paragraphs represent approximately 3 percent of the 2,550 paragraphs.

To convey an idea of how Thomas used theological terms in the categories distinguished above, I shall use the data for the 21 occurrences of the term God (*deus;* 4 of the 21 instances appear as the term "divine" [*divinum*]): 5 in *Category 1;* 7 in *Category 2;* 5 in *Category 3;* none in *Category 4,* which applies to the aftermath of the Condemnation of 1277, and therefore does not apply to Thomas; and 4 that I was unable to classify and therefore place in *Category 5.*

Like Albertus, Thomas also refrained from introducing theological ideas into natural philosophy. Thus in his *Commentary on De caelo,*[50] Thomas follows Albertus Magnus and makes no mention of the empyrean heaven, although both accepted its existence and found occasion to mention it in their theological treatises.[51]

Thomas frequently indicates where Aristotle disagrees with the faith. In 1271, however, near the end of his life, he explained why he did not often mix matters of faith with natural philosophy. In considering a question on the rational soul in man, he seemingly dismisses the question, by asserting that "I don't see what one's interpretation of the text of Aristotle has to do with the teaching of the faith."[52] In Vernon Bourke's judg-

---

50. Thomas's commentary appears in *S. Thomae Aquinatis In Aristotelis libros De caelo et mundo; De generatione et corruptione; Meteorologicorum Expositio cum textus ex recensione leonina,* edited by R. M. Spiazzi (Turin, 1952).

51. Perhaps Thomas refrained from mentioning it in a treatise on natural philosophy, because, as he explains in his commentary on the *Sentences* (bk. 2, distinction 2, qu. 2, art. 1), "the empyrean heaven cannot be investigated by reason because we know about the heavens either by sight or by motion. The empyrean heaven, however, is subject to neither motion nor sight . . . but is held by authority"; cited from E. Grant, *Planets, Stars, and Orbs: The Medieval Cosmos, 1200–1687,* 377, n. 28 (Cambridge, 1994).

52. "Nec video quid pertineat ad doctrinam fidei qualiter Philosophi verba exponatur." The translation and the Latin text are by V. J. Bourke in *St Thomas Aquinas Commentary on Aristotle's "Physics,"* translated by R. J. Blackwell, R. Spath, and W. E. Thirlkel, with an introduction by V. J. Bourke, xxiv (New Haven, Conn., 1963). Bourke does not provide a full reference,

ment, Aquinas did not think he was "required to make Aristotle speak like a Christian" and he undoubtedly "thought that a scholarly commentary on Aristotle was a job by itself, not to be confused with apologetics or theology."[53]

Natural Philosophy in the Fourteenth Century: John Buridan, Nicole Oresme, Themon Judaeus, and Albert of Saxony

It is ironic that the four fourteenth-century natural philosophers whose works are considered here include many more references to God and the faith than did Albertus Magnus and Thomas Aquinas, who were theologians when they wrote their Aristotelian commentaries. This is perhaps partially explicable by the fact that a number of references involved the introduction of counterfactuals by way of references to the absolute power of God to do anything He pleased, short of a logical contradiction, a tactic that is especially noticeable in John Buridan's *Questions on De caelo*.[54] But the effect of the Condemnation of 1277 may have been more pervasive than the introduction of counterfactuals involving God's absolute power. God may also have been invoked to avoid the possible charge of being overly naturalistic at the expense of God's ultimate and underlying role in all events.

The increased invocation of God in the fourteenth century, however, is comparative. It seems more extensive only when compared to the natural philosophical works we have previously examined from Albertus Magnus and Thomas Aquinas. An examination of the 310 questions embedded in the five questions treatises by John Buridan, Nicole Oresme, Themon Judaeus, and Albert of Saxony shows clearly that, like their predecessors in the thirteenth century, most of their texts had little to do with God, the faith, or theology, but were concerned solely with issues in natural philosophy. Of the 310 questions, 217 are free of any entanglement with theol-

---

but the statement occurs in Thomas's *Responsio ad fr. Joannem Vercellensem de articulis* 42 (43), which was printed in Aquinas's *Opera Omnia* (Parma, 1852–1873; reprinted in 25 folio volumes in New York, 1948–1950), 16.167. For a brief description of the treatise, see J. A. Weisheipl, *Friar Thomas d'Aquino: His Life, Thought, and Work*, 390, nr. 65 (Garden City, 1974).

53. The two quotations are from V. J. Bourke's introduction, in *St. Thomas Aquinas Commentary on Aristotle's "Physics,"* xxiii and xxiv.

54. See *Iohannis Buridani Quaestiones super libris quattuor De caelo et mundo*, edited by E. A. Moody (Cambridge, Mass., 1942).

ogy or faith, while 93 (or approximately 29 percent) mention God and the faith. Inspection of any of the 217 questions would not reveal whether the author was Christian, Muslim, Jewish, agnostic, or atheist. Of the 93 with at least a trace of theological sentiment, 53 mention God, or something about the faith, in a cursory manner; of the remaining 40 questions, only 10 have relatively detailed discussions about God or the faith.

The data show that the greatest opportunities for introducing God and the faith occurred in questions and commentaries on Aristotle's *Physics, On the Heavens (De caelo)*, and *On the Soul (De anima)*, with only modest intrusions in *On Generation and Corruption (De generatione et corruptione)* and the *Meteorology (Meteorologica)*. Thus we find that in his *Questions on De caelo*, Buridan discusses God and/or the faith in 27 questions out of 59; that Albert of Saxony discusses such matters in 38 of 107 questions in his *Questions on the Physics;*[55] and Nicole Oresme considers them in 18 of 44 questions in his *Questions on De anima.*[56] By contrast, Albert of Saxony and Themon Judaeus found few occasions for introducing God and faith into their *Questions On Generation and Corruption* and *Meteorologica,* respectively.[57] Albert injected such matters into 6 of his 35 questions and Themon did so in only 4 of 65 questions. In this regard, Albert and Themon were like Thomas Aquinas, who made no mention of God in his commentaries on these two treatises and who defended the faith a total of three times in both together.

It is important to see how the discussions and citations of God and the faith in the 93 questions are classifiable in terms of the five categories distinguished earlier.

55. Albert of Saxony, *Questiones et decisiones physicales insignium virorum. Alberti de Saxoma in octo libros Physicorum; tres libros De celo et mundo; duos libros De generatione et corruptione; Thimonis in quatuor libros Meteororum; Buridani in tres libros De anima; librum De sensu et sensato; librum De memoria et reminiscentia; librum De somno et vigilia; librum De longitudine et brevitate vite; librum De iuventute et senectute Aristotelis. Recognite rursus et emendatae summa accuratione et iudicio Magistri Georgii Lokert Scotia quo sunt tractatus proportionum additis* (Paris, 1518).

56. See Peter Marshall, "Nicholas Oresme's *Questiones super libros Aristotelis De anima:* A Critical Edition with Introduction and Commentary" (Ph.D. diss., Cornell University, 1980).

57. For Albert of Saxony's *Questions on Generation and Corruption,* see *Questiones et decisiones physicales insignium virorem. Alberti de Saxoma in octo libros Physicorum . . .* (Paris, 1518); for Themon's *Questions on the Meteorology,* see ibid., where Themon's name is spelled "Thimon."

Category 1   2
Category 2   12
Category 3   34
Category 4   34
Category 5   16
TOTAL   98[58]

It is obvious that the dominant concern of these fourteenth-century natural philosophers was with *Categories 3* and *4,* which were concerned, respectively, with God used in an illustrative and exemplary sense, and with the absolute power of God. Together, these two categories extend over 68 of the 93 (or 98; see note 58) questions in which something about God and the faith arise in the five treatises. Let us now examine briefly each of the categories in the order 1, 2, 5, 3, and 4, thus reserving the most frequently occurring categories for last. Because of space constraints, I shall include only one example from categories 1, 2, and 5, two from category 3, and more than two from category 4, which is the most interesting categorical use of God in fourteenth-century natural philosophy.

Category 1: The Reaction to Aristotle's Mention of God

In *On Generation and Corruption* (2.10.336b.25–35; Oxford translation), Aristotle declares that God "fulfilled the perfection of the universe by making coming-to-be uninterrupted for the greatest possible coherence would thus be secured to existence, because that coming-to-be should itself come-to-be perpetually is the closest approximation to eternal being." In reaction to this passage, Buridan explains that Aristotle "wishes to declare here and in the second [book] of *De generatione* how such an order is reasonably from God and how all existing things from God, both celestial and inferior, are harmonious with regard to that order that is to be perpetually conserved."[59] Buridan invokes God here in direct response to Aristotle's comments.

---

58. The total is 5 beyond 93 because five questions have been recorded under two different categories. The following duplications occur: Buridan, *De caelo,* bk. 1, qu. 10 appears under *Categories 2* and *4; De caelo,* bk. 2, qu. 10 appears under *Categories 1* and *4;* Albert of Saxony, *Physics,* bk. 2, qu. 10, appears under *Categories 3* and *5; Physics,* bk. 6, qu. 9 appears under *Categories 3* and *4;* and *Physics,* bk. 2, qu. 14 appears under *Categories 1* and *5.*

59. Buridan, *Questions on De caelo,* bk. 2, qu. 10, pp. 171–172.

Category 2: Invocation of God and Faith against the Doctrinal Errors of Aristotle and the Philosophers

In his *Questions on De caelo,* in a question concerned with generable and corruptible things, Buridan declares that

Aristotle says many things that cannot be properly saved. . . . For he holds indeed that nothing corruptible, or having potency for not being, can always exist in the future; and this is in fact false and against the faith because all things except God are corruptible and at some time they are not and are not able to be because they could be annihilated by God. And yet many things are perpetuated and always remain.[60]

Category 5: Mentions of God That Do Not Fit Any of the Other Four Categories

In asking "whether for perfectly understanding something it is necessary to know all its causes,"[61] Albert of Saxony offers eight principal arguments against this proposition. In the second of these, he presents the following proof: "For God is the cause of any whatever thing. Therefore for the perfect understanding of any thing, it is necessary to know God. But since God could not be perfectly known by us, it follows that no thing can be known perfectly by us if, in order to have perfect cognition of anything, it is necessary to know all causes."[62] In his reply to this argument at the end of the question, Albert says: "To the second, I similarly concede that for the perfect cognition of anything, it is absolutely necessary to know God. However, this is not required for perfect cognition in the genus of something."[63] Since this is a discussion about the role of God in understanding things, it does not fit any of the other categories. We should note, however, that Albert argues that we can have perfect cognition of something within a genus, even without perfect knowledge of God.

Category 3: God as Example, Analogy, and Basis of Comparison

In his *Questions on De anima,* Nicole Oresme uses God for illustrative purposes a number of times, occasionally inserting a few within a short

60. Ibid., bk. 1, qu. 26, p. 127.
61. Albert of Saxony, *Questions on the Physics,* bk. 1, qu. 3, fols. 2r, col. 2–3r, col. 1.
62. Ibid., bk. 1, qu. 3, fol. 2r, col. 2.
63. Ibid., *Questions on the Physics,* bk. 1, qu. 3, fol. 3r, col. 1.

space. In a supposition, he declares "that some power makes this or that operation anew without changing itself, just as is obvious with God who continuously produces new effects without any change in Himself."[64] Themon Judaeus employs God in a similar comparative manner, when he assumes that "a pure element is understood [to be] simple, but not simple absolutely, as is God, or an intelligence."[65]

Category 4: God's Absolute Power

In a question inquiring whether every extended thing is a quantity, Albert of Saxony invokes God's absolute power to show that this is not necessarily so. He assumes a given quantity, which he calls a, b, and c, and then argues that "God can separate, without local motion, any whatever thing that is distinct from another thing. Therefore God can separate a quantity from extended thing a, b, and c."[66] Elsewhere, Albert assumes that "God could create another body around this world; and around that body [He could] create another body; and so to infinity. Nevertheless, these bodies are not mutually continuous."[67]

In another question, Albert emphasizes God's power to make a greater magnitude than any given magnitude.[68] In book three, question twelve, Albert mentions God about ten times, all of them relevant to his assumption that God can make an infinite weight by creating a one-foot stone in every proportional part of an hour, and to his assumption that God can annihilate all matter lying between the concave orb of the moon.[69] Albert uses the idea of annihilation again, when he imagines what would happen if God created a vacuum by annihilating all the matter lying between the walls of a pipe (fistula).[70] In another question,[71] Albert invokes God's absolute power in a number of contexts related to local motion. Once again, he assumes that God annihilates matter, this time annihilating all celestial bodies except one, the moon, which rotates from east

64. Nicole Oresme, Questions on De anima, bk. 3, qu. 2, 517 of Marshall's edition.

65. Themon Judaeus, Questions on the Meteorology, bk. 4, qu. 5, fol. 213v, col. 1.

66. Albert of Saxony, Questions on the Physics, bk. 1, qu. 6, fol. 5r, col. 2.

67. Ibid., bk. 3, qu. 11, fol. 39r, col. 2.

68. Ibid., bk. 3, qu. 14, fol. 42r, cols. 1 and 2.

69. Ibid., bk. 3, qu. 12, fols. 39v, col. 1–40v, col. 2.

70. Ibid., bk. 4, qu. 12, fol. 51r, col. 1.

71. What follows in this paragraph appears in Questions on the Physics, bk. 3, qu. 6, fols. 35v, col. 1–36r, col. 1.

to west. The challenge is to explain how a body that has no bodies external to it may be said to be in motion. In the same question Albert also assumes that God could fuse all the celestial orbs, along with all bodies and matter below the moon, into one continuous whole, which He sets into rotation from east to west, or in any way He pleases. Once again, Albert seeks to explain how we are to understand a motion that does not relate in any way to anything outside of itself. In responding to the third principal argument of the same question, Albert places the only limitation one could place on God: not even God could do anything that implies a contradiction.

Similar declarations appear in John Buridan's *Questions on De caelo,* a treatise that offered many opportunities to insert claims about God's absolute power. For example, Buridan argues that "although it is not possible by natural powers that a plurality of motions not exist, it is, nevertheless, not absolutely impossible, because God could make the whole heaven rest."[72] In discussing the empyrean heaven, Buridan says of this immobile heaven beyond the movable heavens that "according to nature [it] does not have any potency or inclination for motion, although it could be moved supernaturally by God Himself, just as all things, except God Himself, could be annihilated by God."[73] Indeed, as an illustration of God's absolute power, adopted directly from Article 49 of the Condemnation of 1277, both Buridan and Nicole Oresme assumed that God could move the entire world with a rectilinear motion.[74]

And going beyond our world, beyond the last immobile empyrean heaven, Buridan declares that "it must be conceded that outside this world, God could easily create a corporeal space and however many corporeal bodies He pleases; but [just] because of this [namely, that He could do it], it should not be assumed that He did it."[75] Among the things that God could make beyond our world are other worlds. Buridan counters Aristotle's arguments against the possibility of other worlds by insisting that "we know from faith that God could make a world, indeed many worlds, and He is also able to destroy them again." Buridan proclaims that "successively different worlds could be made by divine power, but

72. Buridan, *Questions on De caelo,* bk. 2, qu. 10, 170.
73. Ibid., bk. 2, qu. 6, 152.
74. See ibid., bk. 1, qu. 16, 75–76, and Oresme, *Questions on De anima,* bk. 2, qu. 15, 386.
75. Buridan, *Questions on De caelo,* bk. 1, qu. 17, 79.

not by natural power because celestial bodies are not generable or corruptible by natural powers."[76]

God's absolute power had one monumental obstacle that Scholastic ingenuity never surmounted. Could God create an infinite magnitude, or an infinite perfection? Buridan argues that not even God could perform these acts. "I believe," he declares, "that God cannot make a body so great but that He could not make a greater body, nor that He could make a thing so perfect that He could not make one even more perfect."[77] For this reason, Buridan insists that "it is not possible even by the power of God that an infinite body with respect to magnitude [be created], nor [can He make] an effect according to infinite perfection."[78] Thus if God could create an infinite body or perfection, His power would thereafter be limited, since not even God could create something greater than an infinite. But His power is also limited if He cannot create an infinite body or perfection. Scholastics argued this point for centuries without resolution.[79]

The extensive use of the concept of God's absolute power in fourteenth-century natural philosophy gave rise to many counterfactual examples. To avoid even the hint of placing limits on God's omnipotence, natural philosophers assumed that God could create as many other worlds as He pleased, that He could move our world with a rectilinear motion, that He could separate a quantity from its extension, and so on. Indeed, it was always assumed that God could do anything whatever short of a logical contradiction, and therefore that He could do anything that was naturally impossible in Aristotle's world. For anyone who views medieval natural philosophy as about God and the creation, the medieval preoccupation with counterfactuals poses a dilemma: counterfactuals are not about creation. They are about things God could have done, and could do now, but which virtually nobody thought He had done, or would do. Counterfactuals subvert the assertion that natural philosophy is always about God and the creation. Even the part of natural philosophy

---

76. For the Latin text of these two passages, see Buridan, *Questions on De caelo,* bk. 1, qu. 19, 89. Article 34 of the Condemnation of 1277 denounced the opinion that God could not create other worlds. See Grant, *Planets, Stars, and Orbs,* ch. 8 ("The Possibility of Other Worlds") and p. 151 for Article 34.

77. Buridan, *Questions on De caelo,* bk. 1, qu. 17, 79. Buridan makes much the same argument in his *Questions on the Physics,* bk. 3, qu. 15; see Grant, *Planets, Stars, and Orbs,* 111.

78. Buridan, *Questions on De caelo,* bk. 1, qu. 17, 79.

79. See Grant, *Planets, Stars, and Orbs,* 106–10, especially 110.

that is about the world that God created—and which all happily acknowledged that He had created—was not investigated in order to determine
God's nature, to learn about the faith, or to discover religious aspects of
the created world. The paucity of material on these themes in the treatises discussed here is striking, if mute, testimony to these claims.

Conclusion

The evidence presented here reveals that medieval natural philosophers
who explicated the texts of Aristotle's natural books kept their inevitable involvements with God and the faith to a minimum. This is especially true in the thirteenth century, when such famous Dominican
theologians as Albertus Magnus and Thomas Aquinas, who wrote their
commentaries on Aristotle's natural philosophy after they had become
masters of theology, refrained remarkably from intruding theology into
their natural philosophy.

Roger Bacon is even more surprising, because he explicitly advocated
intermingling theology and natural philosophy. But when opportunities
arose to mix the two, he rarely did so.

The Condemnation of 1277 compelled fourteenth-century natural philosophers, most of whom were secular arts masters when they wrote their
relevant treatises, to invoke God's absolute power to do many things that
were naturally impossible in Aristotle's natural philosophy and, at the
same time, to make certain that Aristotle's contrary-to-faith concepts were
plainly identified as errors. Despite these constraints, the overall impact
of specific ideas about God and the faith are quite modest and should not
alter the conception that the content of fourteenth-century natural philosophy was fundamentally about natural phenomena studied in a rational
and secular manner to the fullest extent possible.

No one exemplified this approach better than John Buridan. As a natural philosopher, Buridan was aware that his objective was to describe
and explain nature's operations in terms of natural causes and effects,
and not to explicate God's supernatural actions and miracles. Buridan
had no problems with his faith. He accepted the truths of revelation as
absolute, and acceded to them. But in keeping with the tradition of his
fellow natural philosophers, he acknowledged that his task was to explicate problems about natural actions and phenomena, and not to deal

with the supernatural. In treating a question as to whether every generable thing will be generated, Buridan immmediately acknowledges that one can treat this problem naturally—"as if the opinion of Aristotle were true concerning the eternity of the world, and that something cannot be made from nothing"—or supernaturally, wherein God could prevent a generable thing from generating naturally by simply annihilating it. "But now," Buridan declares, "with Aristotle, we speak in a natural mode, with miracles excluded."[80] Buridan believed that truth was attainable when "a common course of nature [communis cursus nature] is observed in things and in this way it is evident to us that all fire is warm and that the heaven moves, although the contrary is possible by God's power."[81] Natural philosophers like Buridan were usually careful to concede that God could upset the natural order of things by direct intervention. That is why an expression such as the "common course of nature" was so useful. Natural philosophers were primarily interested in natural, not supernatural, powers, for which reason Buridan insisted that "in natural philosophy, we ought to accept actions and dependencies as if they always proceed in a natural way."[82] Although, by His absolute power, God could move an infinite body, Buridan regards it as obvious that Aristotle's arguments "conclude sufficiently with respect to natural powers."[83] If he had to concede that God could use His absolute, unpredictable power to produce any natural impossibilities He wished, Buridan could still save Aristotle and natural philosophy by characterizing Aristotle's arguments as sufficient in the real, natural world, the one he and his fellow natural philosophers sought to understand.

80. See Buridan, Questions on De caelo, bk. 1, qu. 25 (Utrum omne generabile generabitur), 123.

81. From Buridan, In Metaphysicen Aristotelis; Questiones argutissime Magistri Ioannis Buridani in ultima praelectione ab ipso recognitae et emissae ac ad archetypon diligenter repositae cum duplice indicio materiarum videlicet in fronte quaestionum in operis calce (Paris, 1518), bk. 2, qu. 1 ("whether the grasp of truth is possible for us"), fol. 8v, col. 2–9r, col. 1. The translation is by E. Sylla, "Galileo and Probable Arguments," in Nature and Scientific Method, Studies in Philosophy and the History of Science, edited by D. O. Dahlstrom, 22.216 (Washington, D.C., 1991); cited in E. Grant, "Jean Buridan and Nicole Oresme on Natural Knowledge," Vivarium 31 (1993): 88.

82. "Modo in naturali philosophia nos debemus actiones et dependentias accipere ac si semper procederent modo naturali, . . ."; Buridan, Questions on De caelo, bk. 2, qu. 9, 164. Also cited in Grant, "Jean Buridan and Nicole Oresme on Natural Knowledge," 89.

83. "Et sic manifestum est quod rationes Aristotelis sufficienter concludunt quantum ad potentias naturales"; Buridan, Questions on De caelo, bk. 1, qu. 17, 77.

To underscore the fact that medieval natural philosophy was about the natural, not the supernatural, operations of the world, it is important to recognize that in almost any given question *(questio)*, the invocations of religious or theological material usually occupy a small percentage of the total question. Let us recall that of the 310 questions in the five treatises that formed the basis of our investigation of Aristotelian natural philosophy in the fourteenth century, 217 had nothing whatever on God or the faith and only 93 did. Of the 93, however, most had relatively little on theology. For example, in his *Questions on the Physics,* Albert of Saxony asks whether "from the addition of some whole to some whole another whole is made; similarly, [whether] by the removal of some whole from some whole, another whole is made."[84] Of the 201 lines of text in this question, 10 are devoted to the fourth and fifth (of ten) principal arguments in which Albert rejects the proposition as follows:

Fourthly, it would follow that none of us would be baptized. But this is false and the consequence is proved because many particles are added to us. And thus we are greater than when we were baptized. Therefore by addition of some part to the whole there occurs another whole. Therefore it follows that none of us is the same whole which we were in [our] youth, and, consequently, none of us is that [person] which was baptized.

Fifthly, by similar reasoning, it would follow that none of us is the one who was born of his mother, just as Christ was not the same man who was suspended on the cross and who was born of the purest virgin.

Not only do these arguments constitute a small portion of the whole question—slightly less than 5 percent—but the discussions about baptism and Christ are examples, and could have been replaced by other examples of a nonreligious character. Moreover, within the structure of a typical question, the principal arguments and the responses to them at the beginning and end of the question, respectively, represent the least important parts. Between them lies the body of the question in which the author presents the main conclusions and qualifications. In the question we are discussing, the religious component occurs only in the principal arguments (indeed Albert does not even respond to them) and not in the body of the question. Thus they play no significant role in the question.

---

84. Albert of Saxony, *Questions on the Physics,* bk. 1, qu. 8, fols. 7r, col. 2–8r, col. 1. The translation is from Grant, *A Source Book in Medieval Science,* 200 (Cambridge, Mass., 1974).

Because fewer than one-third of the 310 questions considered here had theologically relevant material, and most include much less than 5 percent that pertains to God, the faith, or church doctrine (indeed more than half of the references are little more than passing mentions of God or some aspect of the faith and play insubstantive roles in their respective questions), we may rightly conclude that God and faith played little role in medieval natural philosophy. If natural philosophy was really about God and His creation, why did medieval natural philosophers virtually ignore these themes in their questions? The answer is obvious: because they were irrelevant to their objective, which was to provide natural explanations for natural phenomena. Perhaps, the most important reason why theology did not significantly penetrate natural philosophy is simply that while theology needed natural philosophy, natural philosophy did not need theology.

5 ❧ Medieval Departures from Aristotelian Natural Philosophy

Introduction

Departures from the specific ideas expressed in Aristotle's natural works occurred during the whole span of Aristotle's dominance in European thought. In a recent article, I sought to illustrate the manner in which early modern, or Renaissance, Scholastic natural philosophers abandoned aspects of Aristotelian cosmology that had been accepted in the late Middle Ages. I argued that the departures from medieval cosmology derived from "new external challenges to the Aristotelian system that began to take effect in the sixteenth century."[1] It did not, however, appear to me that early modern, or Renaissance, Scholastics were therefore more innovative or imaginative than their medieval predecessors. Indeed, they may have been less so because medieval Aristotelianism was not significantly challenged by new texts and ideas from outside Western Europe—at least not until the late fifteenth century. Consequently, any important and interesting departures from Aristotle's physical and cosmological texts derived either from ideas already embedded in works that had entered Western Europe during the great age of translation during the twelfth and thirteenth centuries or were the result of innovative ideas developed by medieval natural philosophers themselves, without external stimulation. Whatever the reasons for these departures, it is the purpose of this essay to identify and describe them.

---

1. Edward Grant, "A New Look at Medieval Cosmology, 1260–1687," *Proceedings of the American Philosophical Society* 129 (1985): 417–32.

Departures from Aristotle's Cosmology and Physics

1. Concentrics, Eccentrics, and Epicycles

The first, and probably most significant medieval departure, took place in the thirteenth century in cosmology. It involved a shift from Aristotle's system of concentric spheres, as described in the *Metaphysics*,[2] to a system of solid eccentric orbs as described by Ptolemy in the latter's *Hypotheses of the Planets*.[3] The differences between the systems was obvious to all: concentric spheres were in harmony with the demands of Aristotle's cosmology but could not save the astronomical phenomena, namely, planetary variations in distance and latitude. By contrast, the system of solid eccentrics and epicycles could save the astronomical phenomena but was at variance with important cosmological principles, primarily that of the earth as the sole center for planetary motion.

Ptolemy was apparently well aware of the disparities between the two systems and sought a limited, though important, accommodation wherein he conceded the significance of the earth as primary center by assigning *concentric* convex and concave surfaces to enclose the three or more eccentric orbs that he assigned to each planet.[4] In order to avoid the possibility of vacua within this nested arrangement, Ptolemy also assumed that the outer surface of a planet's total sphere coincided with the inner surface of the immediate superior sphere belonging to the next planet. For example, the convex external surface of Mars's planetary sphere was said to be the same distance from the earth's center, or the center of the world, as the concave external surface of Jupiter's planetary sphere.[5]

2. Translations of the *Metaphysics* were made from Arabic and Greek in the thirteenth century by Michael Scot and William of Moerbeke, respectively (see L. Minio-Paluello, "Aristotle: Tradition and Influence," in *Dictionary of Scientific Biography*, 16 vols., edited by Charles C. Gillispie, 1.267–81 (New York: Scribners, 1970).

3. For J. L. Heiberg's Greek text and German translation of the first book and L. Nix's German translation of the second book from an Arabic version, see J. L. Heiberg, ed., *Claudii Ptolemaei, opera quae exstant omnia, Vol. 2: Opera astronomica minora* (Leipzig: Teubner, 1907), 69–145. The conclusion of bk 1, which is omitted from Heiberg's version, has been published in an Arabic edition with English translation by Bernard Goldstein, in R. B. Goldstein, ed. and trans., *The Arabic Version of Ptolemy's Planetary Hypotheses* (Philadelphia: American Philosophical Society, 1967), 265–302.

4. See Ptolemy's description of Saturn (*Hypotheses of the Planets*, Heiberg, *Claudii Ptolemaei*, 125–31) and O. Pedersen, *A Survey of the Almagest* (Odense: Odense University Press, 1974), 396.

5. Goldstein, *Ptolemy's Planetary Hypotheses*, 7–8; Pedersen, *A Survey of the Almagest*, 393–

Both systems reached Western Europe during the thirteenth century, Aristotle's by way of translations from Arabic and Greek, and Ptolemy's by way of summary accounts, rather than direct translation. The clash of the two systems appears, perhaps for the first time, in Roger Bacon's *Opus tertium* and *Communia naturalium,* works composed in the 1260s.[6] From whatever source he may have derived it, Bacon described the essential features of Ptolemy's account as found in the latter's *Hypotheses of the Planets.*[7] Although Bacon seems ultimately to have rejected Ptolemy's system of solid eccentrics because of its incompatability with Aristotelian cosmology, his description of it became commonplace. But where Bacon rejected the system, most of his Scholastic successors for the next few hundred years embraced it. Indeed, it would still find supporters through much of the seventeenth century.

What did widespread acceptance of Ptolemy's system of material eccentrics and epicycles entail for Aristotelian cosmology and natural philosophy? Numerous questions arose, the answers to which were potentially subversive for Aristotelian cosmology. Did eccentrics imply rectilinear motions for planets? Would heavy bodies fall to the center of the earth? Would vacua occur between successive eccentric orbs which were of necessity of unequal thickness? What was the relationship between successive surfaces of any two successive eccentric or concentric orbs? The answers came in a variety of forms.

In the process of responding to these difficult questions, medieval natural philosophers adopted positions sharply at variance with Aristotle's cosmology or revealed significant difficulties with it. One significant consequence of eccentric orbs was the assumption, contrary to Aristotle, Averroes, and Maimonides, that planets could rotate around a geometric point as well as a physical body (the earth). Celestial motions could thus occur around a multiplicity of centers. Although Pierre d'Ailly saved the idea of an "absolute down" by explaining that heavy bodies would

---

95. In his concept of place, Aristotle identified the innermost, or concave, surface of a containing body with the outermost, or convex, surface of the body it contained (see Aristotle, *Physics* 4.4.211a.31–33).

6. Bacon's discussion of eccentrics and epicycles appears in P. Duhem, *Un fragment inédit de l' "Opus tertium" de Roger Bacon. Précédé d'une étude sur ce fragment* (Quaracchi: Ex typographia Collegii S. Bonaventurae, 1909), 99–137, and in Bacon 1913, 419–56.

7. For a description of the system as Bacon described it, see Edward Grant, "Cosmology," in *Science in the Middle Ages,* edited by D. C. Lindberg, 281–83 (Chicago: University of Chicago Press, 1978).

nonetheless always move toward the center of the earth because the latter is the center of the "total orb," that is the center of the two concentric surfaces that contain each planet's three or more eccentric orbs,[8] he did not attempt to explain away the multiplicity of centers of celestial motion. Perhaps because he agreed with Nicole Oresme, who declared that "whether Averroes likes it or not, we must admit that they [the heavenly bodies] move around various centers, as stated many times before; and this is the truth."[9] On the assumption of the real existence of eccentrics and epicycles, the concept of a plurality of centers became a basic feature of medieval cosmology.

The alleged relationship between the external surfaces of any two successive orbs—that is, the relationship between the convex surface of a contained sphere and the concave surface of its containing sphere—was usually left vague or ignored. Within the context of Aristotelian thought, three possibilities were envisioned: (1) the surfaces are continuous, that is, they coincide; (2) the surfaces are contiguous, that is, they are distinct but in direct contact at every point; or (3) they are wholly or partially distinct and without contact.[10] Although Aristotle did not explicitly declare for any of these three opinions, his conception of place seems to commit him to the view that successive surfaces of celestial orbs are continuous.[11] Thus the concave surface of a containing sphere is identical with the convex surface of the sphere it contains. The two surfaces become one and

8. See D'Ailly's *14 Questions on the Sphere of Sacrobosco*, in Johannes de Sacrobosco 1531, question 13, ff. 163v–164v.

9. Nicole Oresme, *Le Livre du ciel du monde*, in A. D. Menut and A. J. Denomy, eds., *Nicole Oresme "Le Livre du ciel et du monde"* (Madison: University of Wisconsin Press, 1968), bk. 2, ch. 16, p. 463. See also John Buridan's similar attitude in E. A. Moody, ed., *Johannis Buridani, Quaestiones super libris quattuor De caelo et mundo* (Cambridge, Mass.: Mediaeval Academy of America, 1942), bk. 2, question 14, p. 191, lines 19–23.

10. For Aristotle's definitions of contiguous and continuous, see *Physics*, bk. 5, ch. 3. This and what follows on the relationships between the surfaces of celestial orbs is drawn from my article, Edward Grant, "Eccentrics and Epicycles in Medieval Cosmology," in *Mathematics and Its Applications to Science and Natural Philosophy in the Middle Ages: Essays in Honor of Marshall Clagett*, edited by Edward Grant and J. E. Murdoch, 189–214 (New York: Cambridge University Press, 1987).

11. In *Physics* 4.4.212a5–6, Aristotle defined the place of something as the "boundary [or inner surface] of the containing body at which it is in contact with the contained body." He explains further (212a.30) that "place is coincident with the thing, for boundaries are coincident with the bounded." The translation is by R. P. Hardie and R. K. Gaye in J. Barnes, *The Complete Works of Aristotle: The Revised Oxford Translation* (Princeton, N.J.: Princeton University Press, 1984), vol. 1.

the same. Medieval Islamic and Latin astronomers seem to have adopted this interpretation for their system of continuously nested material
spheres. In his widely used *Theorica planetarum*, Campanus of Novara,
for example, assumed that the convex surface of one planetary sphere
was exactly equal to the distance of the concave surface of the next outer
sphere.[12]

Although the concept of continuous surfaces eliminated the fear that
either void spaces or alien matter might intervene between two successive material orbs, it posed formidable problems for natural philosophers.
How could celestial orbs move in different directions if they shared one
and the same surface? As Roger Bacon explained, "it would happen that
these orbs would be moved with equal velocity, even with the same motions, which is contrary to experience."[13] Two successive orbs, and therefore all orbs, would have to move in the same direction.[14] With this in
mind, some authors opted for contiguous surfaces or for their complete
separation. Whatever the problems with contiguous surfaces, the assumption of independent identities and existences allowed the respective
spheres to move in opposite directions. Despite difficulties, both continuous and contiguous surfaces could be made compatible with Aristotle's
cosmology.

But there were those who, like Albertus Magnus and Cecco d'Ascoli,
insisted that the surfaces of the spheres are neither continuous nor contiguous, but without contact and separated from each other. The spaces
between any two such surfaces or orbs are not void, but filled with matter
that is capable of contraction and expansion, as circumstances dictated.[15]
Albertus, who presented the earlier and more detailed account, assumed
that the intervening matter differed from the incorruptible, indivisible

12. F. S. Benjamin Jr. and G. J. Toomer, eds., *Campanus of Novara and Medieval Planetary
Theory: "Theorica Planetarum"* (Madison: University of Wisconsin Press, 1971), 331–37 and 53–55.

13. *Opus tertium*, Duhem, *Un fragment inédit*, 12; *Communia naturalium*, Bacon 1913, 436.

14. Despite his seeming commitment to continuous surfaces, Aristotle did assume that the
celestial orbs could move in different directions and around different axes. See his *Metaphysics*, bk. 12, ch. 8.

15. For Albertus, see *De caelo et mundo*, bk. 1, tract 1, ch. 11, Albertus Magnus, *De coelo et
mundo*, edited by P. Hossfeld, in *Alberti Magni Opera Omnia*, t. V/1, 29–30 (Munster: Aschendorff, 1971). Cecco considered the problem in his *Commentary on the "Sphere" of Sacrobosco*, in
L. Thorndike, *The "Sphere" of Sacrobosco and Its Commentators* (Chicago: University of Chicago Press, 1949), 353. Although both authors give Thebit ibn Qurra as the source of this interpretation, I have not found it in any of Thebit's works that were translated into Latin.

ether, which was incapable of rarefaction and condensation. By contrast, interorbicular celestial matter was divisible and changeable. A number of major and radical divergences from Aristotle's cosmology are represented here. The celestial orbs are no longer in contact. By the assumption of two strikingly different kinds of celestial matter, they would appear to have abandoned the homogeneity of the celestial region. Change had been allowed into the celestial world.

Another major consequence was associated with the lunar epicycle. Because the moon always reveals its same face to us, as can be inferred from the spots on the moon—also perceived as the figure of a "man in the moon"—Aristotle had inferred that all other planets must similarly reveal the same face to us. Maintaining the same face toward us implied that no planet rotated with a proper motion.[16] But if the moon is actually carried around by an epicycle, it would present a different face to us when in the aux of the epicycle than when it is in the opposite of the aux. To avoid this potentially damaging conflict with observation, it was necessary to assume that the moon moved with a rotatory motion contrary to that of the epicycle and with an equal speed. In this way the moon would always reveal the same face to us. For those who accepted the existence of solid eccentrics and epicycles—and most Scholastic natural philosophers did—the moon's rotatory motion was routinely assumed. Here was a significant break with Aristotle: one or more planets was assumed to possess a motion of its own. Not all planets were merely passive bodies carried round by their epicycles. The moon, and perhaps other planets, was now thought to have a capacity for self-motion.[17] Some Scholastics even argued, as did Albert of Saxony, that the moon alone rotated because it differed from the other planets. In this approach, planets were divisible into those that had a capacity for self-rotation and those that did not. Such a radical distinction between planets was no part of Aristotelian cosmology.[18]

16. Aristotle, *De caelo* 2.8.290a25–27.

17. In support of the moon's rotatory motion, see Albert of Saxony, *Questions on De celo,* in Albertus de Saxonia 1518, bk. 2, q. 7, f. 106r, col. 2 [the fifth principal argument]); Pierre d'Ailly, *14 Questions* (Johannes de Sacrobosco 1531, ff. 163v and 164v); and Paul of Venice, *Summa naturalium* (Paulus Venetus, 1476): *Liber celi et mundi,* p. 31, col. 2 (because the work is unfoliated and provided with few signatures, the page numbers have been counted from the beginning of the *Liber celi et mundi*).

18. Unlike Albert of Saxony, John Buridan had argued for uniformity of planetary behav-

Although the system of eccentrics and epicycles was indeed a radical departure from Aristotle's system of concentric spheres, it was relatively easy to ignore because in the frequently asked question on the number of celestial spheres, the response was usually given as anywhere from eight to eleven, which included the attribution of one concentric sphere to each of the seven planets. Natural philosophers were, of course, well aware that eccentric and epicyclic orbs were contained within the concentric sphere assigned to each planet.[19] That the number of cosmic spheres was given in terms of the smaller number of concentric spheres rather than by the more numerous eccentric orbs is perhaps an indication that many Scholastic natural philosophers preferred to think of their cosmological system as Aristotelian rather than Ptolemaic.[20]

The perfect circular motion of the planets, which was a basic ingredient of Aristotle's cosmology, was also challenged, though with little impact. The sun's two simultaneous motions, that is, its daily east to west motion and its annual west to east motion, produced a path that was spiral-like rather than perfectly circular. This phenomenon was already mentioned by Plato in the *Timaeus*[21] and was repeated by numerous subsequent authors. In a remarkable statement, John of Sacrobosco drew attention to the fact that "the sun, moving from the first point of Capricorn through Aries to the first point of Cancer with the sweep of the firmament, describes 182 parallels, to which parallels, although they are not

---

ior and properties: either all planets rotated around their own centers or none did. Buridan adopted the latter alternative. See Edward Grant, *A Source Book in Medieval Science* (Cambridge, Mass.: Harvard University Press, 1974), 524–26, where Buridan's discussion of this issue is translated from his *Questions on the Metaphysics*, bk. 12, q. 11.

19. See Grant, "Cosmology," 280. In his famous encyclopedia of the early sixteenth century, Gregor Reisch distinguished eleven celestial spheres, or heavens, with each planet assigned one sphere (Reisch, *Margarita philosophica* [Basel: Michael Furterius, 1517; reprint, Dusseldorf: Stern-Verlag Janssen & Co., 1973], 245–46). But in his discussion of the motions of the seven planets he assigned at least three eccentric orbs to each planet. On this approach, at least twenty-four orbs were distinguishable. Reisch was merely repeating a common practice.

20. It is ironic that Aristotle assumed between forty-seven and fifty-five concentric spheres and thus had a larger number of spheres than did Ptolemy. Another irony lies in the fact that the assignation of one concentric sphere per planet was derived ultimately from Ptolemy's *Hypotheses of the Planets* rather than from Aristotle's *Metaphysics*.

21. 39 A–B. For a discussion of Plato's meaning and the manner in which the spiral is generated, see T. Heath, *Aristarchus of Samos. The Ancient Copernicus. A History of Greek Astronomy to Aristarchus Together with Aristarchus' Treatise on the Sizes and Distances of the Sun and Moon* (Oxford: Clarendon Press, 1913), 169; for the relevant figure, see 160. This part of the *Timaeus* was known in Latin translation during the Middle Ages.

really circles but spirals, since there is no sensible error in this, no vio-
lence is done if they are called 'circles,' of which number of circles are the
two tropics and the equinoctial."[22] Roger Bacon held that all the planets
as well as the fixed stars moved with spiral motions compounded of two
independent circular motions.[23] Strictly speaking, the planets moved
with circular motions, but the actual resultant motion was not a circle,
but a spiral. Aristotle could hardly have had in mind spirals, or approximate circles, when he described the motions of the planets as circular.
But a number of medieval natural philosophers apparently believed that
the real planetary motions were not circular, but spiral, that is, "nearly"
circular.[24]

2. The Celestial Substance

Although the most significant Scholastic departures from Aristotle's
ideas about the nature of the celestial substance occurred in the seventeenth century, some doubts were already raised in the Middle Ages.
Whether those who accepted spiral motions for the planets were also
aware that this committed them to a denial of Aristotle's claim that the
celestial substance, or ether, moves naturally in circular motions is left
unanswered. Perhaps they assumed with Sacrobosco that there is no sensible difference between spirals and circles. This potentially significant
issue was simply disregarded.

But the relationship of terrestrial matter and celestial matter was of
considerable concern to medieval natural philosophers. Aristotle had
distinguished sharply between the two. The four terrestrial elements
moved with natural nonuniform, rectilinear motions and were subject to
four basic kinds of change: substantial, quantitative, qualitative, and local
motion. By contrast, the celestial substance suffered no change whatever

22. *The Sphere of Sacrobosco,* ch. 3, in Thorndike, *The "Sphere" of Sacrobosco,* 133 (Latin
text, 101).

23. *Opus tertium,* Duhem, *Un fragment inédit,* 118–19; *Communia naturalium,* Bacon 1913,
433 (in line 14, the text has *speras* instead of *spiras*).

24. To the names of Sacrobosco and Bacon, we may add those of Albertus Magnus and
Nicole Oresme. Earlier Theon of Alexandria, Averroes, and al-Bitruji had also described the
spiral motion. See F. J. Carmody, ed., *Al-Bitrûji. De motibus celorum. Critical Edition of the Latin Translation of Michael Scot* (Berkeley and Los Angeles: University of California Press, 1952),
52–54, and Edward Grant, *Nicole Oresme and the Kinematics of Circular Motion: "Tractatus de
commensurabilitate vel incommensurabilitate motum celi"* (Madison: University of Wisconsin
Press, 1971), 31–33, 240, 241.

and moved only with a uniform, natural circular motion that involved rotation in a fixed position. The celestial ether was therefore a fifth element totally different from its four terrestrial counterparts. In brief, the celestial substance was incorruptible and therefore immutable, while the terrestrial substances were corruptible and involved in incessant change.

We must first recognize that a few rather daring medieval Scholastics were prepared to deny that heavenly bodies differed essentially from terrestrial bodies. In this they had at least one significant earlier authority: St. John of Damascus (Damascene), who had declared that "the sun, moon, and stars are composite, and by their very nature subject to corruption."[25] Even earlier similar sentiments had been uttered by John Philoponus[26] and some Church Fathers (St. Basil, for example) who had followed Plato and assumed that the heavens were composed of one or more of the four elements. Following in this tradition, Robert Grosseteste distinguished between a planet and its sphere. The latter was composed of a fifth element, which was immobile and unchangeable; the former was a compound body composed of two or more ordinary terrestrial elements. A planet is therefore a corruptible body![27] In the 1380s and early 1390s, Henry of Hesse also distinguished between planets and the orbs that carry them. The latter were created on the second day, whereas the planets were made on the fourth day from a compound of regular terrestrial elements and were then placed in the heavens by supernatural action.[28]

Although few Scholastics followed Grosseteste and Henry of Hesse, a number of them were prepared to argue that as pure potencies the matter of the celestial and terrestrial regions were identical.[29] Aegidius Romanus, who was usually identified with this interpretation, insisted that

25. See S. John of Damascus, *De fide orthodoxa*, bk. 2, ch. 7, in S. John of Damascus, *On the Orthodox Faith (De fide orthodoxa)*, translated by F. H. Chase Jr., 221 (New York: Fathers of the Church, 1958).

26. Philoponus's opinions were transmitted to the Latin Middle Ages via Simplicius's *Commentary on De caelo* in the Latin translation by William of Moerbeke made around 1260. For Simplicius's discussion, see Simplicus, *Commentaria in quatuor libros De caelo Aristotelis, Guillermo Morbeto interprete* (Venice: Hieronimus Scotus, 1540), ff. 11v, col. 2 and 12r, col. 1.

27. See Grosseteste's *De generatione stellarum* in L. Baur, *Die philosophischen Werke des Robert Grosseteste, Bischofs von Lincoln* (Munster: Achendorff, 1912), 1.32–36.

28. See N. H. Steneck, *Science and Creation in the Middle Ages: Henry of Langenstein (d. 1397) on Genesis* (Notre Dame, Ind.: University of Notre Dame Press, 1976), 61–62.

29. For a discussion, see Edward Grant, "Celestial Matter: A Medieval and Galilean Cosmological Problem," *Journal of Medieval and Renaissance Studies* 13 (1983): 165–71.

if every form were stripped away from celestial and terrestrial matter, the two matters would be identical. And yet Aegidius accepted Aristotle's argument that generation and corruption occur only in the sublunar region. How could this be if both matters were identical? The explanation was based on the doctrine of contraries. Terrestrial qualities have contraries, which cause change. In the heaven, no contrary qualities exist and hence no change can occur. Those who adopted this interpretation insisted on the identity of all matter even as they continued to assume the incorruptiblity of the celestial region.[30]

Despite these departures from Aristotle, few, if any, were prepared to abandon celestial incorruptibility.[31] Not until the seventeenth century did Scholastics begin seriously to deny Aristotle's claim of celestial incorruptibility and follow Tycho Brahe, Galileo, and others in the assumption of a changing and corruptible celestial region. The new star of 1572, the comet of 1577, Galileo's telescopic discoveries described in the *Sidereus Nuncius,* and the fact that a few key Church Fathers had argued for a celestial region composed of fire and water, swayed a number of Aristotelians to abandon not only the idea of celestial incorruptibility but the very notion of hard eccentric and epicyclic spheres.[32]

If medieval Scholastics offered little challenge to celestial incorruptibility, they did, however, begin to weaken the powerful Aristotelian,

30. William Ockham also accepted the identity of the two matters but based his argument on God's absolute power. See ibid., 171–72.

31. Neither William Ockham nor Henry of Hesse, who were positioned to do so, thought the heavens really corruptible (for Ockham, see ibid., 171–72; for Henry, see Steneck, *Science and Creation in the Middle Ages,* 62, where we learn that the elemental forms in the stars or planets are stripped of their contrary qualities when removed to the celestial region and thereby become incorruptible). Possible exceptions are Grosseteste and perhaps Roger Bacon, who, in his *De multiplicatione specierum,* believed that visual species affect the celestial region and that the latter has a certain appetite for those terrestrial virtues that it lacks and needs. Bacon denied, however, that such needs detract from the nobility of the heavens. For Bacon, see D. C. Lindberg, *Roger Bacon's Philosophy of Nature: A Critical Edition, with English Translation, Introduction, and Notes of "De multiplicatione specierum" and "De speculis comburentibus"* (Oxford: Clarendon Press, 1983), 72–75 (text and translation), lxi–lxii (for Lindberg's description).

32. E.g., the three Jesuits, Giovanni Baptista Riccioli (1598–1671) (Riccioli, *Almagestum novum* [Bologna: Typographia Haeredis Victorii Benati, 1651], pars posterior, p. 238, col. 1), Melchior Cornaeus (1598–1665) (Cornaeus, *Curriculum philosophiae peripateticae uti hoc tempore in scholis decurri solet* [Herbipolis: Sumptibus et typis Eliae Michaelis Zink, 1657], p. 489), and George de Rhodes (1597–1671) (de Rhodes, *Philosophia peripatetica ad verum Aristotelis mentem libris quatuor digesta et disputata* [Lyon: I. A. Huguetan et G. Barbier, 1671], pp. 278–81).

Ptolemaic, and Averroistic idea that by their uniform and regular motions, celestial bodies controlled the motion and activities of material things in the terrestrial region. As Averroes put it: "[T]he heaven exists because of its motion; and if celestial motion were destroyed, the motion of all inferior beings would be destroyed and so also would the world."[33]

While a current of Scholastic opinion followed Averroes, numerous others, including Richard of Middleton, Hervaeus Natalis, John Buridan, Albert of Saxony, and Nicole Oresme, challenged the idea of the total dependence of terrestrial changes on the celestial motions. There can be little doubt that the bishop of Paris spurred this reaction by his Condemnation of 1277, where, in Article 156, he condemned the opinion that "if the heaven should stand [still], fire would not act on tow [or flax] because nature would fail to operate."[34] The bishop thus left no doubt of his displeasure with the idea that terrestrial actions were totally dependent on celestial motions.

In commentaries and *questiones* on *De caelo* during the fourteenth century it was not unusual for Scholastics to inquire whether terrestrial elements and bodies could act independently if the celestial motions ceased; or alternatively, they might ask whether a plurality of celestial motions was required for the occurrence of generation and corruption in inferior bodies. In 1377, one hundred years after the condemnation of Article 156, Nicole Oresme took up the problem in his brilliant French commentary on *De caelo*. Straightaway, Oresme denies that a plurality of motions is necessary for the occurrence of sublunar generation, insisting that

if the heavens were at rest, change and growth would still exist, because if fire were at the present moment applied to a matter which it heated and burned, it is unreasonable to suppose that it would stop heating or burning even should celestial motions be stopped. To say the contrary is to support an article condemned at Paris.[35]

---

33. See Averroes's *De substantia orbis*, ch. 4, in Aristoteles, *Opera cum Averrois Commentariis* (Venice: Giunti, 1562–1574), 9, f. 10v, col. 1. For Aristotle's remarks, see *De caelo*, bk. 1, chs. 2, 3; Ptolemy's discussion appears in *Tetrabiblos*, bk. 1, ch. 2, Ptolemy 1948, 5–7. My discussion is based on Grant, "Medieval and Renaissance Scholastic Conceptions of the Influence of the Celestial Region on the Terrestrial," *Journal of Medieval and Renaissance Studies* 17 (1987): 1–23.

34. I have used the text as emended by R. Hissette, *Enquête sur le 219 articles condamnés à Paris le 7 mars 1277* (Louvain-Paris: Publications Universitaires-Vander-Oyez, 1977), 142.

35. Nicole Oresme, *Le Livre du ciel et du monde*, bk. 2, ch. 8, in Menut and Denomy, *Nicole*

To reinforce his position, Oresme introduced the miracle of Joshua at the battle of Gibeon which, in his judgment, and that of others, demonstrated that "generation and destruction did not cease because during the period of cessation the enemies of Gibeon were killed."[36]

Although it was still true that Scholastics acknowledged that without regular celestial motions, the world as we know it would be impossible, they had nonetheless departed from Aristotle and his strict followers by assigning varying degrees of independent action to terrestrial bodies. It was an important step toward the ultimate abandonment of hierarchical distinctions in the universe.

The anomalies cited thus far concerned the very structure of the Aristotelian world. But there were others that not only affected medieval conceptions of the cosmos, but also influenced the manner in which Scholastics did science.

3. Extracosmic Infinite Void Space

Beginning in the fourteenth century, theologians began to assume the real existence of an infinite extracosmic void space beyond our finite cosmos, an assumption that not only violated Aristotle's declaration that neither place, nor void, nor time could exist beyond the world, but also rejected Aristotle's contention that void was impossible.[37] During the sixteenth and seventeenth centuries, numerous Scholastic natural philosophers accepted the existence of extracosmic space, and, like their medieval predecessors, identified it with God's infinite immensity.[38]

4. Hypothetical Conditions "secundum imaginationem"

To these real departures from Aristotle's cosmology, others of a hypothetical, but no less significant, character can be added. Some of these hypothetical assumptions were based on the late medieval concept of God's absolute power to do anything He pleased short of a logical con-

---

Oresme, 375. Oresme had already argued this in his earlier *Quotlibeta*, though without reference to Article 156 (see L. Thorndike, *A History of Magic and Experimental Science* [New York: Columbia University Press, 1923–1958], 3.414).

36. Menut and Denomy, *Nicole Oresme*, 375.

37. See Edward Grant, *Much Ado About Nothing: Theories of Space and Vacuum from the Middle Ages to the Scientific Revolution* (New York: Cambridge University Press, 1981), Part 2 "Infinite Space Beyond the World," ch. 6.

38. Ibid., ch. 7.

tradiction. The Condemnation of 1277 at the University of Paris was an effort to emphasize God's absolute power in defiance of Aristotle's physics and cosmology. The spirit of the Condemnation was emphatically displayed in Article 147, which condemned the opinion "that the absolutely impossible cannot be done by God or another agent," where "impossible is understood according to nature." Physical situations deemed impossible in Aristotle's philosophy were frequently considered and analyzed, occasionally with interesting and important results.

It was in this context that the region beyond our cosmos became the locale for another significant development that was contrary to Aristotle's considered judgment. In *De caelo*[39] Aristotle had argued that our world was unique and that it was impossible for other worlds to exist. As is well known, Article 34 of the Condemnation of 1277 declared it an excommunicable offense to hold that God could not create other worlds.[40] On the mandatory assumption that God could, if he wished, create other worlds, some fourteenth-century Scholastics—including John Buridan and Nicole Oresme—supposed that God did indeed create other worlds and then raised physical problems the solutions to which conflicted with, or departed from, the principles of Aristotelian physics and cosmology. Not only did they find the idea of a multiplicity of equally privileged physical centers theoretically acceptable, but they argued that each world would be self-contained so that heavy bodies would fall to the center of their own world and not to that of another world, as Aristotle had argued. Nicole Oresme even redefined the concepts of up and down independently of a unique geometric center. If God had created such worlds, most were prepared to admit that a void space would lie between them.

Despite the fact that no Scholastic authors proclaimed the actual existence of other worlds, these counterfactual arguments were significant because they insisted on the possibility that other worlds and other world centers might exist and they provided respectable arguments in support of those claims. Where Aristotle thought the existence of other worlds impossible, many Scholastics insisted that at the very least their

39. *De caelo* 1.9.278a.21–279a.14.
40. In what follows for the remainder of this paragraph, I draw upon my article, Edward Grant, "The Condemnation of 1277, God's Absolute Power, and Physical Thought in the Late Middle Ages," *Viator* 10 (1979): 217–25.

existence would not violate the tenets of reason and natural philosophy.

Some of the most interesting and significant departures from Aristotle's physics involved motion in a vacuum. With the exception of Nicholas of Autrecourt,[41] who argued for the existence of interstitial vacua, no Scholastic natural philosophers believed in the actual existence of macro- or microvacua within the cosmos itself, although they regularly conceded that God could create them if He wished. God was thus frequently imagined to annihilate all or part of the matter within the material plenum of our world.[42] But many discussions about the behavior of bodies in vacua did not invoke God's absolute power, but simply assumed the existence of vacua and then analyzed the consequences that might follow for bodies located in such empty spaces.

As is well known, among numerous arguments Aristotle directed against the possible existence of void spaces, one of the most significant was that, without any resistances to oppose it, a body would move in a vacuum instantaneously, or, as Aristotle expressed it, a body "moves through the void with a speed beyond any ratio."[43] Many Scholastic natural philosophers opposed Aristotle on this issue and insisted that if an extended vacuum did exist, bodies would move through it with a finite, temporal motion. The basis for this conviction was the assumption of an analogy between a vacuum and a material plenum.[44] Just as motion through a plenum with distinct and separate termini is successive and temporal, so also would it be successive and temporal in a vacuum if the latter is assumed to possess dimension and extension. Since a material plenum is divisible into parts that must be traversed sequentially, so also must an extended vacuum be potentially divisible into parts that must be traversed in sequence, an activity that necessarily requires time, which, in turn, guarantees that the velocity will be finite. As a further departure from Aristotle, who argued that local motion could only occur if there was an interaction between a motive force and a resistance, the extended void itself was identified as a resistance to the body moving through it. It was the void as resistance that prevented instantaneous motion.

In the fourteenth century, many Scholastics balked at this solution.

41. On Nicholas, see Grant, *Much Ado About Nothing*, 74–77.
42. In what follows, I draw upon my account in Grant, *Much Ado About Nothing*.
43. *Physics* 4.8.215a.24–216a.11; also Grant, *Much Ado About Nothing*, 7.
44. See Grant, *Much Ado About Nothing*, 27–28.

Empty space as a resistance left them uneasy. Instead they chose to locate the resistance to motion in a vacuum in the body itself, a move that led to other significant departures from the physics of Aristotle. In his description of bodies, Aristotle distinguished elemental and mixed, or compound, bodies.[45] The former, earth, water, air, and fire, were hypothetical entities not actually found in nature in their pure state. Bodies actually observed in nature were compounded of two or more of the four elements. Scholastics called them "mixed bodies." Within every mixed body, Aristotle assumed that one of the elements was dominant and would determine the body's natural motion.

For medieval natural philosophers coping with the problem of resistance in a vacuum, the solution was found in the abandonment of Aristotle's conviction that one element was dominant in every mixed body. This step had already been taken by the 1320s when some Scholastics insisted that a mixed body's direction of motion was determined by the relationship between its contrary light and heavy elements. If light elements dominated, the mixed body was characterized as a "light mixed body" wherein the light elements functioned as a motive force that caused the body to rise naturally against the internal resistance of the heavy elements. When the heavy elements dominated, a "heavy mixed body" resulted wherein the aggregate of heavy elements constituted a motive force that caused a natural downward motion which was resisted internally by the natural tendency of the light elements to rise. Every mixed body was thus conceived to possess within itself contrary light and heavy elements. The dominant contrary in that body was assumed to act as motive force while the opposing contrary functioned as an internal resistance. This conjoint action of motive force applied against an internal resistance met the fundamental Aristotelian condition for motion. It followed that a mixed body could fall in a vacuum with a finite speed determined by the ratio of motive force to internal resistance. Here was an instance where Scholastics sought to conform to a basic Aristotelian rule for motion even as they abandoned his fundamental notion that finite motion in a vacuum was impossible.

The doctrine of mixed bodies served as the basis for another de-

---

45. *De caelo* 1.2.268b.27–30. The following is based on Grant, *Much Ado About Nothing*, 44–55.

parture from Aristotle's physics.[46] On the assumption that in a uniform material plenum bodies fall with speeds directly proportional to their weights, Aristotle deduced that in a void, bodies of unequal weight would fall with equal velocities. He reasoned that in a vacuum, no good reasons could be offered to explain differences in speed between bodies of different weights. Without a medium to cleave, all bodies of whatever weight or shape should descend with equal facility and therefore with equal speed. The absurdity of this consequence served as further corroboration for Aristotle that the existence of void space was impossible.

The concept of mixed bodies changed this. Walter Burley, Thomas Bradwardine, and Albert of Saxony argued that mixed bodies of homogeneous composition, whether of equal or unequal weight, would fall with equal speed through a vacuum. As Thomas Bradwardine expressed it in his *Tractatus de proportionibus*, "all mixed bodies of similar composition will move at equal speed in a vacuum." The results and conclusions arrived at in these hypothetical discussions of counterfactual conditions were of considerable importance. Prior to the production of enclosed void spaces by use of air pumps in the seventeenth century, thought-experiments and the assumption of hypothetical conditions were all that could be managed. But these medieval tools of analysis proved powerful. Galileo did not really improve upon them.

5. The Mathematization of Real and Imaginary Phenomena

In conjunction with the proliferation of hypothetical and counterfactual arguments, and to some considerable extent intertwined with it, medieval natural philosophers departed significantly from Aristotle by quantifying real and imaginary physical and nonphysical phenomena to a rather remarkable degree. While Aristotle had placed physics below mathematics in the hierarchy of theoretical knowledge, knowledge of the former was derived independently of mathematics. Indeed the two disciplines had little to do with one another. As Albertus Magnus put it: "we must . . . beware of the error of Plato," who said that "the principles of natural things are mathematical, which is altogether false."[47] Indeed, even more than Aristotle, Albertus weakened the links between geom-

---

46. Grant, *Much Ado About Nothing,* 57–60.
47. From Albertus's *Metaphysics,* bk. 1, tr. 1, ch. 1, as translated in A. G. Molland, "Mathe-

etry and physical nature when he declared that "many of the geometers' figures are in no way found in natural bodies, and many natural figures, and particularly those of animals and plants, are not determinable by the art of geometry."[48]

Although neither was a Platonist, Robert Grosseteste and Roger Bacon greatly enlarged the role of mathematics as envisaged by Aristotle and Albertus. In this, they followed a tradition that stemmed from Boethius.[49] During the fourteenth century dramatic changes occurred, changes that would probably have surprised both Grosseteste and Bacon. Starting in Merton College, Oxford, the application of mathematics and/or logic to real and hypothetical situations became widespread and common. John Murdoch has described this medieval phenomenon as the generation of "measure languages."[50] Among these measure languages were the intension and remission of forms, the language of proportions, and those languages concerned with an analysis of certain continual processes, as in the determination of first and last instants, the beginining and cessation of a process, and the determination of maxima and minima.

It was within the context of these measure languages that a number of significant medieval departures from the content and/or the spirit of Aristotle's physics was effected. The application of proportions to motion led to the formulation of Thomas Bradwardine's "law" and to the further extraordinary extension of it by Nicole Oresme to embrace incommensurable relationships in both terrestrial and celestial motions and changes.[51] Indeed, Oresme would even use his demonstration of the probable

---

matics in the Thought of Albertus Magnus," in *Albertus Magnus and the Sciences: Commemorative Essays, 1980*, edited by J. A. Weishipl, 467 (Toronto: Pontifical Institute of Mediaeval Studies, 1980).

48. From Albertus's *Physics*, bk. 3, tr. 2, ch. 17, as translated by Molland, "Mathematics in the Thought of Albertus Magnus," 469–70.

49. For this assessment, see D. C. Lindberg, "On the Applicability of Mathematics to Nature: Roger Bacon and His Predecessors," *British Journal for the History of Science* 15 (1982): 24–25.

50. See J. E. Murdoch, "From Social into Intellectual Factors: An Aspect of the Unitary Character of Late Medieval Learning," in *The Cultural Context of Medieval Learning. Proceedings of the First International Colloquium on Philosophy, Science, and Theology in the Middle Ages, September 1973*, edited by J. E. Murdoch and E. D. Sylla, 271–339 (Dordrecht and Boston: D. Reidel, 1975).

51. For a discussion of both Bradwardine and Oresme, see Edward Grant, *Nicole Oresme:*

incommensurability of the celestial motions to refute Aristotle's claim that whatever had a beginning must have an end and what has no end cannot have had a beginning.[52]

With the intension and remission of forms or qualities, Scholastics would attempt to quantify the intensities of variable qualities, such as temperatures, sounds, motions, passions, pains, sorrows, joys, and a host of others. To an astonishing degree, these measure languages were also applied within theology to a range of problems that included the relations of God and his creatures, concepts of sin, merit, grace, and so on. Since most—if not all—theologians were also trained as natural philosophers, the quantification of theology is but an extension and illustration of the degree to which medieval Scholastics were enthralled by the quantitative aspects of natural philosophy. As is well known, important achievements emerged from the peculiarly medieval discipline of intension and remission of forms. Not only were graphing techniques developed, but certain important kinematic definitions and theorems were enunciated and demonstrated, as, for example, the definitions of uniform and uniformly accelerated motions and the enunciation and proof of the famous mean speed theorem. Although these important discoveries were not exploited during the Middle Ages, they were nonetheless significant intellectual achievements.

Whatever else may be said of it, the quantification of qualities in the Middle Ages must rank as a major departure from the spirit of Aristotelian natural philosophy. And this holds true despite the hypothetical character of most of the problems to which the quantification was applied.

6. The Earth

Another fundamental tenet of Aristotle's cosmic picture—the earth's absolute immobility—also came under attack in the fourteenth century. The motion accorded the earth, however, was not an axial rotation,[53] but

---

"De proportionibus proportionum" and "Ad pauca respicientes" (Madison: University of Wisconsin Press, 1966).

52. Oresme's argument is described in Edward Grant, "Scientific Thought in Fourteenth-Century Paris: Jean Buridan and Nicole Oresme," in M. Pelner Cosman and B. Chandler, eds., *Machaut's World: Science and Art in the Fourteenth Century* (New York: New York Academy of Sciences, 1978), 112–14.

53. Although there was discussion of the possibility of the earth's daily axial rotation, only Nicole Oresme thought it at least equally plausible as the traditional arguments for its immobility. In the end, however, even he opted for immobility. For the arguments of Buridan and

rather small, though incessant, rectilinear motions by means of which the earth's center of gravity constantly sought to coincide with the center of the world.[54] As a nonhomogeneous body continually suffering change, the earth's center of magnitude differs from its center of gravity. As geologic processes alter the earth's surface, the center of gravity continually shifts to coincide with the center of the world. These minute rectilinear shifts of the earth's center of gravity cause previously submerged parts of the earth to rise above the surface of the seas and oceans. Thus an inert and heavy earth was accorded a degree of rectilinear movement.

The relationship between two of the four terrestrial spheres distinguished by Aristotle was called into question in the sixteenth century.[55] Aristotle had assumed that although the spheres of earth and water were concentric, the concentricity was imperfect because dry land protruded above the water. Following Aristotle, medieval Scholastics always assigned one sphere each to earth and water. It was not until the Portuguese explorations of the southern hemisphere that a new concept was proposed: earth and water formed a single globe called appropriately the "terraqueous sphere." Proposed and adopted first by non-Scholastic authors, it soon won over most Scholastics, especially after Christopher Clavius embraced it in his Commentary on the "Sphere" of Sacrobosco.

Although other medieval departures from Aristotle of greater or lesser significance could be described, enough of them have been presented to convey a good sense of the extent to which Scholastic natural philosophers exercised an unusual spirit of intellectual independence. Under pressure from the emerging and developing new science of the sixteenth and seventeenth centuries, their Scholastic successors in those centuries made further departures as they abandoned what had before seemed like vital elements in Aristotelian cosmology. With Jesuits leading the way, Scholastics not only adopted the terraqueous sphere, as we saw above, but many would reject the incorruptibility of the celestial region and even abandon belief in the celestial orbs, thereby rejecting the existence of eccentric and epicyclic spheres, which had been the foundation of

Oresme, see M. Clagett, *The Science of Mechanics in the Middle Ages* (Madison: University of Wisconsin Press, 1959), 583–609.

54. For references and a brief discussion, see Edward Grant, *In Defense of the Earth's Centrality and Immobility: The Scholastic Reaction to Copernicanism in the Seventeenth Century* (Philadelphia: American Philosophical Society, 1984), 24–25.

55. I have here again drawn from my monograph, ibid., 26–30.

Aristotelian-Ptolemaic cosmology. To replace the traditional spheres, many came to believe that the planets moved by their own power and moved through a fluid celestial medium. They also began the process that eventually subverted the notion of a celestial hierarchy.[56] And, finally, we may conclude by observing that a number of Scholastics would follow Tycho Brahe and assume the existence of a geoheliocentric universe in which all the planets revolved around the sun, which in turn revolved around a stationary earth.

Conclusion

For approximately four hundred years, Scholastics altered the Aristotelian worldview that they had inherited in the twelfth and early thirteenth centuries. By the end of that period, the physical world that Aristotle had fashioned is barely recognizable. What did these departures signify for the history of Aristotelian natural philosophy between the thirteenth and seventeenth centuries?

The changes that we have described were not all of equal intrinsic significance. Cosmological changes in general seem to have been potentially, if not historically, more significant than changes that were associated with, say, the intension and remission of forms, or impetus theory, or Bradwardine's mathematical law of motion. The claim that some celestial orbs could move around empty geometric centers rather than around the earth as physical center altered one of Aristotle's basic structural concepts. The assumption that planets had proper motions of their own in addition to the motions caused by their respective orbs also signified a structural change, as indeed did the conviction that an infinite extracosmic void existed beyond our finite cosmos. The suggested possibility that other worlds might exist each with its own center subverted Aristotle's notion of a single cosmic center. These, and other cosmological changes, altered the physical character of the universe. Although the doctrine of intension and remission of forms, impetus theory, and Bradwardine's law were major departures from Aristotle's thought and of great conceptual importance, they did not alter the basic character of the physical world.

56. See Edward Grant, "Celestial Perfection from the Middle Ages to the Late Seventeenth Century," in *Religion, Science, and Worldview: Essays in Honor of Richard S. Westfall*, edited by M. J. Osler and P. L. Farber, 149–62 (New York: Cambridge University Press, 1985).

Indeed, these ideas were concerned with the intrinsic and descriptive behavior of bodies, whereas many of the cosmological alterations signified an actual physical change in the world—that is, something was added, changed, moved, or extended.

Although some changes involved a degree of physical restructuring of the world and others did not, medieval natural philosophers give no indication of concern about such differences. Nor did they concern themselves about the relative importance of their departures from Aristotle. Where a single scholar may have discussed a number of these departures in a series of separate *questiones*, he was not likely to indicate the greater significance of one over another. Despite the wide range of medieval departures from Aristotle's fundamental ideas, those departures were almost always unrelated. Perhaps this helps explain why Aristotelian natural philosophy endured for so long. No one thought of collecting these departures in order to assess the status of Aristotelian cosmology and physics. Had anyone done so, the chaotic state of Aristotelian natural philosophy would have become evident. An attempt at a new synthesis might have ensued. Of course, this never happened. Despite the departures and changes over the centuries, and despite some significant attempts to accommodate to the new science of the seventeenth century, medieval natural philosophy remained essentially Aristotelian.

If medieval natural philosophy consisted of the physical works of Aristotle and the kinds of alterations of his ideas and principles described above, how does this intellectual mixture affect our understanding of the terms "Aristotelianism" and "Aristotelian"? Are those who upheld most, if not all, of those departures Aristotelians? And is the aggregation of their ideas properly labeled Aristotelianism? With these troubling and difficult questions, which I have attempted to answer elsewhere,[57] I conclude this essay.

---

57. See my article Edward Grant, "Ways to Interpret the Terms 'Aristotelian' and 'Aristotelianism' in Medieval and Renaissance Natural Philosophy," *History of Science* 25 (1987): 335–358.

6 ∞ God and the Medieval Cosmos

It seems appropriate to begin a lecture on "God and the Medieval Cosmos" with the creation of the world. Since modern cosmologists grapple with problems about the formation of our universe—how old it is, how long it took to form our solar system, and similar questions—I shall first describe medieval views about the time it took God to form our world. On this issue, it was apparent early on that Scripture posed a potentially serious dilemma by virtue of two seemingly conflicting statements, one in Genesis, the other in Ecclesiasticus. In Genesis, we are told of a creation that took six days. But in Ecclesiasticus, we are informed that "He who lives forever created all things together [*simul*]"—that is, at the same time, perhaps in an instant. St. Augustine took cognizance of this apparent dilemma when he declared, in his commentary on Genesis [*The Literal Meaning of Genesis*, bk. 4, ch. 33, 1982, 1.142] that "[i]n this narrative of creation Holy Scripture has said of the Creator that He completed His work in six days; and elsewhere, without contradicting this, it has been written of the same Creator that He created all things together." To explain this seeming anomaly, Augustine argued that God did indeed create all things simultaneously but chose to narrate the creation day by day over a period of six days. Thus, for Augustine, God created the world simultaneously *and* in six days. To account for the apparent paradox, Augustine reflected that although God created all things simultaneously there was a "before" and "after" that followed the order of creation in Genesis. To illustrate what he meant, Augustine used the rising sun as an example. Although we see the rising sun in a virtually instantaneous moment, the ray that goes from our eyes to the sun passes over all the intervening spaces in a certain order, nearer things first and then more remote things until it reaches the sun. It was the same with the creation of the world. All things were created in the order described in Genesis, but in an instant, so quickly that "before" and "after" were indistinguishable. God created all things simultaneously by creating them in seeds, or *rationes seminales,* where they came into being in the order described in Genesis.

With a significant exception, Augustine's concept of simultaneous creation received church sanction in the Fourth Lateran Council of 1215, which declared that God is "the creator of all visible and invisible things, spiritual and corporeal, who, by His omnipotent power created each creature, spiritual and corporeal, namely angelic and mundane, at the beginning of time simultaneously from nothing; and then [*deinde*] made man from spirit and body." Creation was thus a twofold process: first, all things except man were created simultaneously, and then man. Although Augustine's interpretation differed from this with regard to the creation of man, his interpretation of simultaneity was the most widely held opinion during the Middle Ages. Nevertheless, variant explanations of simultaneous creation were proposed, along with different interpretations of six days. Some proposed that the world was created in real time over six days, while others, following Augustine, denied that the six days were ordinary, largely because the heavenly bodies were not created until the fourth day, from which one might plausibly infer that the first three days could hardly have been ordinary.

However one might interpret the temporal aspects of the biblical creation, the more complex and difficult question that confronted medieval natural philosophers and theologians was simply this: What kind of cosmos had God created?

Outline of the Medieval Cosmos

From books published in the sixteenth century, it is easy to find woodcuts that give a schematic structural outline of the cosmos as it was understood in the late Middle Ages, say from around 1250 to 1500. Peter Apian included such a diagram in his widely read *Liber cosmographicus* of 1524 (figure 1).

At the center of the cosmos lies a small, dark circle in which earth and water are intermingled; wrapped around earth and water is a sphere, or shell, of air; and around that another sphere, or shell, of fire. Here we have the traditional four elements representing the terrestrial region of Aristotle's natural philosophy, the region where incessant change occurs

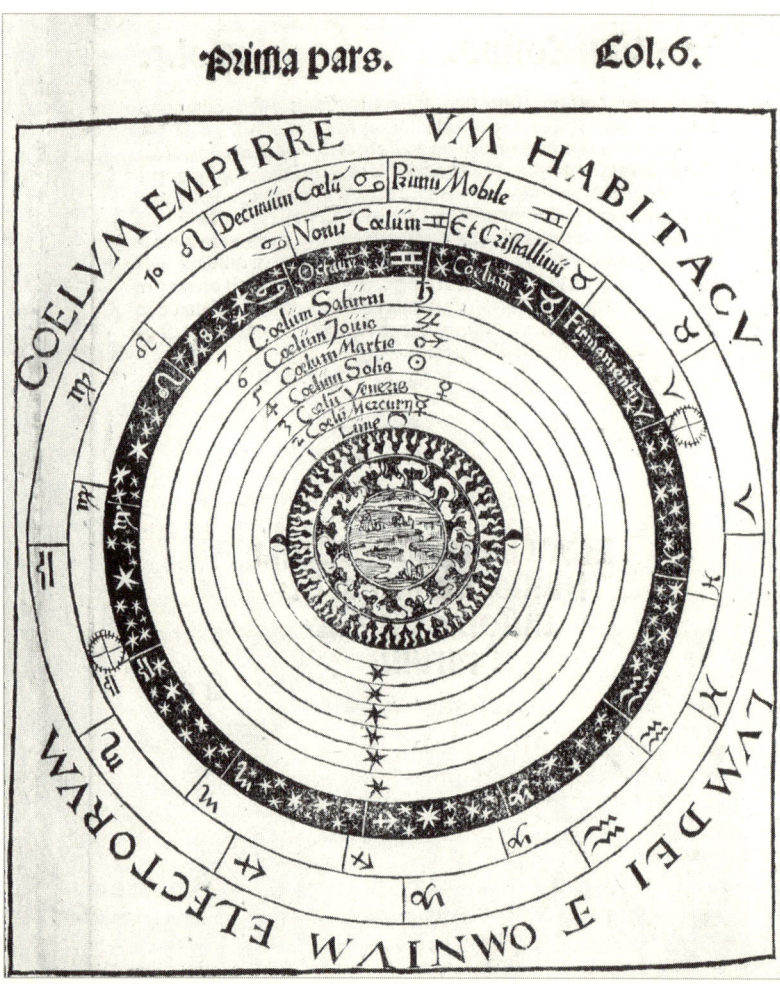

Figure 1. *Movable Celestial Spheres.* Courtesy of Lilly Library, Indiana University, Bloomington.

with things always coming into being and passing away, namely, generation and corruption, as the process was usually called.

Beyond the sphere of fire—and in stark contrast with the ever-changing bodies below—lies the celestial region, which was regarded as uniformly constituted from a special ethereal substance that was incorruptible, unchangeable, and transparent, and capable only of circular motion. All celestial bodies—planets, stars, and orbs—were assumed to be composed of this ether. In his diagram, Peter Apian represents a series of eleven transparent, concentric, celestial shells, or orbs. The first seven represent the seven planets—from the innermost planet, the moon, to the outermost planet, Saturn. Although the celestial ether, and all bodies composed of it, were supposed to move naturally with circular motion, that apparently did not apply to the spherical planetary bodies, which were incapable of self-motion. Each planet was carried around the heavens by the orb within which it was embedded. Although the orb itself was invisible, the planet was sufficiently dense to reflect light and be seen by earthly observers. The eighth sphere encloses all of the fixed stars and was frequently identified as the biblical firmament.

Although Peter's diagram has accounted for all celestial bodies, it has not yet accounted for all celestial motions. An inviolate rule of medieval cosmology, derived from Aristotle, assumed that a celestial orb was capable of only one natural motion. The eighth sphere of the fixed stars was assigned as many as three motions. Since only one motion could be assigned to the orb of the fixed stars, the other two motions required separate spheres, one for each motion. And so it was that ninth and tenth spheres were added, even though they carried no celestial bodies. And beyond the tenth sphere, which was usually described as the first moveable sphere, or the *primum mobile,* Peter Apian located an eleventh orb, the last sphere in our cosmos, an immobile sphere called the empyrean heaven, which was the dwelling place of God and the blessed elect.

The eighth to tenth cosmic orbs not only performed important astronomical functions, but they also came to represent important aspects of the creation account in Genesis. The creation of the firmament on the second day led many Christians to identify it with the sphere of the fixed stars. Since "God made the firmament and divided the waters which were under the firmament from the waters which were above the firmament," Christians sought to identify those waters. And so it happened that the ninth sphere—and occasionally the tenth—was usually identified with the waters above the firmament and was called the crystalline orb. The linkage between ninth sphere and crystalline orb was already made around 1250 by Vincent of Beauvais, in his famous encyclopedia the *Mirror of Nature (Speculum naturale).* Just how the suprafirmamental "waters" were to be interpreted was often argued. Were they fluid or solid? Numerous supporters for each can be found. In general, however, it was not fluidity (softness) or solidity (hardness) that was regarded as the principal attribute of the crystalline orb, but rather its immutability, transparency, and luminosity. As an anonymous author of a French encyclopedia put it around 1400, the crystalline orb is "the heaven that theologians call 'watery,' not because there are waters such as those which are here below, rather they are light waters of a noble nature similar to the heaven [or sky] in clarity and luminosity."[1]

The tenth orb, which also represented a motion usually attributed to the sphere of the stars, was called the "first moveable sphere," or the *primum mobile*. It was the first moveable sphere because beyond it, enveloping all the moving orbs, was the immobile empyrean heaven. The outermost orb of the cosmos—the empyrean heaven—had no astronomical function. And unlike the firmament and crystalline orb, which derived from the second day of creation, the empyrean orb had no obvious biblical sanction, although the heaven created on the first day was sometimes identified as the "empyrean heaven." The empyrean sphere was an essentially theological construct, a product of faith, not science. It emerged as a separate heavenly sphere only in the twelfth century, when theologians such as Anselm of Laon, Peter Lombard, and Hugh of Saint Victor described it as a place of dazzling luminosity. Many theologians regarded it as the dwelling place of God and the angels, as well as the abode of the blessed. Despite its perpetually radiant state, the empyrean heaven did not transmit any of its light to the celestial and terrestrial bodies below. As a sphere, it was assumed transparent, invisible, and incorruptible. Beyond its convex surface, nothing existed. Campanus of Novara, a thirteenth-century astronomer, described the empyrean heaven as "the common and most general 'place' for all things which have position, in that it contains everything and is itself contained by nothing."[2] Although Peter Apian's diagram captured the structural outline of the medieval cosmos, it is a gross oversimplification. Not only does it omit many spheres that Aristotle assigned, but it also ignores a variety of orbs, known as eccentrics and epicycles, that derived from the astronomical

1. See Edward Grant, *Planets, Stars, and Orbs: The Medieval Cosmos, 1200–1687* (Cambridge: Cambridge University Press, 1994), 334.

2. Francis Benjamin Jr. and G. J. Toomer, eds. and trans., *Campanus of Novara and Medieval Planetary Theory: "Theorica planetarum"* (Madison: University of Wisconsin Press, 1971), 183.

works of the great Greek astronomer Claudius Ptolemy, who lived in the second century A.D. Nevertheless, when natural philosophers discussed the cosmos, they had in mind a world akin to Peter Apian's, rather than the astronomical schemas of Aristotle and Ptolemy.

The world depicted by Peter Apian reveals, incidentally, a profound truth about medieval cosmology: its essentially hierarchical nature. As we move up from the earth toward the outermost sphere of the universe, the perfection and nobility of things increases dramatically. Among the terrestrial elements, fire is the most noble because it is located just below the moon and therefore farther from the center of the world than the other three elements. The celestial region beyond, composed of a rare, incorruptible ether, is incomparably better and more perfect than the terrestrial region, where matter is always suffering change. It was axiomatic during the Middle Ages that bodies that are less subject to change and mutation are better and nobler than bodies that are more changeable. And although there were some important objections to the idea that things are more perfect in proportion to their distance from the earth, John Buridan sought to answer them when he declared that "a man is absolutely more perfect than a horse and yet a horse exceeds him in magnitude, speed, and strength. And so, although the three superior planets are absolutely nobler than the Sun, yet it is not absurd that the Sun should exceed them in some properties."[3] And so we can readily understand why Christian theologians placed the empyrean heaven at the edge of the cosmos, at the farthest possible distance from the earth, the most ignoble part of the universe.

Medieval Cosmology beyond the Finite Cosmos

Before I discuss God's role in the cosmos that I have just described, I shall first consider some of the momentous God-related issues that pertain to the cosmos as a whole. Foremost among these was the challenge to the creation account I described at the outset. Was the world created, therefore having a beginning and ultimately coming to an end? Or, is the world eternal, without beginning or end, as Aristotle claimed?

3. Buridan, Paris, 1518, *Questions on the Metaphysics* (*In Metaphysicen Aristotelis quaestiones argutissimae magistri Joannis Buridani*), bk. 12, qu. 12, fol. 75r, col. 1; Grant, *Planets, Stars, and Orbs*, 228.

The Eternity of the World

In light of the importance of the creation account to the Christian faith and to its theology, you will not be surprised to learn that the problem of the eternity of the world was the most controversial issue at the University of Paris during the thirteenth century. It was to the relations between science and religion in the Middle Ages what the Copernican theory was to the sixteenth and seventeenth centuries and the Darwinian theory of evolution to the nineteenth and twentieth centuries. The extent of its significance may be gauged by the Condemnation of 1277 in which the bishop of Paris condemned 219 articles. Assuming the truth of any of these was an excommunicable offense. Among the 219 articles, I count at least twenty-seven that concerned the eternity of the world, a doctrine that masqueraded under many guises.

From Holy Scripture (Genesis 1.1–2; John 1.2–3 and 17.5) Christians learned that the world was created supernaturally and would eventually be destroyed supernaturally. Despite the absence of any explicit statement in Jewish, Christian, or Muslim Scriptures that the world was created "from nothing" *(ex nihilo),* creation from nothing had been widely assumed since the second century A.D. and was made official Christian doctrine by the Fourth Lateran Council in 1215. Its appeal was obvious: a God able to create a world from nothing would seem, prima facie, a more powerful deity than one who, like the Demiurge in Plato's *Timaeus,* did not, and could not. In terms of Aristotelian natural philosophy, however, creation from nothing was impossible, since matter could not come from nothing by natural means.

Although it was forbidden to argue in favor of the eternity of the world, some Scholastics demonstrated that the concept of an eternally existent entity was not self-contradictory, while others sought to prove the absurdity of an eternal world, as did St. Bonaventure (1221–1274). A significant intermediate position seems to have triumphed. Thomas Aquinas (ca. 1225–1274), and others, thought that God might have created the world and yet made it eternal. They believed that logical arguments could neither demonstrate the temporal beginning of the world nor its eternity and absence of a beginning. Thomas argued that it was not a logical contradiction that God could have willed the existence of the world without also causing it to have a temporal beginning. Although this was a popular opinion, Christians, including St. Thomas, knew that they were required to believe, as a matter of faith, that the world had a beginning and was supernaturally created from nothing.

Infinite Void Space Beyond the World

Assuming that our world was created, either in the sense assumed by the Fourth Lateran Council, or in the sense assigned to creation by Thomas Aquinas, one might ask the following questions: Does our created finite cosmos occupy all the space in existence? Does anything lie beyond it? And if so, what? Aristotle's predecessors, the Greek atomists, Democritus and Leucippus, assumed an infinity of other worlds and an infinite space beyond our world. Aristotle found this untenable. In his cosmological treatise, *On the Heavens,* he argued that our world is unique and that nothing could exist beyond it, neither bodies, nor places, nor vacua, nor time (*De caelo* 1.9.279a.10–17). Some, perhaps even many, found Aristotle's restrictions unacceptable. Stoic philosophers and others asked what would happen if a man stood at the extremity of the fixed stars and stretched his hand or a staff beyond. If the hand can be extended, and all believed it could, something must lie beyond, either matter or void. By extending this argument, Stoic philosophers assumed the existence of an infinite three-dimensional void beyond our world. Although such arguments were known in the Middle Ages, they posed no serious challenge to Aristotle's concept of a finite, spherical cosmos beyond which nothing could exist. The assumption of an infinite void space beyond the world as a natural part of the cosmic order would have dealt a devastating blow to Aristotle's natural philosophy. This did not happen.

But if infinite void space was too much for Aristotelian natural philosophers to accept, medieval theologians found occasion to introduce it into their theological deliberations about God's location and the nature of His omnipresence. Stoics and others had assumed the existence of a three-dimensional infinite void beyond the world. But in the fourteenth century, a few theologians assumed the existence of an infinite void space.

The most significant was Thomas Bradwardine (ca. 1290–1349) who first identified God's infinite omnipresent immensity with infinite void space. Bradwardine was an eminent mathematician, natural philosopher, and theologian at Oxford University, who died in 1349 as archbishop of

Canterbury. In a treatise titled *In Defense of God against the Pelagians,* written around 1344, Bradwardine presented five corollaries to show that God is immutable. In the first two corollaries, we learn that God is present everywhere in the world "[a]nd also beyond the real world in a place, or in an imaginary infinite void" (2nd corollary). And in the final corollary, Bradwardine explains that "it also seems obvious that a void can exist without body, but in no manner can it exist without God." Bradwardine was convinced that a precreation void space was necessary in which God could create our world, but it is obvious from the last corollary that the precreation void could not exist independently of God. By identifying infinite void space with God's infinite immensity, Bradwardine avoided two potential pitfalls. First, he did not have to assume the existence of a precreation void space that was independent of God and perhaps coeternal with God, a view that had already been condemned in 1277 at Paris; and second, he did not have to assume that God Himself had created an actual infinite void space in which to create the world anywhere He pleased. This would have been unacceptable because Bradwardine was convinced that God could not create an actually infinite space, or an actual infinite of any kind. Why did Bradwardine think this? Probably for the same reason that a number of other theologians had denied that God could create any actual infinite entity. If God created an actual infinite, He could not create anything greater, since there is nothing greater than an infinite. God would therefore have reached the limit of His power to do anything short of a logical contradiction.

But if God is possessed of an infinite immensity and omnipresence, does this not imply that He is an extended being, spread out over an infinite extension? Since all extended things are divisible, it follows that God would be a divisible being, a consequence that was wholly unacceptable in the Middle Ages. Bradwardine resolved this dilemma by simply declaring that God "is infinitely extended without extension and dimension." Bradwardine's infinite void space was therefore a dimensionless space.

In the seventeenth century, Henry More (1614–1687) rejected the universally accepted medieval conviction that God is dimensionless. Convinced that everything, whether corporeal or incorporeal, is dimensionally extended, More took the logical step and made God a three-dimensionally extended being by assuming that an extended infinite void space was

God's three-dimensional attribute. Though less forthcoming, Isaac Newton appears to have followed Henry More and assumed that infinite, extended, void space is God's attribute. Newton needed an infinitely omnipresent God in an infinite void space to guarantee the lawful operation of
the cosmos.

Medieval Scholastics were the first to divinize infinite void space.
With the major exception of extension, Scholastic theologians assigned
virtually all the properties to it that were conferred upon it by natural
philosophers during the Scientific Revolution, namely, homogeneity, immutability, infinity, and the capacity to coexist with bodies and to receive them without offering resistance. There can be little doubt that medieval ideas about space played a significant role in determining the kind
of space that Isaac Newton fashioned for his new cosmology and physics
in his monumental *Principia* of 1687.

If God created our world in an infinite void space that was His infinite immensity, did He perhaps create other worlds in that infinite immensity? Or is our world unique? The medieval response to such questions was conditioned by the Condemnation of 1277 (mentioned earlier),
which derived from the reaction of medieval theologians at the University of Paris to Aristotle's natural philosophy. Aristotle had shown that
certain occurrences were impossible in nature. Christian theologians insisted, however, that what was impossible in Aristotle's natural philosophy was not impossible to God, who possessed the absolute power to do
anything whatever short of a logical contradiction. And so it happened
that medieval theologians and natural philosophers frequently inquired
about God's ability to do something that was impossible in Aristotle's
philosophy. If no logical contradiction could be found, it was always assumed that God could indeed do it. Among the condemned articles two
are particularly relevant for cosmology: Article 49, which condemned all
who believed that "God could not move the heavens [that is, the world]
with rectilinear motion" because "a vacuum would remain," and Article
34, which condemned the claim "[t]hat the first cause [God] could not
make several worlds." Both condemned articles generated lively discussions.

Article 49 focused attention on the possibility of a void beyond the
world, as well as of other possible worlds. Some of the most astute theologians of the Middle Ages—Thomas Bradwardine, Jean de Ripa, and Ni-

cole Oresme—explicitly cited Article 49 and drew consequences from the assumption that God could move the world with a rectilinear motion. Around 1354, Jean de Ripa, a Franciscan theologian, argued that Article 49 showed that if God did indeed move the whole world in a straight line, there would have to be (imaginary) void places beyond the world that were capable of receiving the whole world, or any part of it. In effect, there would have to be some kind of extracosmic void. De Ripa eventually argued that this extracosmic void was infinite.

Nicole Oresme used Article 49 to demonstrate that the motion of the whole world through empty space was the perfect illustration of an absolute motion, since there was no other body outside the world to which its motion could be compared or related.

Although no one seriously believed that God would actually move our world, more than an echo of this hypothetical manifestation of God's absolute power reverberated through the seventeenth and eighteenth centuries. Pierre Gassendi invoked it when he declared that "it is not the case that if God were to move the world from its present location, that space would follow accordingly and move along with it." In the famous Clarke-Leibniz correspondence, Samuel Clarke, who was the spokesman for Isaac Newton, defended the existence of absolute space when he argued that "if space was nothing but the order of things coexisting [as Leibniz maintained], it would follow that if God should remove the whole material world entire, with any swiftness whatsoever; yet it would still always continue in the same place."[4]

Are There Other Worlds Beyond Our Own?

If the creation account in Genesis strongly suggested a temporal beginning for the world, it also seemed to signify its uniqueness. Here, at least, it seemed that Aristotle and Christianity were in agreement: there is only one world. This apparent unanimity of opinion was, however, deceptive. Although Aristotle's conclusion might be applauded, his derivation of it was offensive because he had argued that the existence of another world was impossible, or, as he put it, "there is not now a plurality of worlds, nor has there been, nor could there be" (De caelo 1.9.279a.7–11). To argue

4. For the quotations from Gassendi and Clarke, see Edward Grant, *The Foundations of Modern Science in the Middle Ages* (Cambridge: Cambridge University Press, 1996), 197; for notes 33 and 34, see p. 214.

that creation of other worlds was impossible, even for God, was viewed as a restriction on God's absolute power to do as He pleased. With a few exceptions, among them Peter Abelard, in the twelfth century, and John Wycliffe, in the fourteenth century, Scholastic theologians and natural philosophers immediately conceded that God could, if He wished, make as many worlds as He pleased. Despite a virtually unanimous conviction that God had not actually created other worlds, the condemnation of Article 34 in 1277 stimulated a significant discussion in which Scholastic theologians and natural philosophers contemplated the consequences of a plurality of worlds for Aristotelian natural philosophy.

They found generally that, contrary to Aristotle, the existence of other simultaneous, identical worlds, the kind that were usually discussed, would not produce absurdities and impossibilities. Aristotle had argued that in a unique world all formally identical elements would move toward one center and circumference, heavy elements, such as earth and water, to the center, and light elements, such as air and fire, toward the circumference of the world. How would such elements behave if there were many worlds? For example, would a heavy particle of earth in another world seek the center of our world? Or would a particle of earth from our world seek the center of another world? If this occurred, the particle of earth would possess two natural motions: one upward toward the circumference of its world and beyond, moving through the intermediate space between the two worlds and entering our world, where a second natural motion occurs, this time straight down toward the center of our world. Such contrary motions for one and the same body would have been regarded as absurd. Scholastic natural philosophers denied that such motions could occur. They argued that the earth of each world would remain at rest in its own world and any parts of the earth that might be separated within that world would return naturally toward the center of that world. The same argument would hold for the other elements. Medieval Scholastics regarded each world as a self-contained system with its own center and circumference. The elements in one world would always remain in their own world and have no effect on the elements of any other world.

Much of Aristotle's cosmological system was built upon the assumption that there could be only one center and circumference toward which light and heavy bodies moved. With the assumption of other possible simultaneously existing worlds, this belief collapses because the centers and circumferences of all worlds are equal, and none is unique. Aristotle's doctrine of natural place also collapses, namely, the concept that each of the four elements, when unimpeded, would move naturally to a single region where it is natural for it to come to rest. With a plurality of worlds, the earths from all worlds will not move toward one exclusive center, but the earth of each world will move toward the center of its own world. The assumption that God could create a plurality of simultaneously existing worlds also compelled natural philosophers to postulate void spaces between those worlds and then to raise numerous problems about the behavior of bodies in these intercosmic empty spaces. Thus it was that the absolute power of God to make as many worlds as He pleased raised physical problems that elicited interesting solutions, most of which conflicted with the principles of Aristotelian physics and cosmology.

In the numerous discussions about other worlds, did anyone consider the possibility that humans might exist on any of them? In the fifteenth century, William Vorilong conjectured (In bk. 1, distinction 44, of his *Sentence Commentary*) that species might exist in other worlds and differ from those of our world. As for the existence of human life on other worlds, William explains that if men did exist in these worlds, they would not exist in sin, because they did not spring from Adam. In answer to the question whether, by dying, Christ also redeemed the inhabitants of these other worlds, William answers in the affirmative. Christ's unique death in our world was sufficient to save the inhabitants of all worlds.

Vorilong barely entered on the subject of extraterrestrial life. But his contemporary, Nicholas of Cusa (1401–1464), presented his ideas in a treatise he called *Of Learned Ignorance (De docta ignorantia),* written in 1440. As the background to his speculations, Cusa repeated a famous definition of God already mentioned in the twelfth century *(Book of the XXIV Philosophers):* "God is an infinite sphere whose center is everywhere and circumference nowhere," a definition that Bradwardine had also cited some hundred years earlier. Despite the definition of God, Cusa denied that the world was infinite. He regarded it, rather, as boundless or unlimited, so much so that it lacked a center or circumference. Within this vast, unlimited, centerless world, Cusa believed that there were numerous earth-like planets—that is, planets composed of terrestrial elements, a daring idea in the fifteenth century. Somewhat more emphatically than Vorilong, Cusa proclaimed the existence of life on these planets and stars. Although Cusa assumed that life differed from planet to planet and star to star, he insisted that all life owed its origins "to God, who is the center and circumference of all stellar regions." In fact, Cusa argued that God filled the universe with material beings, both animate and inanimate, in order to prevent the existence of void spaces.[5] He further believed that the living beings on other planetary worlds are no more noble or perfect than the inhabitants of our earth. Indeed, their worlds are subject to the same kind of corruption and generation as is our world. Cusa's conception of the world represents a total rejection of Aristotle's cosmos. In 1417, Lucretius's *On the Nature of Things* was rediscovered and its atomistic doctrine of an infinity of worlds entered European thought. The doctrine of other worlds with living beings would soon become a major topic of discussion.

Could God Make the World More Perfect?

Scholastics did not confine themselves to the theme of other worlds, but also asked whether God could improve our own world. It was generally agreed that the world could not be absolutely perfect because it would then be equal to God. The world is only relatively perfect. A question frequently asked in theological commentaries (*Sentences*, bk. 1, distinction 44) was whether God could make something better than He had made it. As a special case, it was asked whether God could have made, or could make, a world better or more perfect than the one He made. In the basic theological text of the Middle Ages, Peter Lombard had argued that by virtue of His wisdom, God would have made all things properly the first time and would have no need to improve them. However, viewed from the standpoint of the created thing itself, there is room for improvement, since no created thing can be perfect. Therefore God could make any created thing better than He had made it. A considerable superstructure was built on this assumption.

Scholastic natural philosophers and theologians generally conceded that God could make our world more perfect. Indeed, he could make

5. See Alexandre Koyré, *From the Closed World to the Infinite Universe* (Baltimore: Johns Hopkins University Press, 1957), 22.

better and better worlds ad infinitum. But He could not make a world of infinite perfection, because He could then not improve upon it, which would constitute a limit on His absolute power. To the question why God did not make our world perfect in the first instance, Scholastics responded by conceding that God was under no compulsion to create every possible species of being nor to produce the best version of what He did create. By His omnipotence and inscrutable will, God could do as He pleased. But many Scholastics thought that God had achieved a large degree of perfection and harmony by introducing diversity into our world rather than sameness and repetition. Nicole Oresme, and a few others, held that God had achieved this goal by making the celestial motions incommensurable. In his treatise *On the Commensurability or Incommensurability of the Celestial Motions,* Oresme explains that

[i]t seems more delightful and perfect—and also more appropriate to the deity—that the same event should not be repeated so often, but that [on the contrary] new and dissimilar configurations should emerge from previous ones and always produce different effects. . . . This could not happen, however, without some incommensurability [obtaining] in the celestial motions.[6]

The Medieval Cosmos

It is now time to describe God's role in the operation of the cosmos He created. The most fundamental question that confronted medieval cosmologists was the cause or causes of the celestial motions: What made the orbs go round? Although medieval explanations of celestial motion were based on Aristotle's ideas, they differed significantly, in no small measure because the Christian God was an active God directly involved in the most basic operations of the world, whereas Aristotle's God did not create the world, had no interest in it, and was actually unaware of its existence.

External Movers

In the Christian scheme, God, or the Prime Mover, as He was frequently called, was the ultimate source of all celestial motions. Although He was the Prime Mover, capable of moving all celestial bodies directly, God

6. Grant, *Planets, Stars, and Orbs,* 149.

chose to cause those motions indirectly by creating agents, or movers, for that specific purpose. Medieval cosmologists speculated that God could have achieved his objective by creating movers that were internal or external to the celestial orbs. Internal forces or powers were envisioned as incorporeal entities, either in the form of souls inhering in each celestial orb, or as impressed forces somehow embedded in each celestial orb. By contrast, external movers or causes were always construed as intelligences or angels—the terms were virtually synonymous in the Middle Ages. A popular medieval question asked whether the celestial orbs are moved by external intelligences, or by some internal or innate form or power. With a few notable exceptions, Scholastic natural philosophers chose an external efficient cause in the form of intelligences or angels. They assumed that one angel or intelligence was assigned to each orb and that it had to be in contact with its orb.

Scholastics asked numerous questions about the motions of celestial orbs. Although an angel or intelligence had to be in contact with its orb, where was it actually located: On the surface of the orb? Or within its interior? Did an angel move with its orb, or did it remain motionless as the orb moved? What if the world were eternal? Could a finite being, such as an angel, move its orb with a uniform, finite motion over an infinite time? Would its motive power be eventually exhausted? Indeed, would it become fatigued? Despite the finitude of their respective motive powers, intelligences were not thought to exhaust themselves in moving their orbs, even over an eternal time, because the Prime Mover, God, constantly replaces their depleted energy. Thus the celestial motions will continue until the end of the world.

The extent to which intelligences were deeply embedded in the Western psyche as celestial movers can be seen following the collapse of the idea that orbs carried the planets. This occurred in the late sixteenth and early seventeenth centuries after Tycho Brahe, the great Danish astronomer, demonstrated that the comet of 1577 was a celestial phenomenon, and not a sublunar occurrence, as Aristotle had argued. From this, Tycho inferred that if physical orbs existed, the comet would have smashed into them, or crashed through them. Since no such cataclysmic celestial events had been observed, Tycho abandoned the idea that physical orbs existed in the heavens. At first glance, it seems plausible to assume that the destruction of celestial orbs and their subsequent repudiation by most astronomers and natural philosophers would have resulted in the rejection of angels and intelligences as celestial movers. By no means was this the case. Astronomers and natural philosophers simply transferred the abode of angelic movers from the orbs to the individual planets. An Intelligence now moved each planet. It would take Isaac Newton's theory of universal gravitation to sweep them from the sky.

Internal Movers

In the thirteenth century, a few scholars—John Blund and Robert Kilwardby—argued that the celestial orbs are moved by some kind of innate force, rather than by external intelligences. In the fourteenth century, John Buridan, a master of arts at the University of Paris and perhaps the most famous arts master of the Middle Ages, went beyond the vague innate capacity for circular motion which Blund and Kilwardby attributed to the celestial orbs. Buridan introduced his own impressed force theory of motion, or impetus, which he had earlier invoked as a cause of the motion of terrestrial objects. Before he applied his influential impetus theory to the heavens, Buridan explained that the Bible does not specify that intelligences move celestial bodies. He therefore felt free to suggest that when God created the world, He impressed into each orb a certain force, or impetus, that thereafter served as the cause of the orb's motion. The impetus would always remain constant, because there were no resistances in the heavens to cause its dissipation. Buridan's impetus was thus a permanent quality, which in the absence of external resistances or contrary tendencies, would move each celestial orb with the same uniform velocity forever. Later in the fourteenth century, Albert of Saxony upheld the same essential concept of celestial impetus.

The Role of God in Medieval Cosmology

When we leave the motion of the celestial orbs, God's direct role in the operation of the cosmos ceases. Although God was viewed in the Middle Ages as the creator of the world, it was widely believed that He had structured it to operate by natural causes. The task of natural philosophers was to identify those causes. Virtually all of them believed that they were doing this when they investigated the "common course of nature" (communis cursus nature), by which they understood seeking natural causes. Natural philosophers were usually careful to allow for God to upset the natural order of things by direct intervention. That is why an expression such as the "common course of nature" was so useful. It immediately signified that the investigator is primarily interested in natural, not supernatural, powers, as when John Buridan insisted that "in natural philosophy, we ought to accept actions and dependencies as if they always proceed in a natural way." We may confidently assume that John Buridan's attitude was typical of medieval natural philosophers and most theologians.

Lurking behind this naturalistic approach was the extraordinary concept of God's absolute power. Although natural philosophers used it, the invocation of God's absolute power was especially popular with theologians. No small measure of our interest in medieval cosmology is sparked by the way God's absolute power was invoked to alter the cosmological landscape. It has been said that God created the world from an infinity of possibilities and that once He made His choices, He was bound to honor His own decisions. This is called God's ordained power and it implies a covenant with us that God will uphold the world that He chose to create. Some have inferred from this that God would not employ His absolute power, which enables Him to do anything short of a logical contradiction, to undo anything that He had already ordained.

I believe that those who drew up the articles condemned in 1277 did not share this interpretation of God's absolute power. When they condemned the idea that God could not make other worlds, or could not move the whole spherical cosmos with a rectilinear motion, they were not intending to say that God could not make other worlds in the present, because He had initially decided to make only one world; or that He could not move the world with a rectilinear motion because He had ordained at Creation that it should always lie immobile. It is quite clear from the context within which the condemned articles were drawn that all must concede that God, by His absolute power, could make as many other worlds as He pleased at this very moment, or any time in the future. And that if God wished to move the world with a rectilinear motion, leaving behind a vacuum where the world formerly rested, He could do it immediately, or anytime in the future. I hasten to add, however, that few, if any, thought God had made, or would make, other worlds, or that He would move the world rectilinearly. The vital aspect concerning God's absolute power was not that He would do these hypothetical actions, but that He could do them if He wished. Most appeals to God's absolute power on cosmological matters concerned actions that Aristotle had demonstrated as impossible by natural means. Medieval theologians thought it was essential to have all concede that what was impossible in Aristotle's philosophy was quite possible by supernatural means.

Medieval Scholastics often discussed the possibilities that would arise if God did indeed exercise His absolute power to make other worlds, or move the world with rectilinear motion. I have already discussed these two conjectures. But perhaps an even more popular hypothetical possibility was one in which God was imagined to create vacua by annihilating matter within the boundaries of our cosmos. For example, Albert of Saxony imagined that God created a vacuum by annihilating all matter within the concave surface of the lunar sphere. He then inquired how a body would fall through this vacuum: whether it would fall instantaneously with an infinite speed, because there is no medium to resist it; or whether it would fall in a finite time, however small.[7] Others (for example, John Buridan) assumed the same kind of supernatural annihilation of matter and posed different questions.

The invocation of God's absolute power to annihilate all matter below the moon, or anywhere in the world, proved to be a powerful methodological tool, as is evident by its adoption in the seventeenth century by non-Scholastics who undoubtedly derived it from their Scholastic predecessors, although without acknowledgment. Pierre Gassendi (1592–1655) appealed to repeated supernatural annihilations of parts of our world and of other imagined worlds, in order to demonstrate that an infinite three-dimensional space existed. Gassendi explains that

there is nothing that prevents us from supposing that the entire region contained under the moon or between the heavens is a vacuum, and once this assumption is made, I do not believe that there is anyone who will not easily see things my way.[8]

Thomas Hobbes (1588–1679), an admirer of Gassendi, also made the annihilation of matter a principle of analysis, although he did not invoke

---

7. See Edward Grant, *Much Ado About Nothing: Theories of Space and Vacuum from the Middle Ages to the Scientific Revolution* (Cambridge: Cambridge University Press, 1981), 47, and *A Source Book in Medieval Science* (Cambridge, Mass.: Harvard University Press, 1974), 337–38.

8. Grant, *Much Ado About Nothing*, 390, n. 169.

God as the annihilator, choosing to assume that matter was simply annihilated. But Hobbes, who loathed Scholastics, paid unwitting tribute to them when he declared that

in the teaching of natural philosophy, I cannot begin better (as I have already shewn, than from *privation;* that is, from feigning the world to be annihilated.[9]

By means of this technique, Hobbes formulated his concepts of space and time.

In his famous *Essay Concerning Human Understanding,* John Locke based his argument for the existence of a three-dimensional void space on the assumption that God could annihilate any part of matter, should God do so, a vacuum would remain, "for it is evident that the Space that was filled by the parts of the annihilated body will still remain, and be a Space without Body."[10]

It is obvious that the human imagination could be substituted for God's action. One could simply imagine that other worlds existed; or that the world moves in a straight line; or that this part of the world, or the whole world is suddenly annihilated. We see that while Gassendi and Locke invoked God to annihilate the matter in question, Hobbes did not. He chose to "feign" it. It was easy to eliminate God and simply imagine hypothetical conditions for all "natural impossibilities," as Walter Charleton, an English follower of Gassendi, did when he summarized Gassendi's annihilation argument and explained that "nothing is more usual, nor laudable amongst the noblest order of *Philosophers*" than the assumption of "natural impossibilities."[11] But the inspiration to imagine all manner of "natural impossibilities" was clearly derived from the way medieval Scholastics used God's absolute power to imagine various impossible conditions in order to see how a world would, or could, function under such conditions. We should recognize, however, that medieval appeals to God's absolute power had little, if any, religious motivation or content. Wherever we find it used in Aristotelian treatises, it is almost never intended to make a religious point. It simply became a convenient vehicle for the introduction of highly imaginative questions, the responses to which compelled natural philosophers to apply Aristotelian natural philosophy to situations and conditions that were impossible in

9. Ibid., 390, n. 169.   10. Ibid., 277, n. 77.
11. Ibid., 390, n. 169.

Aristotle's natural philosophy. In the process, some of Aristotle's fundamental principles were challenged. The invocation of God's absolute power made many aware that things might be quite otherwise than were dreamt of in Aristotle's philosophy. By the seventeenth century, it did not much matter whether God's absolute power was made the causal agent of some hypothetical condition, or whether it was the human imagination. The medieval emphasis on the analysis of imaginary conditions had been assimilated into mainstream seventeenth-century philosophy.

Let me now describe, briefly, how God's role in the cosmos was viewed during the late Middle Ages, say from 1200 to 1500. All believed that God was the creator of the universe, and that He had created a world that operated in a rational manner. To the natural philosophers and theologians, this meant that in the "common course of nature" natural causes for all effects must be sought. It was also assumed that God played an indirect, ubiquitous role in the operation of the cosmos, because it was tacitly assumed that God was a coagent in every causal action. Since God, as coagent, was a given for every natural cause, it was unnecessary to mention God, or allude to Him, in the full explanation of a natural cause. To cite God as the cause of any particular natural effect would have been tantamount to offering no explanation at all, because it would have been utterly unilluminating. This generalization applies to all causative explanations in both the terrestrial and celestial regions of the cosmos. Even in the celestial realm, God had delegated the task of turning the spheres to angels or intelligences. The task of the natural philosopher, therefore, was to identify the natural causes of natural effects. Indeed, this was also the task of the theologian when treating cosmological themes. When commenting on the second book of the *Sentences* of Peter Lombard, which was concerned with the creation, theologians discussed numerous questions that were routinely considered in commentaries on Aristotle's works on natural philosophy. Theologians adopted the practice of treating straightforward cosmological questions much as the natural philosophers did. That is, they consciously avoided the intrusion of theology into their deliberations. But even when theologians had occasions to inject remarks about God, or to mention a biblical text, or to refer to a Church Father, the theological elements did not affect their treatment of the main line of argument concerned with natural philosophy.

God's greatest role in medieval cosmology came through the absolute power that was universally attributed to Him. Although, as we saw, these hypothetical situations could have been simply imagined without appealing to God's absolute power, there was a significant difference between the two types of counterfactuals. By invoking God to do this or that naturally impossible action, Scholastics knew they were conjuring up a situation that was possible, since God could do it if He wished, although most were convinced that the probability that God would produce any of these natural impossibilities was virtually nil. Nevertheless, they were dealing with an alternative or option that was a possibility, however remote, a possibility that only God could actualize. By contrast, those, like Hobbes and Charleton, who imagined the annihilation of matter without invoking God as the cause of that annihilation, were using natural impossibilities solely as an analytic methodology. For them the annihilation of matter was utterly impossible.

In medieval cosmology, God is the primary actor, since He created the world. And yet, in the Western approach to science and natural philosophy God had dealt Himself out of natural cosmic operations by creating a rational, self-sufficient world, a world in which He had delegated causal efficacy to natural agents. The most significant role assigned to God lay in the realm of natural impossibilities, within and beyond our world. It is ironic that medieval natural philosophers involved God in cosmic operations that were not in, or about, the world He had created, but were rather in, or about, a world, or worlds, that He had not created, but could have created if He had wished to do so. It was the analysis of divinely possible, though implausible, cosmic acts that captured the Scholastic mind and produced some of the most imaginative and stimulating cosmological discussions, discussions that even influenced some of their most severe critics, among whom were Pierre Gassendi, Thomas Hobbes, John Locke, and Samuel Clarke.

Scholastic natural philosophers sought to understand the physical world at a time when controlled experiments, systematic observations, and the application of mathematics to physical phenomena were rare occurrences. They had to rely on two powerful tools of analysis available to them. The first was *reason,* which they usually applied in a largely a priori manner, based on a minimum of observation and empirical data. They applied reason most effectively in circumstances of their own devising, that is, in the realm of the hypothetical, which enabled them to make use of the second great tool: *the human imagination,* used in ways that I have already described. The conditions they imagined were possible, but only by virtue of God's absolute power, provided that God's action did not imply a logical contradiction. It was by such means that medieval natural philosophers got beyond Aristotle's limited world and came to consider momentous problems about space, vacuum, and other worlds. Many of the non-Scholastics of the sixteenth and seventeenth centuries grappled with the same problems, and profited from the earlier discussions. They were, however, in no mood to acknowledge indebtedness to the defenders of Aristotle and the status quo. After all, they were bringing a new world order into being, a Copernican-Newtonian heliocentric world that would completely replace the Aristotelian-Scholastic geocentric model that had reigned for more than four centuries. They saw no need to be gracious in triumph. Indeed, they had only contempt for their medieval predecessors, whose opinions they mocked and scorned at every opportunity. The image of medieval Scholastic cosmology and science will forever be the hostile one Galileo constructed. Nevertheless, the significant process of "getting beyond Aristotle" had already begun in the Middle Ages, though it would come to spectacular fruition only in the seventeenth century. In retrospect, then, it seems appropriate to accord a small measure of credit to those much-maligned Scholastics, and thereby modestly begin to redress a long-standing injustice inflicted upon the Middle Ages by the victors in the science wars of the sixteenth and seventeenth centuries.

7 ✐ Scientific Imagination in the Middle Ages

> Imagination is more important than knowledge.
> Knowledge is limited. Imagination encircles the world.
> —Albert Einstein

Following Aristotle, medieval natural philosophers believed that knowledge was ultimately based on perception and observation; and like Aristotle, they also believed that observation could not explain the "why" of any perception. To arrive at the "why," natural philosophers offered theoretical explanations that required the use of the imagination. This was, however, only the starting point. Not only did they apply their imaginations to real phenomena, but expended even more intellectual energy on counterfactual phenomena, both extracosmic and intracosmic, extensively discussing, among other themes, the possible existence of other worlds and the possibility of an infinite extracosmic space. The application of the imagination to scientific problems during the Middle Ages was not an empty exercise, but, as I shall show, played a significant role in the development of early modern science.

In any discussion about the scientific imagination in a given historical period and in any civilization in history, a necessary first step requires that the author and readers have a common understanding of what qualified as science in the period in question. This is not a simple matter, largely because some historians of science have denied that proper science existed prior to the seventeenth century, or even later. Indeed, Roger French and Andrew Cunningham have denied the existence of science in the Middle Ages by the very title they assigned to their book: *Before Science: The Invention of the Friars' Natural Philosophy.*[1] In the epilogue to their book, they declare "there was no scientific tradition (in the modern sense of the term 'scientific') of looking at nature in the thirteenth cen-

1. Aldershot, England: Scolar Press, 1996.

tury, only a religio-political way of doing so. Natural philosophy was not the same as modern science." French and Cunningham are quite correct: natural philosophy is not the same as *modern science*, but it undoubtedly comprised a large part of *medieval science*. Indeed, the expression "natural science" was regarded as synonymous with natural philosophy.

During the late Middle Ages, science was composed of two major components: (1) the exact sciences; and (2) natural philosophy, also known as natural science, and occasionally physics. The exact sciences were the mathematical sciences of astronomy, optics, and statics, and, of course, mathematics itself. Aristotle regarded the exact sciences as "mixed sciences," because all of them, except for pure mathematics, involved the application of mathematics to problems in natural philosophy. These sciences were neither purely mathematical nor purely in the realm of natural philosophy. Although some significant contributions were made to the exact sciences during the late Middle Ages, and the scientific imagination was undoubtedly active, especially in optics and statics, the period is perhaps most noteworthy for preserving and studying the exact sciences that had been inherited from Greco-Arabic sources. By preserving and actively engaging in the exact sciences, medieval scholars made it possible for the Copernicuses, Galileos, and Keplers to take that body of scientific knowledge to new heights and fashion early modern science.

During the same centuries, the natural philosophers of the late Middle Ages transformed natural philosophy into something that, while essentially Aristotelian, was nevertheless dramatically different in its imaginative dimensions. They had visions about our world and other worlds that Aristotle had never dreamt of, let alone overtly considered. It is these medieval scientific imaginations that will be the subject of this essay.

Natural philosophy in the late Middle Ages was the study of change and motion in the physical world. The exact sciences, along with medicine, were not regarded as part of natural philosophy. But virtually all other physical changes formed part of its recognized domain. Because of its comprehensiveness, natural philosophy may be rightly regarded as the "Mother of All Sciences." It is the womb from which all the new sciences—physics, chemistry, biology, geology, and all their subdivisions and branches—were born during the seventeenth to nineteenth centuries.[2] Be-

2. Here I draw upon my account in *The Foundations of Modern Science in the Middle Ages: Their Religious, Institutional, and Intellectual Contexts* (Cambridge: Cambridge University Press, 1996), 192–94.

fore their appearances as independent sciences, the ideas and discussions relevant to these different scientific disciplines were scattered in bits and pieces in the commentaries and questions of treatises that were largely composed at medieval universities by members of the arts and theological faculties. The medieval treatises on Aristotelian natural philosophy and the independent tractates on specific themes in natural philosophy form part of the history of science, just as much as seventeenth-century treatises on physics by Galileo and Newton; and eighteenth-century treatises on biology by Carl Linnaeus and Georges-Louis Leclerc, and on chemistry by Antoine-Laurent Lavoisier.

Medieval literature on natural philosophy in the Middle Ages was largely based on Aristotle's so-called natural books—*Physics, On the Heavens, On Generation and Corruption, Meteorology,* and *On the Soul*—which lay at the heart of Aristotle's natural philosophy, to which should be added his treatises on biology and a few brief works known collectively as *The Small Works on Natural Things (Parva naturalia).* The dominant method of explaining and analyzing Aristotle's treatises was by means of a series of questions. It is in these questions that medieval scholars revealed their opinions and judgments about the structure and operations of the physical world, a world that Aristotle had constructed in his natural books. The questions were answered almost exclusively by use of reasoned analysis, with a modicum of empirical data that was, in any event, largely devoid of direct observation.[3] The following questions would have been regarded as proper queries about our physical world, and represent a broad range of scientific inquiry:

whether the whole earth is habitable;
whether spots appearing in the moon arise from differences in
    parts of the moon or from something external;
whether the earth is spherical;
whether by their light the celestial bodies are generative of heat;
whether a compound is possible;
whether there are four elements, no more nor less;
whether any element is pure;

3. For a discussion of the manner in which observation was used in medieval natural philosophy, see Edward Grant, "Medieval Natural Philosophy: Empiricism without Observation," in *The Dynamics of Aristotelian Natural Philosophy from Antiquity to the Seventeenth Century,* edited by Cees Leijenhorst, Christopher Lüthy, and Johannes M. M. H. Thijssen, 141–68 (Leiden: Brill, 2002).

we inquire whether one element could be generated directly from
   another, so that water could be generated directly from air
   without something else being generated from it previously; and
   [the same question can be asked] of the other elements;
whether it is possible for an actual infinite magnitude to exist;
whether the existence of a vacuum is possible;
whether the mass of the whole earth—that is, its quantity or
   magnitude—is much less than certain stars;
whether a comet is of a celestial nature or [whether it is] of an
   elementary nature, say, of a fiery exhalation;
whether lightning is fire descending from a cloud.
On the supposition that a rainbow can occur by reflection of rays, we
   inquire whether such reflection occurs in a cloud or whether it
   occurs in tiny dewdrops or raindrops.[4]

The responses to these varied questions were given in terms of the
current understanding of these phenomena, based largely on Aristotle's
original explanations but often altering those explanations, or adding to
them, in light of subsequent history and contemporary reflections and
conjectures. The scientific imagination that is reflected in responses to
these questions is the basic kind of imagination that drew upon accepted
Aristotelian principles and theories and from which an explanation was
constructed. Rarely was the explanation based on direct observation.
Aristotle himself provides the reason for this when he declares that our
senses "give the most authoritative knowledge of particulars. But they do
not tell us the 'why' of anything—e.g., why fire is hot; they only say that it
is hot" (Aristotle, *Metaphysics* 1.1.981b.10–11). Aristotle and his medieval
followers were primarily interested in the "why" of things, but paid lip
service to the observational basis of knowledge about the physical world.
Certain direct observations were gross and obvious. For example, that
fire heats and the heavens move was observed by all. But how are we to
interpret a statement by Nicole Oresme that "an alteration is when one
thing is changed into another as hotness into coldness, and similarly fire
into air" (*On Generation and Corruption*, bk. 1, qu. 1). All knew that a hot

4. These questions are drawn from Edward Grant, *A Source Book in Medieval Science* (Cambridge, Mass.: Harvard University Press, 1974), 199–210. Almost all are cited in Edward Grant,
God and Reason in the Middle Ages (Cambridge: Cambridge University Press, 2001), 358–59.

thing eventually becomes a cold thing if allowed to cool down. But why should we believe that fire becomes air? It is hardly obvious and is not a common experience. When a fire is extinguished, it disappears. Does it become air? We do not know by observation. But Oresme and all Aristotelian natural philosophers believed it was required by the theory of the four elements. In that theory, air is convertible to fire and fire is convertible to air. Many "observations" in medieval natural philosophy were of this kind: they were driven by theory, or by what had to be. The scientific imagination used in the numerous explanations of phenomena that were not directly observable, such as the mutual conversion of fire and air, was rather straightforward, because it followed accepted principles.

There were however more "imaginative" uses of the imagination that involved both real and hypothetical, or counterfactual, phenomena. Indeed, there were occasional debates as to whether some particular phenomenon was real or imaginary. The most significant of these discussions involved extracosmic existence. It is within this context that Scholastics had occasion to discuss, however cursorily, the role of the imagination. I shall first consider the imagination as it was thought to be involved in the assessment of extracosmic phenomena, and then turn to more mundane, intracosmic uses of the imagination as it was applied to real phenomena.

The Imagination and What Lies Beyond Our World

Aristotle, as is well known, rejected the existence of other worlds and also denied the existence of things beyond our world. In *On the Heavens,* after denying the existence of other worlds, Aristotle declares[5] that

[i]t is therefore evident that there is also no place or void or time outside the heaven. For in every place body can be present; and void is said to be that in which the presence of body, though not actual, is possible; and time is the number of movement. But in the absence of natural body, there is no movement, and outside the heaven, as we have shown, body neither exists nor can come to exist. It is clear then that there is neither place, nor void, nor time, outside the heaven.

5. Aristotle, *On the Heavens,* 1.279a.12–17, translated by J. L. Stocks, in *The Complete Works of Aristotle: The Revised Oxford Translation,* 2 vols., edited by Jonathan Barnes (Princeton: Princeton University Press, 1984). In the lines following, Aristotle explains that what does exist outside the world does not occupy any place nor does time affect it. It is in fact a divine en-

Nevertheless, there are those, Aristotle tells us, who believe that an infinite magnitude exists beyond our world; indeed that there is also an infinite number of worlds. The reason people think that an infinite number of things exist outside our world and that mathematical magnitudes are infinite is "because they never give out in our thought" (*Physics* 3.4.203b.26). Thus the human imagination suggests to some that their world is an infinite magnitude of some kind, or may have an infinite number of worlds, because we can always think of adding another quantity or world.

In the thirteenth century, Scholastic natural philosophers and theologians rejected the existence of an extracosmic infinite void space. They recognized, however, that some people could not accept an end to extension and mistakenly assumed the existence of such a space. Thomas Aquinas, for example, explains: "when we speak of nothing being beyond the heavens, the term 'beyond' betokens merely an imaginary place [*locum imaginatum*] in a picture we can form of other dimensions stretching beyond those of the heavens."[6] In *Questions on the Physics of Aristotle*, falsely ascribed to Siger of Brabant (ca. 1240–ca. 1282), Pseudo-Siger, as we shall call the author, declares that the term "beyond" *(extra)* could signify either a true place or one that is imaginary *(secundum imaginationem)*.[7] In order to imagine a place beyond the world, Pseudo-Siger insists that we must first imagine a place for it, even though that place is not real, but only imaginary. Thus the term "beyond" leads us to assume, or expect, something "out there" beyond the world. And although we may agree with Aristotle that no place, void, or body lies *beyond*, yet by extrapolation from mundane experience, we can imagine spatial dimensions extending ad indefinitum beyond the world, or imagine bodies, and therefore places—for every body must have a place and can only be imagined in a place—beyond the last sphere of our cosmos.

---

tity. In what follows on extracosmic infinite space, I draw on my discussion in *Much Ado About Nothing: Theories of Space and Vacuum from the Middle Ages to the Scientific Revolution* (Cambridge: Cambridge University Press, 1981), 117–21.

6. Translated in Thomas Aquinas, *Summa Theologia, Vol. 8: Creation, Variety and Evil*, Latin text, English translation, introduction, and notes by Thomas Gilby, O.P. (New York: Blackfriars Press, 1967), 75.

7. See Siger de Brabant, *Questions sur la Physique d'Aristote (texte inédit)*, edited by Philippe Delhaye, in *Les Philosophe Belges, Textes et Etudes*, vol. 15 (Louvain: Editions de l'Institut Supérieur de Philosophie, 1941), 179, for Latin text.

Does an Infinite Void Space Exist Beyond Our World?

The extracosmic places and spaces that were regarded as merely imaginary, and therefore nonexistent, by Thomas Aquinas and Pseudo-Siger of Brabant, were made real by others. No one put the case for real extracosmic existents more dramatically than did Nicole Oresme (ca. 1320–1382) in the fourteenth century. Oresme found compelling, and even overriding, the strong intuitive sense that something must exist beyond our finite world. It led him to declare: "[T]he human mind consents naturally, as it were, to the idea that beyond the heavens and outside the world, which is not infinite, there exists some space, whatever it may be, and we cannot easily conceive the contrary."[8] A few lines below, Oresme becomes more specific, explaining that "outside the heavens . . . is an empty incorporeal space quite different from any other plenum or corporeal space. . . ."[9] And although "we cannot comprehend nor conceive this incorporeal space which exists beyond the heavens," Oresme insists that "reason and truth . . . inform us that it exists."[10] Why is it different from any other corporeal space? Because, as Oresme makes explicit, "this space of which we are talking is infinite and indivisible, and is the immensity of God and God Himself, just as the duration of God called eternity is infinite, indivisible, and God Himself, as already stated above."[11] Oresme was here giving utterance to a powerful anti-Aristotelian current that was largely promoted by theologians, the most important of whom was Thomas Bradwardine (ca. 1290–1349), who sometime around 1344 wrote a lengthy theological treatise titled *In Defense of God Against the Pelagians*. In this work, Bradwardine found occasion to examine what God's relationship was, if any, to space. He encapsulated his thoughts in five corollaries that formed part of a chapter titled "That God is not mutable in any way."

(1) First, that essentially and in presence, God is necessarily everywhere in the world, and all its parts;

(2) And also beyond the real world in a place, or in an imaginary infinite void.

8. Nicole Oresme, *Nicole Oresme: Le Livre du ciel et du monde,* edited by Albert D. Menut and Alexander J. Denomy, translated with an introduction by Albert D. Menut, 177 (Madison: University of Wisconsin Press, 1968).

9. Ibid.   10. Ibid.

11. Ibid.

(3) And so truly can He be called immense and unlimited.

(4) And so a reply seems to emerge to the questions of the gentiles and heretics—"Where is your God?" And, "where was God before the creation of the world?"

(5) And so it seems obvious that void can exist without body, but in no manner can it exist without God.[12]

Bradwardine rejected the idea that God could, or did, create an actual infinite void space. He believed instead that God's immaterial, nonextensional, omnipresent infinite immensity was in fact an infinite void space. This was a void without body, but it was not without God, who filled it without extension or dimension. It was Bradwardine's conception of infinite void space that Oresme and a few other theologians adopted in the face of some strong opposition.[13] It was an imaginary infinite void that had to be real because it was identified with God's infinite, omnipresent immensity. Bradwardine's conception of infinite void space was adopted by a number of theologians in the course of the fourteenth to seventeenth centuries. Without perhaps realizing it, they came to equate "imaginary" with real. Bradwardine was probably the first to link God and infinite void space and then to regard that space as real, although without extension. The treatise in which his ideas were embedded was published in 1618 and apparently influenced both Scholastics and non-Scholastics. Thomas Compton-Carleton (ca. 1591–1666), a Jesuit Scholastic, moved dramatically beyond Bradwardine and argued that the infinite space in which God is omnipresent is three-dimensional, although there is no clear evidence that he also assumed that God was therefore also three-dimensional.[14] That step was taken by Henry More (1614–1687), who proclaimed not only the existence of an infinite three-dimensional void space in which God was omnipresent by His infinite immensity, but went beyond and boldly proclaimed that God was omnipresent in infinite void space because He was Himself an infinite three-dimensional being. More made infinite void space an attribute of God. Isaac Newton, who adopted

12. My translation in Grant, *A Source Book in Medieval Science*, 556–57.

13. It was rejected by Thomas Aquinas, Duns Scotus, and Richard of Middleton. See Grant, *A Source Book in Medieval Science*, 146–47.

14. For a description of Compton-Carleton's interpretation, see Edward Grant, *Planets, Stars, and Orbs: The Medieval Cosmos, 1200–1687* (Cambridge: Cambridge University Press, 1994), 183–84.

More's ideas about God and space, assumed the literal three-dimensional omnipresence of God in an infinite space that was God's property and, in effect, his immensity.[15] Thus did the imaginary, extensionless infinite void first postulated by medieval theologian-natural philosophers evolve into Isaac Newton's infinite, three-dimensional space for his new, revolutionary physics.

On the Possible Existence of Other Worlds

One of the most important cosmic ideas that entered Western Europe was embedded in Aristotle's natural philosophy: the powerful belief that only one world exists. Aristotle was unequivocal on this issue, declaring that "there is not, nor do the facts allow there to be, any bodily mass beyond the heaven. The world in its entirety is made up of the whole sum of available matter (for the matter appropriate to it is, as we saw, natural perceptible body), and we may conclude that there is not now a plurality of worlds, nor has there been, nor could there be."[16]

Since Aristotle's position agreed with that of the creation account in Genesis and with the rest of Sacred Scripture, there seemed little reason to believe that any significant conflict would develop over the issue of other possible worlds. Although prior to 1277, some Scholastic natural philosophers allowed that God could create other worlds if He wished to, they chose to present arguments that made it seem unfeasible or contradictory for God to create more than one world. For example, Michael Scot argued that although, by His absolute power, God could create other worlds, nature as a created entity was incapable of receiving other worlds, since it had not been endowed with the capacity to accommodate more than one world. Thomas Aquinas offered similar reasons for supposing that God would not create other worlds.[17] These scholars may also have been motivated by the fact that the existence of other worlds would wreck Aristotle's natural philosophy.

In 1277 in Paris, however, one of 219 articles condemned was Article 34, which rejected the idea "[t]hat the first cause [i.e., God] could not

15. Ibid., 184.
16. Aristotle, *On the Heavens,* translated by W. K. C. Guthrie, Loeb Classical Library, 1.9.279a.7–11 (Cambridge, Mass.: Harvard University Press; London: William Heinemann, 1960).
17. See Grant, *Planets, Stars, and Orbs,* 154–55.

make several worlds."[18] The condemnation signified that if this, or any other article, was defended the defender was subject to excommunication. Although the condemnation was only effective in Paris, the University of Paris was the intellectual center of the Christian world, which made it significant. The condemnation of Article 34 radically altered the subsequent nature of discussions about the possibility of other worlds. The focus of the arguments shifted drastically. No longer was it a question of whether other worlds really existed, but, rather, whether God was capable of creating other worlds. Article 34 made it mandatory for all natural philosophers to concede that God could create other worlds at His pleasure, if He so desired. When commentators now came upon Aristotle's defense of a single, unique world in his famous discussion in *On the Heavens,* they usually felt obligated, if not compelled, to explain that, although Aristotle was right to assume that other worlds could not exist by natural means, God could, if He wished, create other worlds by His absolute, supernatural power. Despite the disclaimer, university natural philosophers did not believe that God did actually create other worlds. This would have destroyed Aristotle's natural philosophy. Thus the situation could have remained in a kind of dead end, where natural philosophers might have routinely and formulaically agreed with Aristotle that while it is naturally possible for only one world to exist, God could, if He wished create as many other worlds as He pleased.

But medieval natural philosophers and theologians chose another more dramatic path. They put God to work to create other worlds, so they might use their imaginations to investigate whether or not those worlds would function as their own Aristotelian world. Immediately following his French translation of Aristotle's discussion (in the first book of *On the Heavens*) demonstrating the natural impossibility of other worlds, Oresme comments: "Now we have finished the chapters in which Aristotle undertook to prove that a plurality of worlds is impossible, it is good to consider the truth of this matter without considering the authority of any human but only that of pure reason."[19] Although Oresme was more intellectually daring and innovative than most, if not all, Scholastic commen-

---

18. For Article 34 and other articles condemned in 1277, see Grant, *A Source Book in Medieval Science,* 48–50. To make certain the reader understands that "the first cause" signifies God, I have added the term within square brackets.

19. Oresme, *Le Livre du ciel et du monde,* 167.

tators, the attitude he expresses here was typical of the medieval approach to the plurality of worlds, as well as to other hypothetical problems.

Indeed, Oresme's reasoned approach to the plurality of worlds question provides an excellent guide to the various types of worlds considered. He imagines the existence of a plurality of worlds in three ways, under essentially two categories: successive and simultaneously existing worlds. Oresme first considers successive worlds, an idea that was derived from the ancient world. According to this theory, the elements of a world come together by love from a confused mass of matter, and after a long period of time it is destroyed by discord and another world will come into being and replace it, ultimately suffering dissolution. "Such a process will take place in the future an infinite number of times, and it has been thus in the past." Oresme immediately adds that he will not discuss this version of a plurality of worlds, because Aristotle rejected it several times. Oresme concludes the topic of successive worlds, by explaining, "It cannot happen in this way naturally, although God could do it and could have done it in the past by His own omnipotence, or He could annihilate this world and create another thereafter. And, according to St. Jerome, Origen used to say that God will do this innumerable times."[20]

Oresme focuses next on the two varieties of simultaneous worlds, the first of which was only occasionally discussed by others, perhaps because of its strange configuration, which Oresme seems to acknowledge by declaring "Another speculation can be offered which I should like to toy with as a mental exercise." Concentric worlds, he suggests, can be imagined to exist in a star or in the moon. For his discussion, Oresme imagines another distinct world that lies concentrically within ours, and even imagines a third world wrapped concentrically around our world. Each world is a replica of ours. Our outer world would have terrestrial and celestial regions, as would inner and outer worlds that are concentric to ours. The major obstacle is size, since "all natural bodies are limited in bigness and smallness, for the size of a man could diminish or grow so much that he would no longer be a man, and the same with all bodies. So, the world we have imagined inside our own world and beneath its circumference would be so small that it would not be a world at all, for our sun would be 2,000 times the size of the other and each of

20. Ibid.

our stars would be larger than this imaginary world." To respond to this problem, Oresme points out that "*large* and *small* are relative, and not absolute, terms used in comparisons," from which it follows that if our world were made "between now and tomorrow 100 or 1,000 times larger or smaller than it is at present, all its parts being enlarged or diminished proportionally, everything would appear tomorrow exactly as now, just as though nothing had been changed."[21] Although the existence of concentric worlds is improbable, Oresme insists that concentric worlds are not impossible "because the contrary cannot be proved by reason nor by evidence from experience, but also I submit that there is no proof from reason or experience or otherwise that such worlds do exist." Oresme concludes his discussion of concentric worlds with this sage advice: "Therefore, we should not guess nor make a statement that something is thus and so for no reason or cause whatsoever against all appearances; nor should we support an opinion whose contrary is probable;[22] however, it is good to have considered whether such opinion is impossible."[23] For Oresme, rejection of any theory or interpretation was only appropriate if it was demonstrated to be impossible.[24]

It is the second kind of simultaneous world that was the major focus of attention by natural philosophers and theologians in the Middle Ages. Is it possible for other worlds to exist simultaneously with ours, but separately and independently from one another? Aristotle had vigorously regarded the existence of more than one world as an utter impossibility. In rejecting these worlds, Aristotle always assumed that they were all identical to our world. Consequently, medieval discussions were almost always about the possible existence of identical Aristotelian worlds. As we shall see, the debate about imaginary, simultaneous, disparate, but identical worlds resulted in extraordinary conjectures and conclusions that actually subverted Aristotle's cosmology and natural philosophy. This resulted from the simple fact that where Aristotle had regarded a plural-

---

21. Ibid., 167–69. For a discussion of concentric worlds in the Middle Ages, see Grant, *Planets, Stars, and Orbs*, 156–57.

22. Here Oresme concedes that the contrary to the existence of concentric worlds is that they do not exist, which he regards as probable.

23. Oresme, *Le Livre du ciel et du monde*, 171.

24. Where Oresme allowed for the possibility of concentric worlds, although he thought them improbable, his predecessors, William of Auvergne and Roger Bacon, dismissed them without hesitation. See Grant, *Planets, Stars, and Orbs*, 156.

ity of identical worlds as impossible, his medieval followers, for religious and rational reasons, regarded them as possible.

It was the condemnation of Article 34 (cited above) that compelled natural philosophers at the University of Paris to reject Aristotle's arguments for the impossibility of other worlds, and to argue that other worlds were certainly possible because God Himself could create as many more worlds as He pleased. Although natural philosophers were thereafter required to concede God's power to make other worlds, in spite of Aristotle's contrary arguments, they were certainly not required to inquire whether those other worlds would function and operate as our own world, or how they might relate to one another. In fact, they were under no compulsion to discuss any aspect of the existence of other worlds. And yet many natural philosophers eagerly chose to investigate such questions, even though none that I know of actually believed that God had created other worlds, or would do so in the future. Why, then, did it become customary for university natural philosophers to include questions about other worlds in their commentaries on Aristotle's books of natural philosophy? Because, it seems, they found questions about other hypothetical worlds a challenge to their imaginations. Could the physical principles that Aristotle had so carefully presented for a world he believed unique be operative in other worlds, if such existed? They obviously thought it important to know this and believed that it would shed light on their knowledge of cosmic possibilities. Moreover, the possibility of other worlds focused their attention on Aristotle's arguments in defense of a unique world, and some of these would be found wanting.

Indeed, Nicole Oresme begins his discussion about simultaneously existing separate worlds by insisting that Aristotle's "arguments are not clearly conclusive." In demonstrating this, Oresme departs significantly from some of Aristotle's fundamental cosmic principles. One of Aristotle's most powerful arguments against a plurality of worlds is, as Oresme expresses it, "that the earth in the other world would tend to be moved to the center of our world, and conversely." To appreciate the force of this argument one must realize that for Aristotle each of the four terrestrial elements—earth, water, air, and fire—had its own absolutely determined natural place, where it would move naturally if unimpeded: earth at the center of the world; fire at the concave surface of the lunar sphere; air right below fire and water below air and just above the earth. These were envisioned as four concentric zones. That each of the four elements had its own exclusive and unique natural place implied the existence of a unique spherical world. If elsewhere in an extracosmic space another spherical world identical to ours existed simultaneously, there would be a second center and circumference, thus precluding the possibility of a unique center and circumference and thus seemingly subverting Aristotle's physics. If all centers are equal and identical, Aristotle's doctrine of absolute, unique natural places for each of the four elements is negated. But Aristotle regarded it as absurd to believe that separate, identical worlds could coexist. If they did, the following impossible consequence would follow: the earth of another world would seek to move toward the center of our world. To do this, it would have to move upward to reach the circumference of its own world and, then move beyond its world toward the center of our world. Thus it would have two natural motions: up and down. Fire would react similarly. It would move naturally upward in its own world and then, upon reaching our world, move naturally downward toward the center of the world.[25] For this, and other reasons, Aristotle rejected the existence of other worlds as impossible and utterly absurd.

Oresme, as we saw, found Aristotle's argument unconvincing. In countering Aristotle, Oresme insisted that no such consequence would follow.[26] To demonstrate this, he dramatically departs from Aristotle's physics by redefining the terms "up," "down," "light" and "heavy." Instead of Aristotle's absolute sense of these terms, Oresme made them completely relative by defining them independently of Aristotle's concept of natural place. In Oresme's new interpretation, a body is judged to be "heavy" and "down" when it is surrounded by light bodies, where the surrounding "light" bodies are conceived as "up." Thus the earth is heavy and down when it comes to rest naturally in the center of lighter bodies that surround it. Such a heavy body would not, therefore, seek a motionless, fixed natural place at the center of the world, as Aristotle believed. Rather, a heavy body would seek to come to rest whenever it is surrounded by lighter bodies, wherever those lighter bodies happen to be. Therefore, a heavy body does not naturally seek to move to the center of the earth, as Aristotle argued.

25. For Aristotle's arguments, see his *On the Heavens*, bk. 1, ch. 8 276a.22–276b.21.
26. I base this on my discussion in *Planets, Stars, and Orbs*, 164–65.

What would happen, however, if a heavy body was not surrounded by lighter bodies? Under such circumstances, that heavy body would be neither up nor down. Oresme then imagines a commonly invoked hypothetical condition in which a void space is assumed to exist between our world and another. If a particle of earth entered that void from the other world seeking to move to our world, it could not do so, because, in the total absence of surrounding lighter bodies there is no up or down, and the body has no inclination to move in any direction. It would immediately come to rest.

Thus did Oresme believe that he had countered Aristotle's major argument against the possible existence of other worlds. Oresme's argument was unique. The important feature of Oresme's position is that he abandoned Aristotle's doctrine of natural place, which was the basis of Aristotle's terrestrial physics. To my knowledge, Oresme's relativistic doctrine of place was fashioned solely to combat Aristotle's arguments against a plurality of worlds. He did not attempt to integrate his new relativistic conception of place into Aristotle's physics, a move that would have required him to restructure and refashion that physics. This was a common tactic among those medieval natural philosophers who diverged from Aristotle's physics. In the late Middle Ages, hypothetical results derived from imaginary situations that were subversive of Aristotle's physical principles were not used to "correct" or challenge Aristotle's physics, which might have proved fatal to it. Perhaps, Oresme, and his fellow natural philosophers, failed to transform Aristotle's physics because, despite their hypothetical conjectures, they really did not believe in the existence of other worlds. Oresme spoke for almost all medieval natural philosophers when he terminated his discussion of possible worlds with the remark: "I conclude that God can and could in His omnipotence make another world besides this one, or several like or unlike it. Nor will Aristotle or anyone else be able to prove completely the contrary. But, of course, there has never been nor will there be more than one corporeal world. . . ."[27]

The main impact of these moves was to show that what Aristotle regarded as impossible was indeed possible. Where Aristotle rejected the existence of other worlds as naturally impossible, many Scholastics ar-

27. Oresme, *Le Livre du ciel et du monde*, 177–79.

gued that if such worlds were created supernaturally, Aristotle's dire predictions would not obtain. I shall now illustrate this with a number of examples.

Aristotle, as we saw, insisted that only one spherical world could exist because, among other arguments, only one circumference and center of the world is possible. By redefining the doctrine of place, Oresme rejected this claim. Other Scholastic authors, however, retained Aristotle's basic doctrine of natural place, but insisted that it could, and would, apply to more than one world. Numerous theologians adopted this position in their theological commentaries. For example, Richard of Middleton (fl second half of thirteenth c.), in a question on "whether God could make another world," insisted that in its own world neither the earth nor any of its parts would move upward contrary to their natural inclination in order to reach the center, or earth, of another world. If for some reason, a heavy earthy body in a particular world was violently removed to a higher elevation, say, in water or air, it would, when unimpeded, always move naturally downward toward the center of its own world. Each world has its own four elements that are completely independent of the set of identical four elements that constitute the terrestrial regions of all other existing worlds. Each world is self-contained and completely free from the operations and influence of any other world.[28] Although a dramatic departure from Aristotle's worldview, and one that would have proved devastating to Aristotle's physics and cosmology, it played no significant role because it was, after all, merely an exercise of reasoned imagination about possible worlds that no one really believed existed, or would exist. God could, of course, create those other worlds, but no one believed that He had done so in the past, or would do so in the future. There was little incentive to tamper with Aristotle's world on the basis of such imaginary discussions.

Motion in a Vacuum within Our Own World

Change and motion lay at the heart of Aristotle's natural philosophy. It is hardly surprising, then, that Aristotle devoted much space to the

28. See Grant, *Planets, Stars, and Orbs*, 159. For a substantial list of advocates of this interpretation, see 160, n. 33.

problem of local motion, both natural and violent. Major challenges to Aristotle came in the latter category, namely, in his explanation of how heavy bodies are forcibly moved out of, or away from, their natural places. Greek and Arabic commentators considered this problem, often rejecting Aristotle's explanations. Thus John Philoponus, the sixth-century Christian Neoplatonist, challenged the role Aristotle had assigned to the external medium, namely, that in air a heavy body could only be moved violently from one place to another if the air in direct contact with the body pushed it along, while the air in front resisted its motion, thereby guaranteeing finite motion. Without an external resistance—that is, in a vacuum—Aristotle believed that a body would move instantaneously, or with infinite speed. When the external air ceased pushing, the heavy body would fall to its natural place.

Philoponus found this implausible, denying the necessity for a resistant medium in local motion and also rejecting the external medium, or air, as the agent or cause of violent, or unnatural, motion. The real cause of violent motions, he insisted, was an incorporeal impressed force, or, as it would be called in the late Middle Ages, impetus. Islamic commentators were often familiar with Philoponus's Aristotelian commentaries and elaborated on his ideas. From Latin translations of some of these works during the twelfth and thirteenth centuries, medieval Latin scholars became familiar with these anti-Aristotelian ideas. Beginning in the late thirteenth century, it became customary for Aristotelian natural philosophers to grapple with two significant questions: Was the existence of a vacuum possible; and (2) If a vacuum did exist, how could bodies move in such a nonresistant medium? All agreed with Aristotle that the existence of a vacuum by natural means was impossible. One therefore had to invoke a hypothetical vacuum, either by assuming that God created one, or simply imagining its existence in order to consider Aristotle's assorted arguments in the fourth book of his *Physics*. Various questions were posed: If a vacuum existed and a real body was placed in it, would that body rise or fall with a natural motion? If it were moved with a violent motion out of its natural place, would it move with a continuous, violent motion? Although Aristotle had denied the possibility of finite motion in a vacuum, most of his medieval followers disagreed with him. In formulating their responses, however, they applied Aristotle's physical principles. They assumed that natural downward motion in a void, just as in a plenum, involves the action of a force operating against a resistance. In a resistance-less vacuum, what could serve as a motive force and what could function as a resistance in heavy bodies moving naturally downward? In the absence of external resistance in a void, Scholastics assumed that compound bodies, which contained both heavy and light elemental bodies, contained within themselves a causative agent, the heavy elements, and a resistive force, the light elements, which moved naturally upward contrary to the downward moving heavy bodies. Thus compound bodies could move in a void because they had within themselves a motive force and a resistance and were therefore independent of their spatial surroundings. But what about the motion of a pure elemental body in a void, say, a particle of earth? There was general agreement that an elemental body could not move with a finite motion in a void because nothing within it could be identified as a resistance that would act against a motive force, even if that motive force were assumed to be the heaviness of the body itself. The void itself, of course, was not regarded as a resistance to motion, although its dimensions had once been thought by a few to serve in that capacity.[29]

But what about violent, or unnatural, motion in a void? The explanations just described for bodies falling with natural motion in a vacuum were of little use to account for the motion of heavy compound bodies that were assumed to be moving away from their natural places. Lightness and heaviness of a body's elements, which were used to explain natural motion downward, were useless to account for violent motion. Here Scholastic natural philosophers resorted to impressed force, or impetus, as it was called. In the history of medieval science, impressed forces played a significant role, exercising an influence on Galileo and Newton, and therefore on early modern physics. Scholastics distinguished two kinds of impressed force: one that is self-exhausting, the other permanent. The difference between them posed a major problem for those who tried to imagine how a heavy body could be moved through an extended vacuum by means of an impressed force. Each kind of impressed force

29. For the manner in which three-dimensional void spaces were assumed to function as external resistances to bodies moving within it, see Grant, *Much Ado About Nothing*, 27–38. John Duns Scotus, for example, declared (31–32) that "the motion of a heavy body in a vacuum would be successive because a prior part of the vacuum would yield first, and the whole heavy body would traverse that part of space and then this [part]." Many objections were raised and it gained little support.

would have dramatically different effects in moving a heavy body through a vacuum. The self-exhausting kind would eventually dissipate itself and the body would come to rest. Nicholas Bonetus, a fourteenth-century theologian, refers to a transitory impetus when he explains "in a violent motion some non-permanent and transient form is impressed in the mobile so that motion in a void is possible as long as this form endures; but when it disappears the motion ceases."[30] But those who regarded impetus as a permanent quality were confronted with an insoluble dilemma if they sought to explain the continuing motion of a body in a vacuum by means of that permanent impetus. How would it come to rest? What in the vacuum would function as a resistance to cause the impressed force to dissipate and gradually lose its capacity to move the body, thus eventually bringing it to rest? They had to conclude that the motion of such a body would never terminate or vary and was therefore endless and impossible. There was nothing in the emptiness of a vacuum to weaken and affect the permanent impetus. For this reason violent motion in the void was usually rejected.

One final noteworthy example of motion in a void involves the fall of two homogeneous bodies of different size and weight. Thomas Bradwardine, Albert of Saxony, and others concluded that two such bodies would fall in a void with equal speed. This conclusion followed because every equal unit of matter in every homogeneous body is identical and therefore has the same ratio of heavy to light elements—that is, the same ratio of motive force to internal resistance, F/R. The fact that the heavier body contained more units of matter did not affect the conclusion because speed was held to be governed by the ratio of force to resistance per unit of matter. This was a rejection of Aristotle's physics in which speed is proportional to heaviness or absolute weight, so that the heavier the body the greater its velocity. In his *De motu* of 1590, Galileo, using specific weight as the determiner of speed, arrived at much the same conclusion when he argued that homogeneous bodies of unequal size, and therefore of unequal weight, would fall with equal speeds in both plenum and void, though their speeds in the latter would be greater than in the former.

Over the centuries, natural philosophers discussed other examples of the behavior of bodies in or around void spaces. From the examples al-

30. See Grant, *Much Ado About Nothing*, 43.

ready described, however, it is obvious that many unusual, purely hypothetical, and imaginary conclusions were seriously considered, virtually all of which were subversive of Aristotle's physics. Although they were never used to transform Aristotle's physics of the real world, they were, nevertheless, important because if God, by His absolute power, chose to create an extended, three-dimensional void space, the behavior of bodies in that void were far more likely to conform to the interpretations of Aristotle's medieval followers than to those of Aristotle. As mere unlikely possibilities, they weakened Aristotle's unqualified claims that they were utterly impossible.

The Imagination Applied to the Real Physical World

The medieval scientific imagination was not confined to imaginary worlds and hypothetical void spaces. It was also applied in interesting ways to the real world filled everywhere with matter. One aspect of medieval imagination consists of "experiences" that are presented as if they had been personally performed or witnessed. They are in fact thought-experiments. In the fourteenth century, John Buridan was the most significant proponent of impetus theory, which he used against two of Aristotle's key explanations, namely, that (1) air, or the external medium, was the cause of continuous projectile motion; and (2) that the attraction of the natural place of a heavy body causes that body to accelerate as it approaches its natural place.

External Medium Is Not the Cause of Continuous Projectile Motion

To refute air as the cause of the continued movement of a projectile, John Buridan, in his *Questions on Aristotle's Physics* presents three "experiences" *(experientie)*. The first and third seem plausibly relatable to direct experience. The second appears to be an imaginary construct, though it too seems plausible. Buridan describes the second experience as follows: "A lance having a conical posterior as sharp as its anterior would be moved after projection just as swiftly as it would be without a sharp conical posterior. But surely the air following could not push a sharp end in this way, because the air would be easily divided by the sharpness."[31]

---

31. Translated by Marshall Clagett in his *The Science of Mechanics in the Middle Ages* (Madison: University of Wisconsin Press, 1959), 533.

The point of the argument is to show that a lance that had conical points at both ends could not be pushed by air, because the pointed end would offer no surface to push against; the air would simply flow by ineffectually. Since he was convinced that impetus, not air, was the cause of continuous projectile motion, Buridan would just as readily have denied that posterior air could push forward a lance with a broad posterior, rather than a conical, surface. But he chose a conical lance for his counterargument against Aristotle, because, with no more surface than a point, it was the most effective and dramatic way to demonstrate the inability of air to move a lance forward after it was hurled into the air.[32]

Proximity to Natural Place Is Not the Cause of Downward Acceleration

Buridan considers Aristotle's second position in his *Questions on Aristotle's On the Heavens* (bk. 2, qu. 12).[33] Aristotle, as we saw, believed that the downward speed of a body accelerates as it approaches its natural place. As part of his refutation of Aristotle, Buridan imagines the following counterexample: he assumes that one stone falls to earth from a high place and another from a low place.[34] If the velocities of the stones are affected only by proximity to their natural place, their speeds should be equal when they are both one foot above the earth, despite the fact that their respective falls began from radically different heights. "Yet," counters Buridan, "it is manifest to the senses that the body which should fall from the high point would be moved much more quickly than that which should fall from the low point, and it would kill a man while the other stone [falling from the lowpoint] would not hurt him."[35] For Buridan, it was obvious that the speed with which the stones fell from radically different heights was determined by those heights, not by their proximity to their natural place. Indeed, he was convinced that it is the continual production of impetus within the falling body that causes its downward acceleration; the greater the height from which it falls, the more impetus

32. For more on this, and the other two experiences, see Grant, "Medieval Natural Philosophy," 146–49.

33. Most of the question is translated by Marshall Clagett, *The Science of Mechanics in the Middle Ages*, 557–62; reprinted in Grant, *A Source Book in Medieval Science*, 280–83.

34. Here I follow my discussion in Grant, "Medieval Natural Philosophy," 153.

35. The translation is by Clagett, *The Science of Mechanics in the Middle Ages*, 533; reprinted in Grant, *A Source Book in Medieval Science*, 281.

will the body produce as it falls, and the greater its speed upon impact with the earth.

Impetus and the Motion of a Body Through a Hole in the Center of the Earth

As an illustration of an imaginary experiment that concerns impetus, Amos Funkenstein cites the *Dialogue Concerning the Two Chief World Systems,* where Galileo compares the following imaginary experiment with the arc of a pendulum:[36]

SALVIATI: From this it seems possible to me . . . to believe that if the terrestrial globe were perforated through the center, a cannon ball descending through the hole would have acquired at the center such an impetus from its speed that it would pass beyond the center and be driven upward through as much space as it had fallen, its velocity beyond the center always diminishing with losses equal to the increments acquired in the descent; and I believe that the time consumed in this second ascending motion would be equal to its time of descent.

Although Funkenstein regarded this imaginary experiment as original with Galileo,[37] it had already been proclaimed in the fourteenth century by Nicole Oresme and Albert of Saxony.[38] After discussing the nature of impetus in the abstract, Albert of Saxony declares:

According to this [theory], it would be said also that if the earth were completely perforated, and through that hole a heavy body were descending quite rapidly toward the center, then when the center of gravity [*medium gravitatis*] of the descending body was at the center of the world, that body would be moved on still further [beyond the center] in the other direction, i.e., toward the heavens, because of the impetus in it not yet corrupted. And, in so ascending, when the impetus would be spent, it would conversely descend. And in such a descent, it would again acquire unto itself a certain small impetus by which it would be

36. Galileo Galilei, "The Second Day," in *Galileo Galilei Dialogue Concerning the Two Chief World Systems—Ptolemaic and Copernican,* translated by Stillman Drake, 227 (Berkeley and Los Angeles: University of California Press, 1962).

37. Amos Funkenstein, *Theology and the Scientific Imagination from the Middle Ages to the Seventeenth Century* (Princeton: Princeton University Press, 1986), 176.

38. Both descriptions have been translated by Clagett, *The Science of Mechanics in the Middle Ages,* 566 (Albert of Saxony) and 570 (Oresme). Albert's discussion appears in his *Questions on the [Four] Books on the Heavens and the World of Aristotle,* bk. 2, qu. 14; Oresme's appears in his *Le Livre du ciel et du monde,* bk. 1, ch. 18. For Menut's version of the passage, see his translation, *Le Livre du ciel et du monde,* 145.

moved again beyond the center. When this impetus was spent, it would descend again. And so it would be moved, oscillating [*titubando*] about the center until there no longer would be any such impetus in it, and then it would come to rest.

Rejection of Aristotle's Moment of Rest

In the eighth book of his *Physics,* Aristotle argues that the motion of a heavy body upward, followed immediately by its downward natural fall, is not a single, continuous motion, but actually two distinct motions— one up and one down—separated by a moment of rest. Most Scholastic natural philosophers rejected Aristotle's moment of rest, largely on the basis of a powerful counterexample that was derived ultimately from Arabic sources.[39] John Buridan, Albert of Saxony, and Marsilius of Inghen were three fourteenth-century Scholastics who rejected the moment of rest in their *Questions on Aristotle's Physics.* "The proof" against a moment of rest, argues Marsilius of Inghen, "is that if a bean [*fabba*] were projected upward against a millstone [*molarem*] which is descending, it does not appear probable that the bean could rest before descending, for if it did rest through some time it would stop the millstone from descending, which seems impossible."[40] This graphic example destroyed Aristotle's moment of rest. But Scholastic natural philosophers did not rest content with this crucial counterargument, inventing additional imaginary experiences to make certain the moment of rest was not resurrected. They conjured up examples involving the contrary motion of ships and the movement of a fly on a lance. As an instance of the latter, John Buridan explains that "if a lance is hanging from a tree [and] a fly ascends on that lance and the cord by which the lance is hanging is broken, and then the lance and the fly fall down, the motion of the fly will be contrary, from up to down, but there will be no moment of rest."[41] Although, in the language of medieval Scholasticism, these examples were "according to the imagination" *(secundum imaginationem),* they were so graphic they may well have functioned as genuine observations.

39. Based on my discussion in Edward Grant, *God and Reason in the Middle Ages* (Cambridge: Cambridge University Press, 2001), 170–72.
40. Translated by Grant, *A Source Book in Medieval Science,* 286–87.
41. Cited from Grant, *God and Reason in the Middle Ages,* 171. Albert of Saxony gives yet another version of the fly on a lance; see ibid.

The Intension and Remission of Forms or Qualities

Although it was arbitrarily applied to phenomena in the real world, the medieval doctrine of intension and remission of forms or qualities was a purely imaginary construct.[42] Based ultimately on Aristotle's discussion in his *Categories,* Scholastics became interested in describing the way qualities vary, becoming either more or less intense. A number of theories were proposed to explain how the intensity of this or that quality varies. In the fourteenth century, qualities were treated as if they were extensive magnitudes that could be quantitatively augmented or diminished by adding quantitative parts or subtracting quantitative parts. This was known as the doctrine of "the intension and remission of forms or qualities." During the early fourteenth century at Merton College (Oxford University), a group of English scholars came to treat variations in velocity, or local motion, in the same manner as variations in the intensity of a quality. Thus, they imagined that the "intensity" of a velocity increased with speed, just as the redness of an apple increased with ripening. As they focused on the quantitative aspects of variations in qualities, the original metaphysical and theological contexts, that had formerly been prominent, were ignored and largely forgotten.

What emerged from all this were treatises in which variations in qualities were presented mathematically, either by means of arithmetic or geometry. They were concerned with qualities that increased at a uniform rate or at a uniformly accelerated rate. When applied to hypothetical bodies in motion, this translated into a successful search for definitions of uniform speed and uniformly accelerated motion. These definitions were then used to derive the famous mean speed theorem, namely that $s = \frac{1}{2}at^2$. An arithmetic proof was given at Oxford[43] and a geometric proof by Nicole Oresme, which appears in his lengthy treatise on intension and remission of forms.[44] But all qualities could be, and many were, treated in a similar manner. In principle, medieval scholars, including

42. For a detailed discussion, see Grant, *The Foundations of Modern Science in the Middle Ages,* 99–104.

43. For the text of the proof by William Heytesbury in his *Rules for Solving Sophisms,* see Grant, *The Foundations of Modern Science in the Middle Ages,* 101.

44. I have described Oresme's geometrical proof of the mean speed theorem in Grant, *The Foundations of Modern Science in the Middle Ages,* 102–3. Oresme's treatise is titled *On The Configurations of Qualities and Motions.*

Oresme, sought to represent variations in virtually all qualities, observable and unobservable. Just as with the mean speed theorem, geometric figures were used to represent the alleged effects of various powers and qualities, such as pains, colors, sounds, tastes, and so forth. For example, Oresme describes how the intensity of a pain can vary. "Let A and B be two pains," he declares, "with A being twice as intensive as B and half as extensive. Then they will be equal simply . . . although pain A is worse than, or more to be shunned than, pain B. For it is more tolerable to be in less pain for two days than in great pain for one day. But these two equal and uniform pains when mutually compared are differently figured . . . so that if pain A is assimilated to a square, then pain B will be assimilated to a rectangle whose longer side will denote extension, and the rectangle and square will be equal."[45]

The highly developed and mathematized medieval doctrine of intension and remission of forms was a wholly imaginary activity. No evidence was offered, nor even demanded, to show that it depicted the behavior of real qualities. The ingenious conclusions and theorems derived within the doctrine of intension and remission of forms were little more than intellectual exercises reflecting the subtle imaginations and logical skills of Scholastic scholars, even though in one place Oresme speaks as if the intensity of every quality has a certain fundamental figuration that represents it. "Thus, for example, the form of a lion demands a different corporeal shape than does the form of an eagle. . . . So, the natural heat of a lion is, in respect to intensity, figurable in a different way than is the heat of an eagle or a falcon; and similarly for others."[46] Although Oresme declares that different shapes or figurations are required for different things and creatures, these shapes are not in any sense real. As Clagett explains, "underlying Oresme's whole treatment of intensities in the *De configurationibus* is the conviction that while intensities may be compared by comparing lines, intensities are not themselves spatial extensions."[47] In the end, they are mental fictions. The mean speed theorem was simply another

45. Cited in Grant, *The Foundations of Modern Science in the Middle Ages,* 172, from the original translation in Marshall Clagett, *Nicole Oresme and the Medieval Geometry of Qualities and Motions: A Treatise on the Uniformity and Difformity of Intensities Known as "Tractatus de configurationibus qualitatum et motuum,"* edited with an introduction, English translation, and commentary by Marshall Clagett (Madison: University of Wisconsin Press), 387.

46. Clagett, *Nicole Oresme and the Medieval Geometry of Qualities,* 233.

47. Ibid., 23.

mental fiction. And yet, by sheer imagination, Scholastic natural philosophers came to formulate one of the seminal relationships of early modern
physics. Not until sometime around 1545 did Domingo Soto, an obscure
Scholastic author, suggest the mean speed theorem be applied to naturally
falling bodies. It remained for Galileo, however, to apply the mean speed
theorem to the motion of real falling bodies and to devise an experiment
to determine if bodies really fall with uniform acceleration. Thus began
the new science of mechanics and the beginnings of modern physics.

The Imagination in Theological Discussions Relevant to Natural Philosophy

During the late Middle Ages, the application of natural philosophy and
logic to theology transformed it into an analytic discipline. The extraordinary nature of this transformation is manifested when we see the kinds
of questions that were routinely discussed in the average theological
treatise—that is in the average *Commentary on the "Sentences" of Peter
Lombard*. By inspection, it seems apparent that logic and natural philosophy were required to answer them, as is true for the following questions
about God and angels.

God

Whether God could make a creature exist for only an instant.
(Hugolin of Orvieto, *Sentences*, bk. 2, dist. 2, unique question, art. 3.)

Whether God could do contradictory things simultaneously.
(Richard of Middleton, *Sentences*, bk. 1, dist. 42, unique article, qu. 4.)

Whether God, by His infinite power, could produce an actually
infinite effect.
(Gregory of Rimini, *Sentences*, bk. 1, dist. 42–44.)

Angels

Does an angel exist in a place?
(Thomas Aquinas, *Summa theologiae*, vol. 9, Angels, qu. 52 ("On the Relationship of
Angels to Places"), article 1, pp. 44–45.)

Can an angel be in several places at once?
(*Summa theologiae*, vol. 9, Angels, qu. 52, article 2.)

Can several angels be in the same place at once?
(*Summa theologiae*, vol. 9, Angels, qu. 52, article 3.)

Can an angel move from place to place?

(*Summa theologiae*, vol. 9, Angels, qu. 53 ("On the Local Motion of Angels"), article 1.)

Does an angel, moving locally, pass through an intermediate place?

(*Summa theologiae*, vol. 9, Angels, qu. 53 ("On the Local Motion of Angels"), article 2.)

Whether an angel exists in a divisible or indivisible place.

(Gregory of Rimini, *Sentences*, bk. 2, dist. 2, qu. 2.)

Whether an angel could be moved from place to place successively in some time.

(Gregory of Rimini, question in *Sentences*, bk. 2, dist. 6.)

Whether an angel could be moved from place to place in an instant.

(Gregory of Rimini, question in *Sentences*, bk. 2, dist. 6.)

Whether an angel could sin or be meritorious in the first instant of his existence.

(Gregory of Rimini, bk. 2, dist. 3–5, qu. 1.)

To respond to the questions about angels, it is obvious that, at the very least, Scholastics had to utilize their knowledge of the Aristotelian concepts of place and local motion. Also involved were questions about the meaning of terms like "succession" and "instant." Although they were spiritual beings, the behavior of angels with respect to their places and motions was dependent on the analogous behavior of material bodies. Therefore only natural philosophy could provide the answers.

The same may be said for questions about God. The questions and the responses to them had little religious content or significance. Most of the questions concerned what God could or could not do, and what He could know or not know. They were questions invented by the imaginations of medieval theologians, who seem to have concocted the questions in order to apply their knowledge of logicomathematical analytic techniques acquired during their courses of instruction at the medieval universities. There seems to be no other explanation for the strange, and even bizarre, questions they raised. Why would a theologian think it important to know "whether God could make a creature exist for only an instant"; or "whether God could make the future not to be"; or "whether God could cause a past thing [or event] to have never occurred"; or "whether God could know something that He does not know";[48] and so on. The answers to questions of what God could or could not do were

---

48. For these questions, see Grant, *God and Reason in the Middle Ages*, 277.

based on the law of noncontradiction. If, in performing this or that action, God was involved in a logical contradiction, the near unanimous conclusion was that not even God could do it. If no logical contradiction was involved, God could do anything whatever. One problem that was never resolved was whether God could create an actual infinite dimension or infinite multitude. Richard of Middleton and John Buridan argued that if God created an infinitely great magnitude, it would follow that He could not create anything greater, since there is nothing greater than an infinite magnitude. But if God could not create anything greater than an infinite He had already created, this would constitute a denial of His absolute power to do anything whatever, which was regarded as a logical contradiction. But there were numerous other Scholastics, known as "infinitists," who argued that God could indeed create infinites, some, like Gregory of Rimini arguing that He could create three kinds of infinites, namely, infinites of magnitude, multitude, and intensity.[49]

To illustrate how the scientific imagination could perform in such questions, we need only mention Gregory of Rimini's extraordinary discovery in the course of discussing the question "whether God, by His infinite power, could produce an actually infinite effect." Gregory found it necessary to discuss the meaning of such terms as "part," "whole," "greater than," and "less than." He concluded that these terms were also applicable to infinites in a special sense. In describing how these terms relate to infinites, Gregory arrived at a momentous idea about the relationship between infinites, an idea that lies at the heart of the modern theory of infinite sets.[50] One infinite can be part of another infinite, Gregory explains, but the infinite that is part is nevertheless equal to the infinite of which it is a part. This can happen, according to Gregory, when "some infinite is less than some [other] infinite, because the infinite which is the part does not contain all the things which the infinite that is the whole contains." Gregory gives no example, but the relationship of the natural numbers to the subset of even numbers provides an excellent illustration of Gregory's intent. Despite the fact that the infinite subset of even numbers is part of the infinite set of natural numbers, the two infinite sets are equal, or in modern terminology, they have the same cardinality. Greg-

49. Ibid., 243–45.
50. For the full argument, see my discussion in ibid., 245–48.

ory had discovered the counterintuitive idea that in the domain of the infinite, a part can equal the whole. In evaluating Gregory of Rimini's contribution, John Murdoch declares[51] "Since the 'equality' of an infinite whole with one or more of its parts is one of the most challenging and, as we now realize most crucial aspects of the infinite, the failure to absorb and refine Gregory's contentions stopped other medieval thinkers short of the hitherto unprecedented comprehension of the mathematics of infinity which easily could have been theirs." Despite this failure, medieval Scholastic theologians grappled with problems of the infinite in a number of contexts, prompting Murdoch to add that, nonetheless, "their deliberations over this particular paradox as involved in problems like the eternity of the world and the continuum again and again led them, as it seldom did their ancient predecessors, to the heart of the mathematics of the infinite. The fact that they seemed to realize that it was the heart, and that in treating it they fared as well as, and at times better than, anyone else before, it appears, the nineteenth century, is unquestionably to their credit." Perhaps the great fascination with the infinite in the late Middle Ages occurred because it was an exercise of the imagination and a keen test of analytic skills of the logicomathematical variety.

Conclusion

In his study of the scientific imagination from the Middle Ages to the seventeenth century, Amos Funkenstein discussed a number of counterfactuals that were analyzed by medieval theologians and natural philosophers, including the possibility of a plurality of worlds, the role of impetus theory, and motion in void space. Funkenstein explains

[S]eventeenth-century science articulated some basic laws of nature as counterfactual conditionals that do not describe any natural state but function as heuristic limiting cases to a series of phenomena, for example, the principle of inertia. Medieval schoolmen never did so; their counterfactual yet possible orders of nature were conceived as incommensurable with the actual structure of the universe, incommensurable either in principle or because none of their entities

51. The two passages below are from John E. Murdoch, "*Mathesis in Philosophiam Scholasticam Introducta:* The Rise and Development of the Application of Mathematics in Fourteenth-Century Philosophy and Theology," in *Arts Libéraux et Philosophie au Moyen Age* (Montreal: Institut d'Etude Médiévales, 1969); for the two passages, see 224.

can be given a concrete measure. But in considering them vigorously, the theological imagination prepared for the scientific.[52]

This assessment is essentially correct. But it fails to mention the important fact that Scholastic natural philosophers had no desire to make their counterfactuals about our world, and what might lie beyond, commensurable with the Aristotelian physical cosmos, which almost all regarded as the best representation of the universe and its operations. Rather, they only sought to show that what Aristotle had deemed impossible in our world was in fact possible, even if only by God's absolute power, that is, by means of supernatural action. For example, they showed that, contrary to Aristotle, if a void space existed, bodies would move in it with finite, successive motions; or if other worlds existed they could be self-sufficient worlds that would not interfere with our world, and so on. But these hypothetical conclusions were never intended to apply to the real, physical world. However, some conclusions derived from thought-experiments assumed to occur in the physical world were exceptions. Thus we can learn something about the properties of impetus from the assumption of a body falling through an imaginary hole in the center of the earth; and we may infer from the impact of a falling millstone on an ascending bean that, contrary to Aristotle, there is no moment of rest separating the upward and downward motions of that bean.

As I think has been reasonably demonstrated, the medieval imagination in natural philosophy and theology was a formidable instrument. It produced a number of significant discoveries and theories, as well as numerous interesting conjectures. Although it transformed theology, it did not alter the basic Aristotelian worldview. No meaningful effort was made to convert hypothetical arguments into real knowledge about the physical world. The sense that there was any real need to do so was simply absent, largely because medieval natural philosophers were convinced that they already had significant knowledge about what the world is and is not. Many of them undoubtedly believed that conclusions derived from reasoned arguments about physical phenomena are probably true. But they seem to have been as much interested, if not more so, in treating questions about hypothetical and imaginary phenomena

52. Funkenstein, *Theology and the Scientific Imagination from the Middle Ages to the Seventeenth Century* (Princeton, N.J.: Princeton University Press, 1986), 11.

to which they applied concepts and ideas from Aristotelian physics and cosmology that were often found to be inapplicable. However, they did not hesitate to use their imaginations to challenge, and reject, as we saw, basic Aristotelian ideas such as the cause of projectile motion and the moment of rest.

Thought-experiments have played a significant role in the history of science and have been widely discussed and analyzed.[53] The thought-experiment, in Peter King's judgment, "is peculiarly well-suited for uncovering conceptual incoherencies and inadequacies; it demands a high degree of rigor as well as logical sophistication; it is as precise an analytical tool as can be found."[54] King rightly observes that "[t]hought-experiments in their mediaeval use, support theories which have no check or control, no way to test their correctness or incorrectness, as opposed to the modern experimental method."[55] It is important to recognize, however, that medieval Scholastic natural philosophers and theologians did not think they had to test their thought-experiments. They were convinced that their thought-experiments would prove true if the hypothetical conditions they described were actually brought into being.

Before science could have reached the stage it did in the seventeenth century, there had to be a widespread use of reason and reasoned analysis. The medieval universities supplied the intellectual context for all of Western Europe. They developed an approach to nature that I have characterized as "probing and poking around." Indeed, the very questioning method they used as the foundation of their natural philosophy clearly reveals such an approach. This spirit of inquiry—this constant "probing and poking around"—was institutionalized within the medieval university where it was practiced for some four centuries. Nothing like it had ever been seen before in previous or contemporary civilizations and cultures. In retrospect, we can see that all this was an indispensable background for the emergence of early modern science. Although most of the responses to questions about an Aristotelian world, or other worlds re-

53. See Tamara Horowitz and Gerald J. Massey, eds., *Thought Experiments in Science and Philosophy* (Savage, Md.: Rowman & Littlefield, 1991) and Roy A. Sorenson, *Thought Experiments* (New York/Oxford: Oxford University Press, 1992).

54. Peter King, "Mediaeval Thought-Experiments: The Metamethodology of Mediaeval Science," in *Thought Experiments in Science and Philosophy,* edited by Tamara Horowitz and Gerald J. Massey, 56 (Savage, Md.: Rowman & Littlefield, 1991).

55. Ibid.

lated to it, have been rejected along with the Aristotelian worldview, that worldview served for many centuries to render intelligible a world that would otherwise have been largely magical and enigmatic. That intelligibility was acquired almost wholly by a questioning, analytic reasoning inspired by the medieval imagination. It was the methodology and spirit of inquiry—not the actual content of medieval natural philosophy—that makes the medieval contributions to the emergence of early modern science vital and lasting. Medieval natural philosophy was rejected because by the seventeenth century, natural philosophers had learned that reason alone was inadequate to the task of describing the world's operations. It had to be supplemented by observation, experiments, and the application of mathematics. But science has remained an enterprise of resolving one question after another, without end. This indispensable activity, which is the heart of modern science, did not emerge from out of the void. Nor did it originate with the great scientific minds of the sixteenth and seventeenth centuries, from the likes of Copernicus, Galileo, Kepler, and Newton. It came out of the Middle Ages from many faceless Scholastic logicians, natural philosophers, and theologians. It was a lasting contribution from the beginning centuries of the new civilization that had emerged in Western Europe around 1000 A.D. and endured until the rejection of the Aristotelian worldview in the seventeenth century.

8 ∞ Medieval Natural Philosophy
Empiricism without Observation

In his splendid book *The Scientific Revolution: A Historiographical Inquiry,* H. Floris Cohen identifies the major issues that historians of science have emphasized in distinguishing between natural philosophy in the Middle Ages and natural philosophy in the seventeenth century.[1] During the eighteenth and nineteenth centuries, it was customary "to contrast the 'empiricism' of the new science with the sterile *a priori* reasoning taken to be characteristic of previous philosophies of nature, in particular Aristotle's." This interpretation was largely abandoned in the twentieth century when "it was discovered how deeply the empirical fact had gone into the making, not only of early modern science, but also of its apparent antithesis, Aristotelianism." But what kind of empiricism do we find in the Middle Ages? "By and large," Cohen explains,

natural philosophies prior to the Scientific Revolution were concerned with such phenomena of nature as present themselves immediately to the senses (e.g. trees burning, stars shining, stones being thrown). In contrast, the Scientific Revolution may be characterized by the *expansion* into, as well as the *exploration* of, a new phenomenal domain previously hidden to the senses (e.g. infusoria, Jupiter's satellites, an apparently void space on top of a mercury column in a glass tube deliberately erected in a dish of the same fluid).[2]

The empiricism of the Middle Ages differs even further from that of the seventeenth century by the emphasis that was placed on "the commonsense reliance on daily experience" and "'naïve,' more or less unguided observation." This is contrasted with the seventeenth-century approach

---

1. H. Floris Cohen, *The Scientific Revolution: A Historiographical Inquiry* (University of Chicago Press, 1994).

2. All quotations from Cohen's book are from 183, 184.

in which the idea of *experiment* takes hold, a concept that involves "the deliberate subjection of nature to searching questioning" and "active interrogation."[3] In this essay, my objective is to examine some of the ways in which medieval natural philosophers used observation and experience in their questions on Aristotle's natural books.

It is almost a truism that Aristotelian natural philosophy is, in sharp contrast to Platonic philosophy, rooted in sense perception. For Aristotle, scientific knowledge is ultimately based on perception. As Jonathan Barnes explains, Aristotle's scientific treatises "are scientific, in the sense that they are based on empirical research, and attempt to organise and explain the observed phenomena."[4] For Aristotle, "knowledge is bred by generalisation out of perception."[5] Thus it is hardly surprising that medieval natural philosophers laid heavy emphasis on sense perception and observation as the foundation of knowledge and science. Already in the twelfth century, even before the impact of Aristotle's thought had taken hold, A. C. Crombie sees an emphasis on sense perception when, he explains, "the saying *nihil est in intellectu quod non prius fuerit in sensu*, became a commonplace."[6]

With the dissemination of Aristotle's natural philosophy in the thirteenth century, it was not unusual for Scholastic natural philosophers to emphasize experience, as did, for example, Albertus Magnus and Roger Bacon. It would be difficult to find a more emphatic statement in favor of observation and experience than Albert the Great's declaration that

[a]nything that is taken on the evidence of the senses is superior to that which is opposed to sense observations; a conclusion that is inconsistent with the evidence of the sense is not to be believed; and a principle that does not accord with the experimental knowledge of the senses [*experimentali cognitioni in sensu*] is not a principle but rather its opposite.[7]

3. For these ideas and brief quotations, see Cohen, *The Scientific Revolution,* 184.

4. Jonathan Barnes, *Aristotle* (Oxford: Oxford University Press), 61.

5. Ibid., 59.

6. A. C. Crombie, *Medieval and Early Modern Science,* 2 vols. (Garden City, N.Y.: Doubleday Anchor Books, 1959), 2.10.

7. Translated by Wallace, *Causality and Scientific Explanation,* 2 vols. (Ann Arbor: University of Michigan Press, vol. 1: 1972, vol. 2: 1974), 1.70. Wallace's translation is from the Borgnet edition of the Latin text of Albertus's *Commentary on Aristotle's Physics* (Albert the Great, *Liber VIII Physicorum* [Borgnet], tract. 2, cap. 2, p. 564). The Latin text in the fourth volume of the modern edition of Albert's *Opera omnia* is as follows: "Omnis enim acceptio, quae firmatur

In a similar vein, Roger Bacon asserts that "there are two modes of acquiring knowledge, namely by reasoning and experience." He then goes on to exalt experience over reasoning when he explains that

[r]easoning draws a conclusion and makes us grant the conclusion, but does not make the conclusion certain, nor does it remove doubt so that the mind may rest on the intuition of truth, unless the mind discovers it by the path of experience; . . . For if a man who has never seen fire should prove by adequate reasoning that fire burns and injures things and destroys them, his mind would not be satisfied thereby, nor would he avoid fire, until he placed his hand or some combustible substance in the fire, so that he might prove by experience that which reasoning taught. But when he has had the actual experience of combustion his mind is made certain and rests in the full light of truth. Therefore reasoning does not suffice, but experience does.[8]

In the fourteenth century, John Buridan emphasized the importance of induction when, in a question on the possible existence of a vacuum, he declared that

every universal proposition in natural science [*in scientia naturali*] ought to be conceded as a principle which can be proved by experimental induction [*per experimentalem inductionem*], just as in many particular [occurrences] of it, it would be manifestly found to be so and in no instances does it fail to appear.

---

sensu, melior est quam illa quae sensui contradicit, et conclusio, quae sensui contradicit, est incredibilis, principium autem, quod experimentali cognitioni in sensu non concordat, non est principium, sed potius contrarium principio." See Albert the Great, *Alberti Magni Opera omnia, Physica*, part 2, bks. 5–8, ed. Paul Hossfeld (Aschendorff: Monasterii Westfalorum, 1993) [Hossfeld], bk. 8, tract. 2, cap. 2, p. 587, col. 2.

8. Roger Bacon, "On Experimental Science," *The Opus Majus of Roger Bacon*, 2 vols., translated by Robert Belle Burke (New York: Russell & Russell, 1962), vol. 1, part 6, chap. 1, p. 583. Reference is made to this passage in Bacon by N. W. Fisher and Sabetai Unguru, "Experimental Science and Mathematics in Roger Bacon's Thought," *Traditio* 27 (1971): 358. For an account of Bacon's ideas about experiment, see Jeremiah Hackett, "Roger Bacon on *Scientia experimentalis*," in *Roger Bacon and the Sciences: Commemorative Essays* (Leiden: Brill, 1997), 277–315. Whatever Bacon's "experimental science" (*scientia experimentalis*) may have been— and its meaning is unclear—he employed it largely in optics, or perspective (*perspectiva*). Bacon does not appear to have applied it to his own *questiones* on Aristotle's natural books, such as his questions on *Physics* and *De caelo*. Nor did it play any part in natural philosophy at the universities, if we judge by Bacon's assertion in his *Opus majus* that "this Experimental Science is *wholly unknown to the rank and file of University students . . .*" (Hackett, "Roger Bacon on *Scientia experimentalis*," 294). We may be confident that whatever Roger Bacon had in mind by "experimental science" played no role in the natural philosophy produced at the University of Paris, or any other medieval university.

For Aristotle puts it very well [when he says] that many principles must be accepted and known by sense, memory, and experience.[9] Indeed, at some time or
other, we could not know that every fire is hot [except in this way].[10]

These are powerful statements in favor of experience and observation.
Although few medieval natural philosophers have matched the enthusiasm for experience exhibited in the above passages, we may plausibly assume that most of them were empiricists in the Aristotelian sense and in
the sense expressed by Albert the Great, Roger Bacon, and John Buridan.
That is, ideally, they thought it desirable to begin with an observation or
sense perception as the basis of a generalization or conclusion.

In the *Posterior Analytics*, however, Aristotle presented another picture
of science. Here it was not observation that was crucial, but a deductive Eu-
clidean-type methodology that began from first principles that were themselves indemonstrable. But there is also an empirical and inductive component to Aristotle's method, since he believes that we get to know universal
premises—axioms, definitions, and hypotheses—by induction.[11]

Aristotle himself did not follow his own advice and did not apply the
rigorous methodology of the *Posterior Analytics* to his own natural and
biological books. G. E. R. Lloyd conveys an idea of the confusion in Aristotle about the relationship between the particular of observation and
the universal. Aristotle, says Lloyd,

stresses that we cannot *know* the particular, only the universal: this is stated explicitly at 81b6f., for instance, and again in *Metaphysics*, for example at B, ch. 4,
999a26ff. And yet he also recognises on four separate occasions that we arrive at
knowledge of the universal only from an examination of particulars.[12]

9. *Posterior Analytics* 2.19.100a4–9.
10. "Item omnis propositio universalis in scientia naturali debet concedi tanquam principium que potest probari per experimentalem inductionem sic quod in pluris singularibus
ipsius manifeste inveniatur [corrected from *inveniaur*] ita esse et in nullo nunquam apparet
instantia, sicut enim bene dicit Aristoteles quod oportet multa principia esse accepta et scita
sensu, memoria et experientia; immo aliquando non potuimus scire quod omnis ignis est cali-
dus." From John Buridan, *Questions on the Physics*, bk. 4, qu. 7, in *Acutissimi philosophi reverendi Magistri Johannis Buridani subtilissime Questiones super octo phisicorum libros Aristotelis
diligenter recognite et revise a Magistro Johanne Dullaert de Gandavo antea nusquam impresse*
(Paris, 1509; reprinted Frankfurt, 1968), fol. 73v, col. 1. The translation is mine (slightly altered)
from Edward Grant, *A Source Book in Medieval Science* (Cambridge, Mass.: Harvard University Press, 1974), 326.
11. See G. E. R. Lloyd, *Aristotle: The Growth and Structure of His Thought* (Cambridge:
Cambridge University Press, 1968), 125.
12. Ibid., 126.

Aristotle believed that our senses "give the most authoritative knowledge of particulars. But they do not," he explains, "tell us the 'why' of anything—e.g. why fire is hot; they only say that it is hot."[13] Despite an emphasis on sense perception and observation of the particular, Aristotle's natural books—*Physics, On Generation and Corruption, Meteorology, On the Heavens, On the Soul,* and the *Parva naturalia*—are largely theoretical accounts of their subject matters. They attempt to tell us the "why" of things. The picture of the cosmos that Aristotle constructed in these works seems far removed from its observational foundation. It seems rather an account of a world that was made to conform to Aristotle's preconceived ideas of what the universe had to be like in order to function in a manner worthy of a divine cosmos.

Because of his own ambiguities and confusions, Aristotle's legacy to the Latin Middle Ages was a natural philosophy that presented a rather confused relationship between the theoretical and the empirical. Just what role did observation play in medieval natural philosophy? Most, if not all, historians of medieval science and natural philosophy do not regard the Middle Ages as a period in which the habit of observation was well developed or much cultivated. Thus although we might concede that scholars in the Middle Ages were philosophically committed to Aristotelian empiricism, it would be helpful to know what role observation really played in their resolution of physical questions. Toward this purpose, I have chosen a number of examples to discuss, many drawn from problems about motion. They represent only a small fraction of the possible domain of examples from the whole range of natural philosophy. Any one else who chose to write a similar essay could easily select a wholly different set of examples. But to a certain extent, I trade on the fact that medieval natural philosophers probably shared a common outlook and attitude toward the role of experience, and the way it should be used in questions on Aristotle's natural books. To gain some insight into the way observations and experience were used in medieval natural philosophy, I have drawn from the works of five famous fourteenth-century natural philosophers: John of Jandun, John Buridan, Nicole Oresme, Albert of Saxony, and Marsilius of Inghen.

In medieval natural philosophy, observations are often identified by some form of the term *experientia*, rather than by any form of the term

13. Aristotle, *Metaphysics* 1.1.981b10–11.

observatio, although the latter does occasionally appear.[14] In many instances, however, no term is used to characterize the observation or experience. It is just given.

In his *Questions on Generation and Corruption*, book 1, question 1, Nicole Oresme offers some distinctions in one of which he declares that "another [distinction] is evident in principles known by experience, just as that fire can heat."[15] Later, in the same question, he explains that "an alteration is when one thing is changed into another, as hotness into coldness, and similarly fire into air." Buridan presents similar observations when he declares that we can attain truth based on evidence that occurs when "a common course of nature [*communis cursus nature*] is observed in things," as when "it is evident to us that all fire is warm and that the heaven moves, although the contrary is possible by God's power."[16] What do these rather common observations in medieval natural philosophy signify? At first glance, the instances just cited seem to represent experiences of a most fundamental and obvious kind, the kind that "present themselves immediately to the senses," as Floris Cohen expressed it. All fire is warm, or fire heats; all hot things will eventually cool and become cold; and the heavens move. But what kind of an experience is the conversion of fire into air? It is hardly an obvious or common experience. Indeed, it is not clear that Oresme really intended it as an observation. Why, then, did medieval natural philosophers believe that fire is transformed into air? Largely, I think, because it was assumed by all Aristotelian natural philosophers that the four elements were transformable into each other. Just as, for example, water is transformable into air and vice versa, so is air convertible into fire and vice versa. Thus, the conversion of

14. St. Thomas Aquinas used the term *observatio* in such expressions as "the observation of celestial bodies" (*per observationem caelestium corporum*) and "astronomical" [or "astrological"] observations" (*per astrologicas observationes*). For these and other usages of *observatio*, see Roy J. Deferrari, Sister M. Inviolata Barry, and Ignatius McGuiness, *A Lexicon of St. Thomas Aquinas Based on the "Summa Theologica" and Selected Passages of His Other Works* (Baltimore: The Catholic University of America Press, 1948), 758.

15. "Alia [distinctio] est evidentia in principiis notis per experientiam, sicut quod ignis potest calefacere." From Nicole Oresme, *Nicole Oresme Quaestiones super De generatione et corruptione*, edited by Stefano Caroti (Munich: Verlag der Bayerischen Akademie der Wissenschaften, 1996), bk. 1, qu. 1., p. 4, ll. 45–46.

16. The translation is by Edith Sylla, which I quote from my article "Jean Buridan and Nicole Oresme on Natural Knowledge," *Vivarium* 31, no. 1 (1993): 88. The passage occurs in Buridan's *Questions on the Metaphysics*. See John Buridan, *In Metaphysicen questiones Aristotelis: Questiones argutissime Magistri Ioannis Buridani . . .* (Paris, 1518), bk. 2, qu. 1, fols. 8v, col. 2–9r, col. 1.

fire into air does not appear to be a proper observation, but seems rather an "observation" that is required by theory, the theory of the four elements and their mutual conversions. Many "observations" are of this kind.

1. The Impetus Theory of Projectile Motion

But let us now turn to less immediate kinds of observations. Few, if any, medieval natural philosophers were more empirically minded than John Buridan. We see this in his well-known question on impetus theory, where he asks "whether a projectile after leaving the hand of the projector is moved by the air, or by what it is moved."[17] To refute Aristotle's theory of *antiperistasis,* the judgment that air was the cause of the continued movement of a projectile after leaving the hand of the projector, Buridan counters with three experiences *(experientie).* "The first experience" that he invokes

concerns the top [*trocus*] and the smith's mill [i.e. wheel—*mola fabri*] which are moved for a long time and yet do not leave their places. Hence, it is not necessary for the air to follow along to fill up the place of departure of a top of this kind and a smith's mill. So it cannot be said [that the top and smith's mill are moved by the air] in this manner.[18]

The second experience is this: A lance having a conical posterior as sharp as its anterior would be moved after projection just as swiftly as it would be without a

17. The translation appears in Marshall Clagett, *The Science of Mechanics in the Middle Ages* (Madison: University of Wisconsin Press, 1959), 532–38. It is reprinted in Grant, *A Source Book,* 275–80.

18. In the final question of his *Questiones in libros Physicorum,* bk. 8, qu. 13, Albert of Saxony cites the same experience involving the smith's wheel and the top in a question "by what is a projectile moved upward after its separation from the projector?" ("Ultimo quaeritur a qua movcatur projectum sursum post separationem illius a qua proiicit"). Here is what Albert says: "Similiter ista opinio non habet locum in motu mole fabri; similiter in motu troci. Vidimus enim quod trocus post exitum eius a manu proiicientis diu movetur circulariter absque hoc quod aliquis aer ipsum insequatur, movet enim super eodem puncto spatii." See Albert of Saxony, *Questiones et decisiones physicales insignium virorum. Alberti de Saxonia in octo libros Physicorum; tres libros De celo et mundo; duos libros De generatione et corruptione; Thimonis in quator libros Meteorum; Buridani in tres libros De anima; librum De sensu et sensato; librum De memoria et reminiscentia; librum De somno et vigilia; librum De longitudine et brevitate vite; librum De juventute et senectute Aristotelis. Recognitae rursus et emendatae summa accuratione et judicio Magistri Georgii Lokert Scotia quo sunt tractatus proportionum additis* [Lokert] (Paris, 1518), fol. 83v, col. 1.

sharp conical posterior. But surely the air following could not push a sharp end in this way, because the air would be easily divided by the sharpness.

The third experience is this: a ship drawn swiftly in the river even against the flow of the river, after the drawing has ceased, cannot be stopped quickly, but continues to move for a long time. And yet a sailor on deck does not feel any air from behind pushing him. He feels only the air from the front resisting [him]. Again, suppose that the said ship were loaded with grain or wood and a man were situated to the rear of the cargo. Then if the air were of such an impetus that it could push the ship along so strongly, the man would be pressed very violently between that cargo and the air following it. Experience shows this to be false. Or, at least, if the ship were loaded with grain or straw, the air following and pushing would fold over [plico] the stalks which were in the rear. This is all false.[19]

The experiences Buridan presents are offered as counterinstances to Aristotle's claim that air pushes projectiles along after the projectile has lost contact with the projecting agent. It is very likely that Buridan actually observed the rotatory motions of the mill wheel and the top. These were experiences that he probably recalled, realizing that they were useful in his argument against Aristotle. Although it is, of course, possible that Buridan specifically observed these two phenomena in connection with his treatment of the question on impetus, he gives no indication of such actions and it is more than likely that he did no such thing, but relied rather on his earlier experiences with such phenomena.

The second experience is a "reasoned" experience. In the absence of any statement to the contrary, we ought not to assume that Buridan tested this with two lances, one with a conical posterior and one without. It is far more plausible to assume that he reasoned that a conical posterior would readily divide the air that sought to push it. Consequently a lance with a conical posterior would not be moved by the air, as contrasted to a lance that had a broad posterior surface against which the air could push. Although this is presented as an experience, it is extremely unlikely that Buridan hurled a conical lance and a nonconical lance to see which was carried further. Even if he did, it is not likely to have resolved the issue. What we have here is a hypothetical "experience" wherein Buridan rea-

19. Clagett, *The Science of Mechanics,* 533; Grant, *A Source Book,* 275–76. For the Latin text, see John Buridan, *Acutissimi philosophi reverendi Magistri Johannis Buridani subtilissime Questiones super octo physicorum libros Aristotelis* (Paris, 1509), bk. 8, qu. 12, fols. 120r, col. 2–120v, col. 1.

soned about the way a conical posterior would divide the pushing air as opposed to the way a nonconical lance would react.

The third experiential example, in which a ship continues to move even after the shiphaulers cease to pull it, is a phenomenon that, at some point, or points, in his life, Buridan probably observed directly. It was a common experience. He trades on this experience to show that air could not plausibly be assumed to push the ship. Here Buridan again resorts to hypothetical conditions to argue that if air were the cause of the ship's motion, a sailor would be "pressed very violently between that cargo and the air following it"; or, if "the ship were loaded with grain or straw, the air following and pushing would fold over the stalks which were in the rear," all of which is false.

Later in the same question, Buridan adds some hypothetical elements to his direct observations. He argues that "if you cut off the air on all sides near the smith's mill by a cloth [*linteamine*], the mill does not on this account stop but continues to move for a long time. Therefore it is not moved by the air." It is highly unlikely that Buridan ever carried out this "experiment." He simply assumed its truth. Moreover, if air moved a ship, and if the ship were carrying straw or grain, then "the air ought to blow the exterior stalks toward the front. But the contrary is evident, for the stalks are blown rather to the rear because of the resisting ambient air."[20] Here again, Buridan rightly deduces this consequence from the conditions of the ship's motion and the nature of straw and grain. We do not observe such a consequence therefore, the air is not the motive force causing the ship's motion. If air were truly a motive force, "it follows," Buridan argues,

that you would throw a feather farther than a stone and something less heavy farther than something heavier, assuming equal magnitudes and shapes. Experience shows this to be false. The consequence is manifest, for the air having been moved ought to sustain or carry or move a feather more easily than something heavier.[21]

Done with his arguments against Aristotle's explanation, Buridan presents his own impetus theory and offers a few more appeals to experience that he regarded as supportive of that theory. In one appeal, he

20. Clagett, *The Science of Mechanics*, 534; Grant, *A Source Book*, 276.
21. Clagett, *The Science of Mechanics*; Grant, *A Source Book*.

declares that impetus theory explains why "one who wishes to jump a long distance drops back a way in order to run faster, so that by running he might acquire an impetus which would carry him a longer distance in the jump. Whence the person so running and jumping does not feel the air moving him, but [rather] feels the air in front strongly resisting him."[22] Here is a common phenomenon that Buridan probably experienced. Moreover, his impetus theory gave a consistent explanation for it, whereas Aristotle's theory did not.

[1] If light wood and heavy iron of the same volume and of the same shape are moved equally fast by a projector, the iron will be moved farther because there is impressed in it a more intense impetus, which is not so quickly corrupted as the lesser impetus would be corrupted. [2] This also is the reason why it is more difficult to bring to rest a large smith's mill which is moving swiftly than a small one, evidently because in the large one, other things being equal, there is more impetus. And for this reason [3] you could throw a stone of one-half or one pound weight farther than you could a thousandth part of it. For the impetus in that thousandth part is so small that it is overcome immediately by the resisting air.[23]

In some form or other, Buridan had surely observed phenomena similar to those he cites here. But he gives no indication that he ever actually threw the objects he mentions and then compared the results; or that he sought to halt the motions of a large and small mill. And yet these experiences ring true and served as powerful evidence in favor of his impetus theory. Indeed it is not implausible to assume that he actually threw a one-half or one pound stone and a tiny stone and compared the distances traversed.

2. Is There a Moment of Rest in Contrary Motions?

In the eighth book of his *Physics* (8.8.264a.7–35), Aristotle argued that a continuous motion up and down is not a single motion, but really two separate motions separated by a moment of rest. This problem was frequently discussed in medieval questions on the *Physics*, as it was by

22. Clagett, *The Science of Mechanics*, 536; Grant, *A Source Book*, 277.
23. Clagett, *The Science of Mechanics*, 535; Grant, *A Source Book*. I have added the numbers in brackets.

Buridan,[24] Albert of Saxony,[25] and in the *Questions on the Physics* (Lyons, 1518) attributed to Johannis Marcilius Inguen, who is presumably not Marsilius of Inghen. The three authors just mentioned all appeal to a particular experience, although that "experience" is best characterized as a thought experiment, one, however, that was, in principle, capable of being carried out, but which, for these Scholastic authors, was undoubtedly confined to the realm of the imagination.

Marsilius argues that a moment of rest need not occur

between any motions that turn back [*reflexos*] [over the same path]. The proof is that if a bean [*faba*] were projected upward against a millstone [*molarem*] which is descending, it does not appear probable that the bean could rest before descending, for if it did rest, through some time it would stop the millstone from descending, which seems impossible.[26]

This frequently reported "observation" reveals another feature of the medieval attitude toward experience-like phenomena. An experience that was drawn from another author, or authors, seems to have had as much validity as one that was directly experienced or carried out. Indeed, there were very few of the latter and many of the former. Moreover, medieval natural philosophers rarely distinguished between experiences they had personally observed and those that they devised for the occasion.

Many, if not most, of the important and interesting experiences introduced into medieval *questiones* were *secundum imaginationem*. The question on the moment of rest seems to have stimulated the inventive juices, if not genius, of medieval Scholastics. In addition to the millstone striking the bean, they devised a ship experience in which Marsilius assumes that

Socrates [*Sortes*] is moved toward the west in a ship that is at rest. Then it is possible that Socrates might cease moving in any instant. Now let it be assumed that in the [very] same instant in which Socrates should cease to be moved [to-

24. John Buridan, *Quaestiones super octo physicorum libros*, bk. 8, qu. 8, fol. 116r, col. 1.

25. Albert of Saxony, *Quaestiones in octo libros Physicorum*, bk. 8, qu. 12, fol. 82v, cols. 1–2.

26. Translation by Edward Grant in Grant, *A Source Book*, 286–87. For the Latin text, see Marsilius of Inghen (?), *Questiones subtilissime Johannis Marcilii Inquen super octo libros Physicorum secundum nominalium viam . . .* (Lyon, 1518), fol. 84r, col. 1. This argument against Aristotle derives from Islamic sources, appearing in Abu'l Barakat al-Baghdadi's (ca. 1080–ca. 1165) counterargument against Avicenna's defense of Aristotle's position. For others in the West who cited this argument, see Grant, *A Source Book*, 287, n. 18.

ward the west], the ship with all its contents, begins to be moved toward the east. Hence, immediately before, Socrates was moved to the west, and immediately after, will be moved toward the east. Therefore, previously he was moved with one motion and afterward with another, and contrary, motion without a moment of rest.[27]

Buridan achieved the same objective by citing an analogous experience based on a fly. He explains that

if a lance is hanging from a tree [and] a fly [*musca*] ascends [in lieu of descends] on that lance and the cord by which the lance is hanging is broken, and then the lance and the fly fall down, the motion of the fly will be contrary, from up to down, but there will be no moment of rest.[28]

Albert of Saxony also found the fly and lance an attractive example, perhaps drawing it from Buridan. Albert abandons the tree and the cord and simply assumes that the fly ascends the lance quicker than the lance descends. Albert then assumes that the upward speed of the fly diminishes until it becomes less than the descent of the lance. But at the instant in which the speed of ascent of the fly and the speed of descent of the lance are equal, "it is true to say that immediately before [the speeds were equal] this fly was ascending; and it is [also] true to say that immediately after [the speeds were equal] it descends, because immediately after this the descent of the lance will be quicker, from which it again follows that between the ascent and descent of the fly there is no moment of rest."[29]

The ascending fly fits better in Albert of Saxony's version than in Buridan's.[30] Buridan's example raises an interesting question. Why did

27. Grant, *A Source Book,* 287.

28. "Similiter si lancea pendente ad trabem musca ascendat [in place of *descendat*] per illam lanceam et rumpatur corda ad quam pendebat lancea, et tunc cadat lancea cum musca deorsum. Motus musce erit reflexus de sursum ad deorsum; et non erit quies media." From Buridan, *Questions on the Physics,* bk. 8, qu. 8, fols. 116r, col. 2–116v., col. 1

29. I give the full text of Albert's example: "Tertio ponatur quod aliqua musca ascendat super aliquam lanceam velocius quam illa lancea descendat. Et remittatur velocitas ascensus illius musce donec fiat minor quam velocitas descensus illius lancee. Tunc in instanti in quo velocitas musce et velocitas lancee sunt equales verum est dicere quod immediate ante hoc musca ascendebat; et verum est dicere quod immediate post hoc descendet quia immediate post hoc descensus lancee erit velocior. Ex quo iterum sequitur quod inter ascendere et descendere ipsius musce non sit quies media." From Albert of Saxony, *Questions on the Physics,* bk. 8, qu. 12, fol. 82v, col. 2.

30. If Marsilius of Inghen knew about the fly and lance example, he chose to ignore it in his question.

he choose a fly to ascend the lance? He could hardly have been unaware of the fact that at the very moment when the cord broke and the lance fell, the fly would have flown off the lance and moved off in almost any direction, except straight down with the lance. Although it was Buridan's illustration, and he was free to present it in any manner he chose, · his example loses much of its impact by the implausible choice of a fly to ascend and then descend. In Albert's example, the fly is continually ascending, although after its velocity becomes less than that of the lance, it is actually descending.

Nevertheless, the experiences of Socrates on the ship and the fly on the lance served a useful purpose. They conjured up situations in which a moment of rest could not occur, thus presenting significant counterinstances, which were then generalized, to subvert Aristotle's argument in favor of a moment of rest. It is worth noting that, although the experiences cited above against the moment of rest are *secundum imaginationem*, they are not counterfactual. They could conceivably have occurred.

3. On the Cause of the Acceleration of Freely Falling Bodies

That natural motion is accelerated was assumed as a fact by all, based not only upon Aristotle's affirmation,[31] but presumably also on the basis of direct observation. Thus, when he considered the question "whether natural motion ought to be swifter in the end than in the beginning,"[32] John Buridan declared that "the great difficulty in this question is why this [acceleration] is so." In the process of determining the cause of such acceleration, Buridan devised experiences to contradict the theories he rejected, as well as to support the true interpretation.

One rejected theory, attributed to Aristotle in the latter's *De caelo*, assumes that as a body falls it heats and rarefies the air. Therefore, the air

31. See *On the Heavens* 1.8.277a28–30, where Aristotle says that "earth moves more quickly the nearer it is to the centre," and *On the Heavens* 2.6.288a.21, where he declares that "we expect natural motion to reach its maximum at the goal, unnatural motion at the starting-point, and missiles midway between the two." I have used Aristotle, *The Complete Works of Aristotle: The Revised Oxford Translation*, edited by Jonathan Barnes, 2 vols. (Princeton: Princeton University Press, 1984).

32. Translated by Clagett, *The Science of Mechanics*, 562–64, from Buridan's *Questions on De caelo*, bk. 2, qu. 12. The translation is from the edition of E. A. Moody, *Johannis Buridani Quaestiones super De caelo el mundo* (Cambridge, Mass.: Mediaevel Academy of America, 1942), 176–81.

offers less resistance to the body, which consequently increases its speed. In opposition to this explanation, Buridan offers a rather odd "experience" in which he claims that

a man moves his hand just as swiftly as a stone falls toward the beginning of its movement. This is apparent, because striking another person hurts him more than the falling stone, even if the stone is harder. And yet a man so moving his hand does not heat the air sensibly, since he would perceive that heating. Therefore in the same way the stone, at least from the beginning of the case, does not thus sensibly heat the air to the extent that it ought to produce so manifest an acceleration [*velocitatio*] as is apparent at the end of the movement.[33]

The second theory Buridan rejects is that heavy bodies are attracted by natural place so that "the heavy body is moved more swiftly by the same amount that it is nearer to its natural place." To refute this opinion, which he attributes to Aristotle and Averroes in the fourth book of the *Physics*, Buridan invents three experiences that appeal to reason and common sense. I shall mention only one of them. Buridan assumes that one stone falls to earth from a high place and another from a low place. Now if the velocities of the stones are effected only by proximity to their natural place, when they are one foot from the earth their speeds should be equal, despite the fact that they began their respective falls from radically different heights.

Yet it is manifest to the senses that the body which should fall from the high point would be moved much more quickly than that which should fall from the low point, and it would kill a man while the other stone [falling from the low point] would not hurt him.[34]

From this argument, it is evident that the height of the stones determines their speed and not their proximity to their natural place.[35]

To account for a body's acceleration in natural fall, Buridan invokes his impetus theory. He sets the stage for its introduction by a bold move in which, by a series of assumptions, he standardizes all the material conditions that might affect a body's downward velocity: (1) "the stone is found to be equally heavy after the movement as it was before it"; (2) "the

---

33. Clagett, *The Science of Mechanics*, 558; Grant, *A Source Book*, 280–81.
34. Clagett, *The Science of Mechanics*, 559; Grant, *A Source Book*, 281.
35. Buridan also mentions, and rejects, a third opinion to account for natural acceleration.

resistance which arises from the medium remains the same or is similar, since, as I have said, it does not appear to me that the air lower and near to the earth should be less resistant than the superior air." Thus "if the moving body is the same, the total mover is the same, and the resistance also is the same or similar, the movement will remain equally swift, since the proportion of mover to moving body and to the resistance will remain [the same]."[36] From such conditions, Buridan rightly infers a uniform downward motion. But he knows from Aristotle and observation that a body accelerates in its natural downward motion. To explain this Buridan resorts to his impetus theory, concluding that it is the continual production of impetus within the body that causes the downward acceleration.

As evidence that impetus is manufactured within moving bodies, and in support of his own theory, Buridan cites the smith's mill, or wheel, which, we saw earlier, he also employed in his *Questions on the Physics* (book 8, question 12; see above). He informs his audience that

you have an experiment [*experimentum*] [to support this position]: if you cause a large and very heavy smith's mill [i.e., a wheel] to rotate and you then cease to move it, it will still move a while longer by this impetus it has acquired. Nay, you cannot immediately bring it to rest, but on account of the resistance from the gravity of the mill, the impetus would be continually diminished until the mill would cease to move.[37]

In this question, Buridan needed an example, or experience, to show how impetus functioned. The smith's mill served that function admirably. It is one of the few instances where an experience played an essential role in giving credibility to a theoretical argument.

Nicole Oresme also included the same question in his *Questions on De caelo*.[38] Like Buridan, he accepts impetus as the explanation for the downward acceleration of heavy bodies. Although Oresme seems to rely less on experiential claims than did Buridan, he does include some interesting examples. In his refutation of the claim that "heavy things are

---

36. Clagett, *The Science of Mechanics*, 560; Grant, *A Source Book*, 282.

37. Clagett, *The Science of Mechanics*, 561; Grant, *A Source Book*, 282.

38. See Claudia Kren, ed. and trans., *The "Questiones super De celo of Nicole Oresme* (Ph.D. diss., University of Wisconsin), bk. 2, qu. 7: "Consequently, it is sought whether natural motion is faster in the end than in the beginning," 526–76.

moved faster at the end because of the longing they have for the *terminus* and the end of the motion itself,"[39] Oresme counters that

> if there were two stones, A and B, wholly equal and similar and equally distant from the center, and A was moved previously and B would now begin to be moved, then by experience [*per experientiam*], it is clear that A would be moved faster and yet B would be equally near that natural place or that end.[40]

In the second paragraph following this "experience," Oresme refutes another opinion that claims that "the center has a certain attractive force for the heavy thing itself."[41] The refutation involves a reference to the "experience" just cited, which Oresme now calls "the imagination previously posited [*per ymaginationem prius positam*],"

> namely, let A and B be heavy things, wholly similar and an equal distance from the center and other things being equal, assume that A was moved previously; then it is evident to the senses [*ad sensum patel*] that A is moved faster, and yet they ought to be drawn [down] equally [fast].[42]

By being "evident to the senses," Oresme presumably means that he, or anyone witnessing these motions, would observe that A strikes the ground before B. Of course, we do not know if Oresme personally witnessed the fall of the two stones. He may have been guided rather by the logic of the situation, which determines that B, which fell from a greater height than B, would necessarily fall faster than B. This was a quite common procedure. In this instance, we might say that Oresme's experience is an "imagination" masquerading as an "experience."

To disprove the interpretation that air, or any medium, causes the downward acceleration of a body, Oresme presents a number of experiences. In one of them, a descending material sphere, say wood or lead, "passes through an opening which is immediately reclosed, so that the air following is kept back." Although the air is cut off from the descending body, the latter will, nonetheless, accelerate, thus proving that air is not the cause of the acceleration.[43] In this truly imaginary experience, it is difficult to understand the basis of Oresme's confidence that the body would continue to accelerate after the air has been cut off. How could he know this?

---

39. Oresme (Kren), *De celo,* 540.
40. Ibid., 540. I have slightly altered Kren's translation.
41. Ibid. Oresme attributes this opinion to the Commentator, Averroes.
42. Ibid., 542. I have slightly altered the translation.
43. See ibid., 548.

Like Buridan, Oresme not only accepts the truth of the impetus theory, but he also uses the observation of a wheel as evidence for the existence of impetus. Oresme explains that a wheel that is set in motion, "cannot be stopped immediately without great difficulty, but is moved by a certain impetus, though separated from the first moving agent and [is not moved] by the surrounding air."[44]

4. The Vacuum and Sense Experience

Discussions about the vacuum in commentaries on the fourth book of Aristotle's *Physics* were another significant source of experiences based on a mixture of real and imaginary observational data. Of the many experiences reported and discussed, only a few can be mentioned here. Perhaps the most notable was the extraordinary experience introduced by John Buridan in his discussion of the popular question: "whether it is possible that a vacuum exist."[45] It is in this question that Buridan's statement about experimental induction, quoted above, appears. Indeed immediately after that passage, Buridan cites the nonexistence of the vacuum as an instance of experimental induction. The "experimental induction" is simply that "everywhere we find some natural body, namely air, or water, or some other [body]."[46] By implication, of course, we never find a vacuum, which therefore does not exist. Indeed, Buridan, agreeing with Aristotle, goes all the way: a vacuum cannot possibly exist or be made to exist naturally. In support of this claim, Buridan seeks to "show by experience that we cannot separate one body from another unless another body intervenes." As evidence, Buridan offers two experiences, the first of which is quite dramatic:

If all the holes of a bellows [*follis*][47] were perfectly stopped up so that no air could enter, we could never separate their surfaces. Not even twenty horses could do it if ten were to pull on one side and ten on the other; they would nev-

---

44. Ibid., 560–62.

45. "Queritur septimo utrum possibile est vacuum esse." From Buridan, *Questions on the Physics,* bk. 4, qu. 7, fols. 72v, col. 2–73v, col. 2.

46. The Latin text for these sentiments is: "Sed per talem inductionem experimentalem apparet nobis quod nullus locus est vacuus quia ubique invenimus aliquod corpus naturale, scilicet vel aerem vel aquam vel aliud." From Buridan, *Questions on the Physics,* fol. 73v, col. 1.

47. I have translated *follis* as "bellows." But Buridan might have had in mind a leather ball or pouch, or something capable of being inflated with air and then deflated by removing the stop, or stops.

er separate the surfaces of the bellows unless something were forced or pierced through and another body could come between the surfaces.[48]

Buridan may have seen how difficult it is to part the sides of a bellows after most of the air has been squeezed out; or, he may been made aware of it by others. But, in the absence of any claim to the contrary, we ought not to assume that he ever personally tried to separate the sides of a bellows himself. And we can assume with virtual certainty that he never harnessed ten horses on each side of a bellows and witnessed their failure to pull the sides apart.

Buridan was not the first to use a bellows to argue against the existence of vacua (see the section on John of Jandun, below). This so-called experience is driven by Buridan's unqualified conviction that it is impossible for a vacuum to be made by natural means. Consequently, he had to conclude that nothing whatever could pull the sides of the bellows apart: not ten horses on each side, not even ten thousand!

Buridan's experience bears a striking resemblance to the famous experiment carried out by Otto von Guericke (1602–1686) in Magdeburg in 1657, approximately three hundred years after Buridan. He built a large copper sphere formed from two half spheres. After evacuating the sphere by means of an air pump, von Guericke showed that because of air pressure on the spheres and the void within, not even two teams of eight horses, one team harnessed to each side, could pull them apart. Thus where Buridan's team of horses sought to show the impossibility of vacuum, Von Guericke's horses labored mightily to demonstrate the existence of a vacuum and the force of air pressure.[49]

Some years later, Albert of Saxony treated the same question and invoked the same bellows experience, without horses, however.[50] Did he de-

---

48. "Et iterum nos experimur quod non possumus unum corpus ab alio separare quin interveniat aliud corpus. Unde si perfecte obstruerentur omnia foramina follis adinvicem ita quod non posset aer sub intrare, nunquam possemus latera follis ab invicem elevare. Imo nec viginti equi hoc possent si decem traherent ad unam partem et decerm ad aliam nunquam enim separarent latera follis ab invicem nisi aliquid rumperentur vel perforaretur per quid aliud corpus posset intercidere." From Buridan, ibid., fol. 73v, col. 1.

49. See Grant, A Source Book, 563, n. 54. For a brief sketch of Von Guericke's life, see Fritz Krafft's article on "Guericke (Gericke), Otto Von," in Gillispie, ed., Dictionary of Scientific Biography, 5.574–76.

50. "Secunda conclusio per nullam potentiam naturalem possibile est esse vacuum . . . probatur quibusdam experientiis. Primo si omnia foramina alicuius follis obstruerentur, nulla potentia posset elevare unum asserem ab alio nisi fieret alicubi ruptura per quam subintraret

rive it from Buridan? Since Albert derived some of his physical thoughts and ideas from Buridan, it would not be surprising if he took this experience as well. If he did, he obviously decided to abandon the horses. But even before Buridan, John of Jandun sought to demonstrate the impossibility of a vacuum by using a bellows in conjunction with a firm and sturdy vessel. He assumed that the mouth, or nozzle, of the bellows was inserted into a single opening of a vessel and that the bellows-vessel system was completely sealed off from the air outside. He further assumed that the sides of the bellows were separated to form a vacuum, and that there was air in the vessel part of this closed system. From this configuration, Jandun saw two possibilities. In the first, all the air from the vessel enters the space between the sides of the bellows, thus leaving a vacuum in the space of the vessel; in the second possibility, the air remains in the vessel and the space within the bellows remains void. Because formation of a vacuum is impossible, neither of these two alternatives is possible. Jandun resolves the issue by appeal to a "universal nature" (natura universalis), the function of which is to prevent formation of a vacuum by any means necessary. In this case, the universal nature avoids development of a vacuum by preventing the separation of the sides of the bellows.[51]

Why did John of Jandun, John Buridan, and Albert of Saxony believe that the sides of a bellows would remain in contact to avoid formation of a vacuum unless a body, or air, intervened between those sides?[52] When the sides of a bellows meet after forcing out the air within, the sides are

---

aer. Quo facto faciliter unus istorum asserem leveratur ab alio, nam tunc esset aliquid quod posset recipi inter latera ipsius follis. Hoc videtur esse signum naturam abhorrere vacuum." From Albeit of Saxony, Questions on the Physics, bk. 4, qu. 8, fol. 48v, col. 1. The question Albert is discussing is "Utrum vacuum esse sit possibile" (fol. 48r, col. 2). For a translation of Albert's bellows experiment, see Grant, A Source Book, 325.

51. My discussion of John of Jandun is based on my earlier discussion in Grant, Much Ado About Nothing: Theories of Space and Vacuum from the Middle Ages to the Scientific Revolution (Cambridge: Cambridge University Press, 1981), 82–83. For the Latin text, see 312, nn. 92, 93. Jandun treats this problem in his Questions on the Physics, bk. 4, qu. 10. For the full title of his 1519 edition, see my bibliography on p. 426. On the "universal nature," see Grant, Much Ado About Nothing, 69–70.

52. Nicole Oresme also discussed this question in his Questions on the Physics, bk. 4, qu. 7 ("Consequenter queritur utrum naturaliter possit esse vacuum in hoc mundo"), in Stefan Kirschner, ed., Nicolaus Oresme Kommentar zur Physik des Aristoteles: Kommentar mit edition de Quaestionen zu Buch 3 unde 4 der Aristotelischen Physik sowie von vier Quaestionen zu Buch 5 (Stuttgart: Franz Steiner Verlag, 1997), 325–28.

difficult to pull apart, because the air pressure pressing on the external surfaces is greater than the pressure of the air on the interior surfaces within the bellows. As air enters, the sides can be readily parted. By extrapolation, it would have been easy to believe that if no material body could enter the interior of the bellows, its sides would remain in contact to prevent formation of a vacuum. But that extrapolation was plausible only because Aristotelian natural philosophers were already convinced that nature abhors a vacuum and would never permit its formation. Aristotelian physics depended on that belief. The bellows experience as presented by Buridan, Albert, and John of Jandun is an artificial construct based upon a genuine artifact, the bellows. John of Jandun's version is the most extreme, since he has two compartments within a single closed system, in one of which a vacuum is assumed.

Medieval natural philosophers employed other experiences to demonstrate, or illustrate, what they already knew, namely, that nature abhorred a vacuum and would never permit its existence. So great was nature's aversion to the vacuum that to prevent it, it would cause a disruption in its own common course, as, for example, compelling heavy bodies to rise or light ones to fall. A number of these experiences were derived from ancient Greek sources, especially from Philo of Byzantium. They involved a burning candle in an enclosed vessel the bottom of which was filled with water, siphons, and elepsydras.[53]

5. The Relationship between Imagination and Empiricism in Medieval Natural Philosophy

Earlier, I had occasion to mention the *secundum imaginationem* aspect of medieval natural philosophy. A number of the experiences cited thus far were really imaginary and hypothetical, but were presented as if they were genuine empirical phenomena. But I shall close my account of medieval empiricism by showing how the *secundum imaginationem* approach manifested itself in the most unempirical of circumstances. Perhaps the most famous imaginative case in the history of medieval natural philosophy is the derivation of the mean speed theorem from a concern

---

53. For descriptions of these instruments and experiments, see Grant, *Much Ado About Nothing*, 77–100.

with the variation of qualities. The derivation of the mean speed theorem appears in the medieval subject area known as the "configuration of qualities," or the intension and remission of forms or qualities. It was an attempt to compare the variations of all kinds of qualities.

The instrument employed for these comparisons was mathematics: arithmetic at Merton College, Oxford, and geometry at the University of Paris, where Nicole Oresme composed the major treatise on the subject. Observed qualities could be compared in the way they varied and in the effects they had. Although the Mertonians and Oresme were generally relating and representing variations in virtually all qualities, both visible and invisible, the observational component in their approach is quite strange. Oresme, for example, used geometrical figures to represent the alleged effects of various powers and qualities, such as pains, colors, joys, sounds, and so forth. He shows how two equal qualities might have different effects. One way this can occur is if the qualities though equal are of unequal intensity. For example,

let A and B be two pains, with A being twice as intensive as B and half as extensive. Then they will be equal simply, although pain A is worse than, or more to be shunned than, pain B. For it is more tolerable to be in less pain for two days than in great pain for one day. But these two equal and uniform pains when mutually compared are differently figured, . . . ; so that if pain A is assimilated to a square, then pain B will be assimilated to a rectangle whose longer side will denote extension, and the rectangle and square will be equal.[54]

It is on the basis of such comparisons that Oresme took the great step and geometrized the Merton mean speed theorem, which he described as "Every quality, if it is uniformly difform, is of the same quantity as would be the quality of the same or equal subject that is uniform according to the degree of the middle point of the same subject."[55] This cumbersome verbalization expresses the modern relationship $s = (\frac{1}{2})at^2$, where s is distance, a is acceleration, and t is the time of acceleration. Oresme's proof

54. For the passage, see Oresme's *Tractatus de configurationibus qualitatum et motuum*, part 2, ch. 39, in Marshall Clagett, ed. and trans., *Nicole Oresme and the Medieval Geometry of Qualities and Motions: A Treatise on the Uniformity and Difformity of Intensities Known as "Tractatus de configurationibus qualitatum et motuum,"* 387.

55. See Oresme, *De configurationibus*, part 3, ch. 7, in Clagett, *Nicole Oresme and the Medieval Geometry of Qualities and Motions*, 409. For William Heytesbury's version of the mean speed theorem, see Clagett, *The Science of Mechanics*, 270.

of the theorem was done geometrically by equating the area of a triangle, which represents the distance traversed in a certain time interval by a uniformly accelerated quality, or velocity, with the area of a rectangle,[56] which represents the distance traversed by a uniform velocity moving with a speed equal to the instantaneous speed acquired at the middle instant of the time of its uniform acceleration. By equating a uniformly accelerated motion with a uniform motion, it was possible to express the former by the latter.

The mean speed theorem applied to qualities as well as velocities. It was assumed to apply to any qualities that changed in the same way, that is, where one quality changed in a manner analogous to a uniform acceleration and the other quality changed uniformly with the mean speed of the quality that was uniformly increasing. But Oresme and his medieval colleagues never sought to determine experimentally or observationally whether qualities and speeds really changed in the manner described in the mean speed theorem, or in accordance with numerous other relationships that were attributed to the range of qualities. Until Domingo de Soto in the sixteenth century, no one is known to have suggested that uniformly accelerated motion might apply to naturally falling bodies. But it was not until Galileo that the mean speed theorem was not only applied to naturally falling bodies, but an experiment was devised to determine if bodies really fell with uniform acceleration. In the *Two New Sciences,* Galileo has Simplicio raise the question by declaring that "I am still doubtful whether this is the acceleration employed by nature in the motion of her falling bodies." Simplicio urges his colleagues to devise an experiment that agrees with the conclusions. Salviati replies to Simplicio declaring

Like a true scientist, you make a very reasonable demand, for this is usual and necessary in those sciences which apply mathematical demonstrations to physical conclusions, as may be seen among writers on optics, astronomers, mechanics, musicians, and others who confirm their principles with sensory experiences that are the foundations of all the resulting structure.[57]

56. For the relationship of the triangle and rectangle, see Fig. 21(a) in Clagett, *Nicole Oresme and the Medieval Geometry of Qualities and Motions,* 409.

57. Galileo, *Two New Sciences,* a new translation with introduction and notes by Stillman Drake (Madison: University of Wisconsin Press, 1974), Third Day, p. 169.

With regard to the experiments, Salviati says that "the Author has not failed to make them, and in order to be assured that the acceleration of heavy bodies falling naturally does follow the ratio expounded above, I have often made the test [*prova*] in the following manner, and in his company." Galileo then presents his famous inclined plane experiment.[58]

Although the configuration of qualities, or intension and remission of forms, was concerned with representing the variation of qualities mathematically, and therefore seems connected to sense perception, it was really an abstract and hypothetical application of mathematics to imaginary qualitative changes that were connected to the real world only in the sense that most of the qualities existed in the real world. When we turn to conjectures about the possible movement of bodies in a vacuum we find ourselves as far removed from the medieval world of experience as one could get. And yet, despite their unanimous view that nature abhorred an extended vacuum, medieval natural philosophers thought it important to answer questions about hypothetical activities of observational entities in hypothetical vacua.

In one of his questions on vacuum, Albert of Saxony declares that

we have never experienced the existence of a vacuum, and so we do not readily know what would happen if a vacuum did exist. Nevertheless, we must inquire what might happen if it existed, for we see that natural beings undergo extraordinarily violent actions to prevent a vacuum.[59]

And then to demonstrate how nature extends itself to prevent formation of a vacuum, Albert mentions the siphon, or tube argument, where, as he puts it, "we see that if some tube [*fistula*] is put in water, and air is drawn

---

58. See ibid., 169–70.

59. Albert of Saxony, *Questions on the Physics*, bk. 4, qu. 12 ("Utrum si vacuum esset aliquid posset moveri in ipso velocitate finita seu motu locali seu motu alterationis"), fols. 51r, col. 1–51v, col. 1. The translation is from Grant, *A Source Book,* 339. The translation of the question is (p. 338): "whether, if a vacuum existed, something could be moved in it with a finite velocity or local motion, or with a motion of alteration." Marsilius of Inghen made a similar statement in his question "Whether a motion could occur in a vacuum, if one existed" ("Utrum in vacuo si esset posset fieri motus"; *Questions on the Physics,* bk. 4, qu. 12, fol. 54r, col. 2), when he declares that "because we have never experienced what happens in a vacuum, no one could know what would follow if the existence of a vacuum were assumed. Thus the conclusions stated previously are probable and conjectural." From ibid., fol. 54v, col. 1. The Latin text of this passage is given by Henri Hugonnard-Roche, "L'hypothétique et la nature dans la physique parisienne du XIVe siècle," in *La nouvelle physique du XIVe siècle,* edited by Stefano Caroti and Pierre Souffrin (Florence: Leo S. Olschki, 1997), 175.

from the tube, the water follows by ascending, striving to remain contiguous with the air lest a vacuum be formed."[60]

The key to understanding medieval interpretations of motion in hypothetical void space is to realize that medieval natural philosophers imported into the void, and analyzed, the same bodies they discussed in the plenum of their ordinary world.[61] They then imagined how such bodies would behave in a milieu that was devoid of material resistance. They considered both elemental and mixed bodies, the latter consisting of two or more elements forming a compound body. Although some allowed that an elemental body could fall with a finite speed in a void, most rejected such motion because the void lacked a medium that could serve as an external resistance to an elemental body. But most assumed that mixed bodies could fall in a void because they possessed a motive force and an internal resistance, the two essential requirements for finite motion, even in a vacuum.

To render the situation in a void more analogous to that in a plenum, some Scholastics assumed a vacuum that was produced by the annihilation of all matter within the concave surface of the lunar sphere, or occasionally all matter below the sphere of fire. Some also assumed that the former natural place of each element, now void, nevertheless retained the properties it had when it functioned as an elemental plenum. Thus one could speak of the "vacuum of fire" *(vacuum aeris),* or the "vacuum of air" *(vacuum aeris),* or the "vacuum of water" *(vacuum aque).* In the examples used, it became customary to assign degrees of heaviness and lightness to the elements in the compound. One conclusion that was generally reached was that a mixed body could descend with a successive motion in a vacuum, as Albert of Saxony indicates in the question "whether if a vacuum did exist, a heavy body could move in it":

60. Albert of Saxony, *Questions on the Physics,* fol. 54v, col. 1. Both Buridan and Marsilius of Inghen cite the same experience. Buridan replaces the water with wine (Grant, *A Source Book,* 326; for Marsilius's version, see ibid., 327).

61. Henri Hugonnard-Roche discusses the role of hypothetical physics in medieval natural philosophy and declares that "[l]e domaine de cette physique des cas imaginaires, ou physique 'hypothétique,' a été construit à l'aide de critères sémantiques touchant les conditions de vérité des propositions du domaine, et d'instruments conceptuels tirés de la physique 'naturelle,' ou physique de la *ratio generalis corporum.* Mais dan le meme temps qu'elle s'étendait *ad imaginabilia,* cette physique de cas impossibles *secundum quid* s'éloignait de la 'nature' en la dépouillant d'une partie de ses attributs, pour devenir imaginaire." See Hugonnard-Roche, "L'hypothétique et la nature," 177.

A second conclusion [is this]: By taking "vacuum" in the first way, as it is commonly taken in this question,[62] a heavy mixed body is easily moved in it successively. This is clear, for let there be a heavy mixed body whose heaviness [*gravitas*] is as 2 and lightness [*levitas*] as 1. And let it reach the concave [surface] of air and descend successively until its center of gravity [*medium gravitalis*] is the middle [or center] of the world [*medium mundi*]. [This will happen] because it has an internal resistance, for it has one degree of lightness inclining [or tending] upward and two degrees of heaviness inclining downwards.[63]

Because a mixed body can have varying relationships between its constituent elements, the fall of a mixed body in a void can produce results at variance with the fall of the same body in a plenum. Albert gives two significant instances of such differences. In a third conclusion of the question as to whether a heavy body could move in a vacuum, he shows that, under the right circumstances, "a heavy mixed [or compound] body [*mixtum*] could be moved quicker in a plenum than in a vacuum." Albert then draws another startling consequence in the same third conclusion, arguing that "the natural motion of some heavy body can be quicker in the beginning than in the end." For example,

If a mixed [or compound] body of four elements should have one degree of fire, one of air, one of water, and four of earth and if everything were annihilated within the sides of the sky except this mixed body, and if the mixed body were placed where the fire was, then this mixed body would descend more quickly through the vacuum of fire [*vacuum ignis*] than through the vacuum of air [*vacuum aeris*], and so on, as can easily be deduced from this case. But you [now] say, what should be said, therefore, about the common assertion that natural motion is quicker in the end than at the beginning? One can say that is universally true of the motion of heavy and light [elemental] bodies but not of the motion of heavy and light mixed [or compound] bodies.[64]

Although Aristotle had distinguished between elemental and mixed bodies, he had made no use of the distinction in his arguments against the vacuum.[65] Albert shows how Scholastic natural philosophers used it to make finite, successive motion in a vacuum seem possible and intel-

---

62. That is, as a separate, extended space devoid of body.
63. Grant, *A Source Book*, 336.
64. Ibid., 337. I have added ["elemental"].
65. On the distinction between elemental and mixed bodies, see Grant, *Much Ado About Nothing*, 44–45.

ligible. We see this in the two conclusions just described, namely that "a heavy mixed [or compound] body [*mixtum*] could be moved quicker in a plenum than in a vacuum" and "the natural motion of some heavy body can be quicker in the beginning than in the end." To these two significant deviations from Aristotelian physics, we must add one more, the concept that two mixed bodies of homogeneous composition, but of unequal weight, would fall with equal speeds in a vacuum. Thus Albert declares that

Mixed [or compound] bodies of homogeneous composition [*consimilis compositionis*] are moved with equal velocity in a vacuum but not in a plenum. The first part [concerned with fall in a vacuum] is obvious, because they are of homogeneous mixture. The ratio of motive power to total resistance in one body is the same as in another homogeneous [*consimilis*] body, because they both have only internal resistance.[66]

More than two centuries later, Galileo arrived at the same conclusion in a quite different manner.[67]

Observation did not contribute to the analysis of motion, other than the fact that natural philosophers could see that a body was faster at the end of its motion than at the beginning, a fact that convinced them that falling bodies accelerated. The analysis of motion was, however, largely a rationalistic, rather than an empirical, process. Late medieval discussions of motion in a void are significant because whatever observational component there was in discussions about motion in a material plenum was also applied to motion in a void. The only difference lay in the fact that external resistance did not exist in the void. Internal resistance, however, operated just as it did when bodies fell in a material medium. Indeed, they treated fall in the void with as much seriousness as they did motion in a plenum. Even though void was naturally impossible, they treated it as the limiting case for motion in a plenum.

---

66. See Grant, *A Source Book*, 341. This conclusion is the eighth in Albert's twelfth question of the fourth book of his *Questions on the Physics*. Thomas Bradwardine had already asserted this conclusion in 1328 in his *Tractatus de proportionibus*, ch. 3, theorem 12. See Grant, *A Source Book*, 305. Albert goes on to show, in the same conclusion, that the same two homogeneous bodies would fall with different speeds in a plenum. This is so because in a plenum there is also an external resistance, in addition to the internal resistance.

67. See Grant, *Much Ado About Nothing*, 61–66.

6. Conclusion

In fact, there is a great anomaly in medieval natural philosophy. Aristotelianism was empirical and rooted in sense perception. Near the beginning of this essay, I cited passages from Albert the Great, Roger Bacon, and John Buridan that emphasized the importance of empiricism, induction, and observation. In a chapter on physics in the late Middle Ages, A. C. Crombie emphasized theoretical discussions about the nature of induction in the fourteenth century.[68] But theoretical interest in induction by those already mentioned, and by the likes of John Duns Scouts and William of Ockham, did little to encourage and incite other Scholastics to emphasize observation and experience. We see very little direct observation in the questions literature on Aristotle's natural books. Very few questions were decided by appeals to observation. Despite the emphasis he placed on experience and induction, Crombie recognized that

[f]rom the beginning of the 14th century to the beginning of the 16th there was a tendency for the best minds to become increasingly interested in problems of pure logic divorced from experimental practice, just as in another field they became more interested in making purely theoretical, though also necessary, criticisms of Aristotle's physics without bothering to make observations.[69]

It is no exaggeration to characterize medieval Aristotelianism as empiricism without observation. This is true despite the fact that one finds, as we saw above, many empirical elements in medieval questions. But the authors who report them, or use them to support or refute an argument, did not directly observe them. The kinds of observations Floris Cohen mentioned, the "phenomena of nature as present themselves immediately to the senses (e.g. trees burning, stars shining, stones being thrown)," are too general and ubiquitous to serve any significant function. Medieval empiricism is far more complex. Empirical elements and observations appear in numerous examples and illustrations, but the examples were not observed by those who report them. This is not surprising when one realizes that questions on Aristotle's natural books could contain anywhere from forty to one hundred and twenty questions. It is implausible to think that a medieval natural philosopher would observe directly

68. See Crombie, *Medieval and Early Modern Science*, 2.28–35.
69. Ibid., 2.22–23.

the empirical elements that went into all the questions on a given work of Aristotle. There were simply too many of them. Some authors, however, might, on occasion, have experienced, or been familiar with, some of the observations they reported. But they were under no compulsion to check out any of the observations they included in a given question. No one expected it. In many, if not most, instances observations that appear in medieval questions were drawn from another source, either from within the Latin medieval tradition, or from Greek or Arabic sources.

Medieval observations were not introduced for their own sake, namely, to learn more about the world. They were intended rather to uphold an a priori view of the world or to serve as an example or illustration. The idea of observation was important in the Middle Ages because it was the basis of the medieval Aristotelian epistemology, which was founded on sense perception. But it was clearly not enough, as Aristotle understood and as we see in the crucial passage cited earlier, where Aristotle declares that although the senses "surely give the most authoritative knowledge of particulars they do not tell us the 'why' of anything—e.g., why fire is hot; they only say that it is hot."[70] Medieval natural philosophers were in agreement with Aristotle on this major point, which accounts for why, during the late Middle Ages, empiricism was, and remained, the servant of the analytic and a priori, which provided the "why" of things to explain and interpret the empirical world. John Murdoch has perceptively argued that although it is true that

empiricist *epistemology* was dominant in the fourteenth century this did not mean that natural philosophy then proceeded by a dramatic increase in attention being paid to experience and observation (let alone anything like experiment) or was suddenly overwrought with concern about testing or matching its results with nature. On the contrary, its procedures were increasingly *secundum imaginationem* (to use an increasingly frequently occurring phrase) and when some "natural confirmation" of a result is brought forth, more often than not it too was an "imaginative construct."[71]

The most powerful tool medieval natural philosophers possessed was not empiricism as manifested by observation *per se,* but rather experience

<hr />

70. Aristotle, *Metaphysics* 1.1.981b10–11.

71. John E. Murdoch, "The Analytic Character of Late Medieval Learning: Natural Philosophy without Nature," in *Approaches to Nature in the Middle Ages,* edited by Lawrence D. Roberts (Binghampton, N.Y.: Center for Medieval and Early Renaissance Studies, 1982), 174.

as adapted for use in thought experiments *(secundum imaginationem)*. Most of the experiences cited in this essay are really thought experiments designed to refute or uphold a theory. But the "experiences" were not actually performed—in most cases, they had not even been experienced by the author—although they were usually examined and analyzed with great seriousness. They only had to appear plausibly appropriate and relevant to be accepted and then utilized as part of an overall argument for or against some real or imagined position.

It was one thing to write about induction and observation, and to uphold their importance, as did Roger Bacon, Albertus Magnus, John Buridan, and others, but it was quite another to come to the realization that it was essential to make observations in the real world on a routine basis, and to design experiments to learn things about the world that could not be derived from raw observation and experience. And to make all this a routine and regular feature of natural inquiry. This stage of development was not reached in the Middle Ages. It had to await the seventeenth century, the century of Newton. But if Scholastic natural philosophers developed an empiricism without observation, and focused attention on hypothetical, rather than real and direct, observations, they did, at least, recognize that experience and observation were important ingredients in doing science and natural philosophy.

Because they failed to realize the importance of regular and direct observation and of the need for devising experiments to yield nature's patterns of behavior, medieval natural philosophers did the next best thing. They sought to uphold the laws of Aristotle's world as well as they could. Where they found it at variance with reason and observation, they changed those laws and perceptions. But they did this in the way Aristotle had taught them, and also by means of a new tool that they had devised for themselves. That is, they used observation and sense perception, guided by reason, to support the positions they believed true. But they relied most heavily on their imaginations, which were guided by reason in the form of analytic techniques and logical analysis. It was in this manner that they concocted thought experiments for the real world, as well as for the world Aristotle had regarded as naturally impossible, the world of imaginary void space. By these methods, they arrived at some rather startling theories and conclusions, such as the mean speed theorem, impetus theory, the possibility of finite motion in a vacuum, and claims for the existence of extracosmic void space. They achieved all this with a "natural philosophy without nature," to use John Murdoch's perceptive and felicitous phrase, and, perhaps not surprisingly, by employing an "empiricism without observation."

In light of all this, one is inexorably driven to ask: Did medieval Scholastic natural philosophers believe that their responses to the multitude of questions they posed about the workings of nature provided them with truths about the structure and operations of the physical world? To this question, we must, I believe, respond in the affirmative, since we have no evidence to the contrary. To reply in the negative is to assume that they knowingly and willingly labored to no purpose, an untenable assumption.

When we realize that the contributions described above, and others, were made without the sophisticated methodologies that would become a routine part of scientific inquiry in the seventeenth century, we should recognize that medieval natural philosophers deserve a much greater measure of respect than has hitherto been accorded them.

9 ∾ Science and Theology in the Middle Ages

Science and theology were never more closely interrelated than during the Latin Middle Ages in Western Europe. In this occasionally stormy relationship, theology was clearly the dominant partner. Limited challenges to that dominance occurred only when a sufficiently powerful natural philosophy was available to offer alternative interpretations of cosmic structure and operation. Conflict between science and theology rarely arose in the technical sciences, but developed in that part of natural philosophy concerned with the larger principles of cosmic operation, especially where theology and science sought to explain the same phenomena.[1] Prior to the twelfth century, when the scientific fare of Latin Christendom was meager, science lacked powerful metaphysical foundations and consisted of little more than a few of Aristotle's logical treatises, some medical works, two-thirds of Plato's *Timaeus,* a few astrological books, and, especially, a series of Latin encyclopedic handbooks written by Pliny, Solinus, Calcidius, Macrobius, Martianus Capella, Boethius, Cassiodorus, Isidore of Seville, and the Venerable Bede. An important feature of this body of secular learning was the famous seven liberal arts

1. Although there are significant differences between the modern term *science* and the medieval term *natural philosophy,* the two will be used here interchangeably. In practical terms, natural philosophy (or "natural science," as it was occasionally called) was generally identified with Aristotle's "natural books" *(libri naturales),* which treated themes in cosmology, physics, and matter theory. As one of the three major subdivisions of speculative philosophy, natural philosophy was concerned exclusively with mobile bodies and their changes. Although natural philosophy was distinct from mathematics, sciences that used mathematics, such as optics and astronomy, but were also concerned with mobile bodies could also fall under the consideration of natural philosophy. For the place of natural philosophy in the medieval division of the sciences, see Robert Kilwardby, O.P., *De Ortu Scientiarum,* edited by Albert G. Judy, O.P. (Toronto: British Academy and the Pontifical Institute of Mediaeval Studies, 1976), 15–29, and Domingo Gundisalvo, *On the Division of Philosophy,* partially translated by M. Clagett and E. Grant in *A Source Book in Medieval Science,* edited by Edward Grant, 62–65 (Cambridge, Mass.: Harvard University Press, 1974).

with their twofold division into language and mathematics or mathematical science, the former consisting of grammar, rhetoric, and dialectic (or logic) and designated by the term *trivium,* the latter embracing arithmetic, geometry, astronomy, and music, collectively known as the *quadrivium.*

The narrow conception of science embodied in the *quadrivium* of the seven liberal arts was expanded into the broader sense of natural philosophy during the twelfth and thirteenth centuries when, in the course of an unparalleled period of translating activity from Arabic and Greek into Latin, the bulk of Greek science and natural philosophy was finally introduced into Latin Christendom, some eleven hundred years after the birth of Christianity.

Science as Handmaiden to Theology

Through much of the Middle Ages, science was assigned the status of a "handmaiden to theology" *(philosophia ancilla theologiae),* a role first envisaged for it by Philo Judaeus in the first century A.D., subsequently adopted by Clement of Alexandria (ca. 150–ca. 215) and St. Augustine (354–430) in late antiquity, and fully reinforced centuries later by Hugh of Saint-Victor (d. 1141) and St. Bonaventure (1221–1274). According to the handmaiden concept, science was not pursued for its own sake but only for the aid it could provide in the interpretation of Holy Scripture. St. Bonaventure even devoted a special treatise to the ancillary and subsidiary role of the arts to theology. In this work, which he titled *On Retracing the Arts to Theology (De reductione artium ad theologiam),* Bonaventure interpreted the "arts" *(artes)* as almost synonymous with philosophy and science and believed that he had demonstrated "how all divisions of knowledge are handmaids of theology," for which reason "theology makes use of illustrations and terms pertaining to every branch of knowledge." It was the purpose or "fruit of all sciences, that in all, faith may be strengthened, [and] *God may be honored.*"[2] The glo-

---

2. See Sister Emma Therese Healy, ed. and trans., *Saint Bonaventure's "De reductione artium ad theologiam": A Commentary with an Introduction and Translation* (Saint Bonaventure, N.Y.: Franciscan Institute, 1955), 41. I have added the bracketed word. For the manner in which Roger Bacon subordinated mathematics to theology, see David C. Lindberg, "On the Applicability of Mathematics to Nature: Roger Bacon and His Predecessors," *British Journal for the*

rification of God was the ultimate goal of the scientific study of nature. Some two centuries earlier, Peter Damian (1007–1072) also reflected patristic and early medieval attitudes toward the relationship of God and nature. Because God created the world from chaos, Damian considered Him the direct and immediate cause of nature's laws and its ordered beauty.[3] God encourages the study of the external, visible world with a twofold purpose: to provide in us the contemplation of its invisible, spiritual nature, so that we should better love and adore Him, and to enable us to gain dominion over it as described in Psalms 8:6–9.[4] Achievement of these goals is made possible by the sciences of number and measure in the *quadrivium*.[5] For Peter Damian, as for Bonaventure later, the study of nature and its laws was not an end in itself, pursued merely for the sake of knowledge. It had to serve the higher needs of religion and theology. Under these circumstances the secular sciences could hardly avoid the status of handmaidens.

Revolt of the Handmaidens: Natural Philosophy Challenges Theology

The subservience of science to theology, however, was always relative. It was more complete during the early Middle Ages than later, a condition attributable, in no small measure, to the enfeebled state of natural philosophy in the five or six formative centuries of the early medieval period. The bulk of Greek science and philosophy was simply absent from the corpus of secular learning that passed for science. So low was the level of science in this period that it posed no threat whatever to Christian tradition and doctrine. With the exception of Plato's *Timaeus,* most of it was encyclopedic, unintegrated, and frequently confused or contradictory. Devoid of cohesion or guiding principles, it could inspire little by

History of Science 15 (1982): 3–26. The most frequently cited biblical passages in support of the handmaiden idea were Exodus 3:22 and 12:36, which spoke of despoiling the Egyptians of their treasures. In 1231, when he sought to justify the expurgation of Aristotle's physical treatises, Pope Gregory IX referred to the despoiling of the Egyptians by the Hebrews (for the passage, see Grant, *Source Book*, 43).

3. André Cantin, *Les sciences séculières et la foi: Les deux voies de la science au jugement de S. Pierre Damien (1007–1072)* (Spoleto: Centro Italiano di Studi sull'Alto Medioevo, 1975), 557, 578.

4. Ibid., 580.   5. Ibid., 536 ff.

way of new interpretations or insights about the nature of the world that might prove subversive of Christianity.

By the twelfth century, significant changes were under way that would eventually challenge theology's interpretation of the cosmos and the God who created it. The threat to theology and the church did not derive from astrology or magic, which, though potentially dangerous, were successfully contained in the Middle Ages. It came from Greek natural philosophy and science, initially in its benign Platonic and Neoplatonic forms in the twelfth century and then in its powerful and truly menacing Aristotelian form in the thirteenth century. The beginnings of this momentous process are already apparent in the enthusiastic study of Plato's *Timaeus* in the twelfth century. Evidence of significant change is readily available. Inspired perhaps by Honorius of Autun's (fl. 1122) joyous sentiment that "all of God's creation gives great delight to anyone looking upon it,"[6] a sentiment shared by his contemporary Thierry of Chartres (d. ca. 1155),[7] and by earlier authors such as Peter Damian, scholars came to investigate nature for its own sake. William of Conches (ca. 1080–ca. 1154), for whom physical laws took precedence over ecclesiastical authority, reflected the new attitude when he denounced those who, "ignorant themselves of the forces of nature and wanting to have company in their ignorance ... don't want people to look into anything; they want us to believe like peasants and not to ask the reason behind things."[8] To explain causes and phenomena by mere appeal to God's omnipotence or a biblical passage was now tantamount to a confession of ignorance.[9] It was the obligation of philosophy, not Holy Scripture, to teach about nature and its regular causes and events. A newfound confidence in human reason and sensory experience had emerged. Even the Bible, especially the creation account of Genesis, had to conform to the demands of physi-

6. From Honorius's *Elucidarium* 1.12, as translated by M. D. Chenu, O.P., *Nature, Man, and Society in the Twelfth Century: Essays on New Theological Perspectives in the Latin West,* preface by Etienne Gilson, edited and translated by Jerome Taylor and Lester K. Little, 8 n. 15 (Chicago: University of Chicago Press, 1968; original French version published in 1957).

7. See Tina Stiefel, "The Heresy of Science: A Twelfth-Century Conceptual Revolution," *Isis* 68 (1977): 350.

8. From William's *Philosophia mundi* 1.23, as translated in Chenu, *Nature, Man, and Society,* 11. For William's attitude toward the relationship between physical law and the exegetical tradition on Genesis, see Helen R. Lemay, "Science and Theology at Chartres: The Case of the Supracelestial Waters," *British Journal for the History of Science* 10 (1977): 229–33.

9. Chenu, *Nature, Man and Society,* 12.

cal science. The bold new emphasis on rational inquiry, with which the names of Adelard of Bath, Peter Abelard, William of Conches, Bernard Silvester, Clarenbaldus of Arras, and others in the twelfth century were associated, marked the beginning of an unsuccessful, though vigorous, attempt to separate science from theology. Separation, however, did not signify that science was to be pursued solely for its own sake. On the contrary, its application to the exegesis of Holy Scripture and to the elucidation of theological problems would produce a role reversal: science began to encroach upon theology. Thus were the seeds of a science-theology confrontation planted, the bitter fruits of which would grow to maturity in the thirteenth century following upon the introduction of Aristotle's scientific works, which formed the crucial core of the new Greco-Arabic science that entered Western Europe. By the early thirteenth century, Latin translations (from Arabic) of Aristotle's scientific, logical, and metaphysical works had taken Europe by storm. No match for the depth and diversity of the Aristotelian treatises with their elaborate scientific methodology and foundational principles, Plato's *Timaeus*, which had formed the basis and inspiration of the twelfth-century worldview, soon fell into abeyance.

Aristotle's treatises on physics, metaphysics, logic, cosmology, the elements, epistemology, and the nature of change furnished the Middle Ages with its conception of the structure and operation of the physical world. They assumed this fundamental role because their introduction into Western Europe coincided with, and probably contributed toward, the establishment of that uniquely medieval institution, the university. For approximately four hundred and fifty years, from 1200 to 1650, the universities of Western Europe emphasized a philosophical and scientific curriculum based on the works of Aristotle, whose logic and natural philosophy were studied by all who received the master of arts degree. Since the latter was usually a prerequisite for entry into the higher faculty of theology, most theologians were well acquainted with contemporary science.

The impact of Aristotle's thought on the late Middle Ages cannot be overestimated. For the first time in the history of Latin Christendom, a comprehensive body of secular learning, rich in metaphysics, methodology, and reasoned argumentation, posed a threat to theology and its traditional interpretations. Where Plato's creation account in the *Timaeus*, which featured a creator God who sought to share his goodness by fabricating a world from preexistent and coeternal matter and form, was reasonably compatible with Christianity, Aristotle's cosmic system, which assumed a world without beginning or end and a deity who had no knowledge of that world, was not. When to these difficulties were added those concerning the soul (it apparently perished with the body) and a strong tendency to employ naturalistic and even deterministic modes of explanation, it becomes obvious that the Aristotelian world system was not readily reducible to the status of a theological handmaiden. While numerous theologians and almost all arts masters eagerly embraced the new Aristotelian learning at the University of Paris, which possessed the most prestigious and powerful faculty of theology in all of Christendom, there was a growing uneasiness among more traditionally minded theologians, as evidenced by a ban on Aristotle's natural books issued in 1210 and 1215 and an abortive attempt to expurgate them in 1231.[10] All such attempts were in vain, and by 1255 Aristotle's works were not only officially sanctioned but constituted the core of the arts curriculum.[11]

Those who had hoped for a harmonious relationship between theology and philosophy were to be bitterly disappointed. During the 1260s and early 1270s a fundamental split developed. On the one side were radical arts masters and liberal theologians who found Aristotle's philosophy essential to a proper understanding of God and his creation. Opposed to them were traditional theologians for whom significant aspects of Greek philosophy were dangerously subversive to the Christian faith.[12] Typified by the likes of Siger of Brabant and Boethius of Dacia, the more radical arts masters perceived Aristotle's natural philosophy as the indispensable key to a proper interpretation of the cosmos and concluded that philosophy was not only independent of theology but at least its equal and perhaps its superior. Although they would surely have denied it equality, many theologians regarded philosophy as worthy of independent study and assigned it a central role. The most illustrious member of this group

10. The bans of 1210 and 1215 were issued by the provincial synod of Sens, which included the bishop of Paris. The order to expurgate the books of Aristotle in 1231 came from Pope Gregory IX, who appointed a three-member committee for the purpose. Whatever the reasons, the committee never carried out its assignment. For a translation of the documents of 1210 and 1231, see Grant, *Source Book,* 42–43.

11. For the document of 1255, see Grant, *Source Book,* 43–44.

12. The different reactions to pagan philosophy are described by John Wippel, "The Condemnations of 1270 and 1277 at Paris," *Journal of Medieval and Renaissance Studies* 7 (1977): 195.

was undoubtedly Thomas Aquinas (ca. 1225–1274), who considered the-ology the highest science because of its reliance on revelation. Without revelation the truth of the metaphysics that philosophers might devise would be incomplete and imperfect.[13] Yet Aquinas not only embraced philosophy with enthusiasm but regarded Aristotle as the greatest of philosophers, one who had achieved the highest level of human thought without the aid of revelation.[14] Rightly understood, philosophy, which included secular science, could not contradict theology or faith.[15]

Suspicious of the emphasis on philosophy and secular learning that had occurred during the 1260s and fearful of the application of Aristote-lian philosophy to theology, traditional and conservative theologians, in-spired by St. Bonaventure, sought to stem the tide by outright condemna-tion of ideas they considered subversive. Since repeated warnings of the inherent dangers of secular philosophy and the perils of its application to theology had been of little avail,[16] the traditional theologians, many of whom were neo-Augustinian Franciscans, appealed to the bishop of Paris, Etienne Tempier, who responded in 1270 with a condemnation of 13 propositions, which was followed in 1277 by a massive condemnation of 219 propositions, any one of which was held at the price of excommu-nication.

The Impact of Theology on Science

Controversial and difficult to assess, the Condemnation of 1277 looms large in the relations between theology and science. Except for articles directed specifically against Thomas Aquinas, which were nullified in

13. See Frederick Copleston, S.J., *A History of Philosophy*, 9 vols., 2:318–319 (Westminster, Md.: Newman Press, 1946–1975).

14. Ibid., 319. Edith Sylla observes ("Autonomous and Handmaiden Science: St. Thom-as Aquinas and William of Ockham on the Physics of the Eucharist," in *The Cultural Context of Medieval Learning*, edited by John E. Murdoch and Edith D. Sylla, 354, 363 [Dordrecht and Boston: D. Reidel, 1979]) that despite Aquinas's acknowledgment of the autonomy of philoso-phy (which includes natural philosophy) from theology, he often subordinated the former to the latter, as exemplified in his discussion of the Eucharist. By contrast, William of Ockham (ca. 1285–1349) refused to bend physics and natural philosophy to the needs of theology, choos-ing rather to explain physically inexplicable religious phenomena by God's direct intervention.

15. Wippel, "Condemnations of 1270 and 1277," 175.

16. For some of these warnings, see Leo Elders, S.V.D., *Faith and Science: An Introduction to St. Thomas' "Expositio in Boethii De Trinitate"* (Rome: Herder, 1974), 51 and nn. 42, 43.

1325, the condemnation remained in effect during the fourteenth century and made an impact even beyond the region of Paris, where its legal force was confined. Hastily compiled from a wide variety of written and oral sources, the 219 condemned errors were without apparent order, repetitious, and even contradictory.[17] Orthodox and heterodox opinions were mingled indiscriminately.[18] A number of the errors were relevant to science. Of these, many were condemned in order to preserve God's absolute power *(potentia Dei absoluta)*, a power that natural philosophers were thought to have unduly restricted as they eagerly sought to interpret the world in accordance with Aristotelian principles.[19] If the condemned errors accurately reflect contemporary opinion, some natural philosophers were prepared to deny the divine creation of the world; that God could create more than one world; that he could move the world in a straight line, leaving behind a void space; that he could create an accident without a subject; and so on. In denying to God the capacity to perform these and other actions that were impossible in the physical world as conceived by Aristotle and his followers, philosophers were severely constraining God's power. The theologians who compiled the list of condemned errors sought to curb the pretensions of Aristotelian natural philosophy by emphasizing the absolute power of God to do whatever He pleased short of a logical contradiction. Indeed, Article 147 made this quite explicit by rejecting the claim "that the absolutely impossible

17. The Latin text of the 219 articles, in their original order, appears in Heinrich Denifle and Emile Chatelain, *Chartularium Universitatis Parisiensis*, 4 vols., 1:543–555 (Paris: Ex typis Fratrum Delalain, 1889–1897); for a methodical regrouping of the articles aimed at facilitating their use, see Pierre F. Mandonnet, O.P., *Siger de Brabant et l'Averroisme latin au XIIIme siècle, IIme partie: Textes inédits*, 2d ed. (Louvain: Institut supérieur de philosophic de l'Université, 1908), 175–191. Using Mandonnet's reorganized version, Ernest L. Fortin and Peter D. O'Neill translated the articles into English in *Medieval Political Philosophy: A Sourcebook*, edited by Ralph Lerner and Muhsin Mahdi, 337–354 (New York: Free Press of Glencoe, 1963). Their translation was reprinted in *Philosophy in the Middle Ages: The Christian, Islamic, and Jewish Traditions*, edited by Arthur Hyman and James J. Walsh, 540–49 (Indianapolis: Hackett Publishing Co., 1973). Selected articles relevant to science have been translated in Grant, *Source Book*, 45–50. For a discussion of each article, including its sources, see Roland Hisette, *Enquête sur les 219 articles condamnés à Paris le 7 mars 1277*, Philosophes médiévaux vol. 22 (Louvain: Publications Universitaires; Paris: Vander-Oyez, 1977).

18. Wippel, "Condemnations of 1270 and 1277," 186.

19. Here and in what follows on the Condemnation of 1277 I follow my article "The Condemnation of 1277, God's Absolute Power, and Physical Thought in the Late Middle Ages," *Viator* 10 (1979): 211–44. For the distinction between God's absolute power and his ordained power *(potentia ordinata)*, see 215.

cannot be done by God or another agent," which is judged "an error, if impossible is understood according to nature."[20] With respect to nature, then, all had to concede that God could do things that were contrary to prevailing scientific opinion about the structure and operations of the cosmos. In short, God could produce actions that were naturally impossible in the Aristotelian worldview. It was thus Aristotelian natural philosophy on which the Condemnation of 1277 pressed most heavily. If we can judge from those condemned errors that asserted that "theological discussions are based on fables," that "nothing is known better because of knowing theology," that "the only wise men of the world are philosophers,"[21] and that "there is no more excellent state than the study of philosophy,"[22] the Condemnation of 1277 may have served as a vehicle of sweet revenge for the theologians who compiled it. It offered an opportunity to humble the professional Aristotelian natural philosophers from whom those hostile sentiments derived.

The Paris condemnation of 219 diverse errors in theology and natural philosophy was a major event in the history of medieval natural philosophy. Whatever the doctrinal and philosophical disputes, or personal and group animosities that produced it, emphasis on God's absolute power was its most potent feature. Although the doctrine of God's absolute power was hardly new in the thirteenth century,[23] the challenge from Aristotelian natural philosophy and physics, and Greco-Arabic thought generally, conferred on it a new significance. The growing tendency prior to 1277 was to interpret cosmic phenomena in accordance with natural causes and explanatory principles derived from Aristotelian physics and cosmology. After 1277, appeals to God's absolute power were frequently introduced into physical and cosmological discussions. Whether by implication or explicit statement, many of the articles of the condemnation proclaimed God's infinite and absolute creative and caus-

20. Grant, *Source Book,* 49.

21. Articles 152, 153, and 154 as translated in Grant, *Source Book,* 50.

22. Article 40 as translated in Wippel, "Condemnations of 1270 and 1277," 187.

23. It had already been proclaimed by St. Peter Damian in the eleventh century (for a translation of the relevant sections from Damian's *On Divine Ominipotence* [*De divina omnipotentia*], see *Medieval Philosophy from St. Augustine to Nicholas of Cusa,* edited by John F. Wippel and Allan Wolter, O.F.M., 143–52, esp. 148–49 [New York: Free Press, 1969]) and by Peter Lombard in the twelfth century (the passage from Peter's *Sentences* is translated in Grant, "Condemnation of 1277," 214 n. 10).

ative power against those who would circumscribe it by the principles of Aristotelian natural philosophy. As a consequence, natural impossibilities, usually cast in the form of "thought-experiments," were hereafter entertained with increasing frequency and occasionally with startling consequences. The supernatural alternatives considered in the aftermath of the condemnation of 1277 conditioned Scholastics to contemplate physical possibilities outside the ken of Aristotelian natural philosophy, and frequently in direct conflict with it. As the means of achieving these hypothetical possibilities, God's absolute power was usually invoked. Indeed, hypothetical possibilities based upon supernatural actions became a characteristic feature of late medieval Scholastic thought. To illustrate these tendencies we need only consider two articles concerned with the possibility of other worlds and the movement of our own.

Both Aristotle and the Bible agreed that only one world existed. With a variety of arguments Aristotle had demonstrated the impossibility of other worlds. For some of his enthusiastic medieval followers it was an easy inference that God could not create other worlds even had He wished to do so. Thus a limitation was placed upon divine power, a limitation that was condemned in Article 34, which threatened excommunication to any who held that God could not possibly create more than one world. Although it was in no way required to believe that God had created a plurality of worlds—indeed, no one in the Middle Ages did so believe—but only that He could do so, the effect of Article 34 was to encourage examination of the conditions and circumstances that would obtain if God had indeed created other worlds. In this spirit a number of Scholastic authors formulated arguments that sought to make the possible existence of other worlds intelligible. Sometime around 1295 Richard of Middleton (d. ca. 1300) argued in his commentary on Peter Lombard's *Sentences* that if God created other worlds identical with ours, the very kind Aristotle had discussed, each of them would behave just as ours does, since no good reasons could be adduced for supposing otherwise. Hence each world would be a self-contained, closed system with its own center and circumference.[24] It surely followed that if God did indeed create more than one world, no unique and privileged center would exist, an inference that subverted the foundation of Aristotle's cosmology,

24. For the references and further discussion, see Grant, "Condemnation of 1277," 220–23.

namely, that the center and circumference of our world are unique. This extraordinary result, which would be repeated in the fourteenth century by the likes of William of Ockham, John Buridan, and Nicole Oresme, was achieved merely by considering possible, hypothetical worlds, not real ones.

Consideration of other worlds immediately posed the problem of what might lie between them. Prior to 1277 the possibility that a vacuum might intervene was rejected because Aristotle had demonstrated the impossibility of void space within and beyond our unique world. In light of Article 34 of the Condemnation of 1277, however, Nicole Oresme and others now boldly proclaimed the existence of intercosmic void space. Indeed, the necessity of conceding the existence of void space beyond our world—and therefore the possibility that void space could intervene between our world and other possible worlds—could be directly inferred from another article (no. 49), which made it mandatory to concede that God could, if He wished, move the last heaven, or the world itself, with a rectilinear motion even if a vacuum were left behind.[25]

A few fourteenth-century Scholastics moved beyond the merely hypothetical and boldly proclaimed the real existence of an infinite, extracosmic void space, which they identified with God's immensity. Late medieval Scholastics introduced God into space in a more explicit manner than that suggested by the vague metaphors found in earlier patristic, cabalistic, and hermetic traditions.[26] In the fourteenth century Thomas Bradwardine, Jean de Ripa, and Nicole Oresme proclaimed the existence of a real, extracosmic, infinite void space filled by an omnipresent deity. Oresme explicitly identified infinite, indivisible space with God's immensity. These ideas were developed further by Scholastic authors of the sixteenth and seventeenth centuries. The medieval Scholastic idea that God must bear an intimate relationship to space remained a viable concept well into the eighteenth century and played a role in the scientif-

---

25. A detailed discussion appears in ibid., 226–32.

26. By "hermetic" tradition is meant the approximately fifteen anonymous Greek treatises written sometime between A.D. 100 and 300 and ascribed to the Egyptian god Hermes Trismegistus ("Thrice-Great Hermes"). A diverse collection of mystical and spiritual works that incorporated popular Greek philosophy along with Jewish and Persian elements, the hermetic treatises exercised some influence during the Middle Ages but had their greatest impact during the Renaissance. For an account of their significant role in Western thought, see Frances A. Yates, *Giordano Bruno and the Hermetic Tradition* (Chicago: University of Chicago Press, 1964).

ic and theological thought of Isaac Newton himself. From the assumption that infinite space is God's immensity, Scholastics derived most of the same properties for space that non-Scholastics did subsequently. As God's immensity, space was necessarily assigned divine properties, such as homogeneity, immutability, infinity, lack of extension, and the capacity to coexist with bodies to which it offered no resistance. Except for extension, the divinization of space in Scholastic thought produced virtually all the properties that would be attributed to space during the course of the Scientific Revolution.[27]

Although no articles of the Condemnation of 1277 concerned vacua within the cosmos, it followed inexorably that if God could create a vacuum beyond our world and between possible worlds, He could surely create one or more within our world. Throughout the fourteenth century and later, God was frequently imagined to annihilate all or part of the matter within the material plenum of our world.[28] After 1277 all sorts of situations were hypothesized within such wholly or partially empty spaces. The questions raised became an integral part of a large literature on the nature of vacuum and the imagined behavior of bodies therein. Would the surrounding celestial spheres collapse inward instantaneously as nature sought to prevent formation of the abhorred vacuum? Indeed, could an utterly empty interval, or nothingness, be a vacuum or space? Would a stone placed in such a void be capable of rectilinear motion? Would people placed in such vacua see and hear each other? Analyses of these and similar thought-experiments in the late Middle Ages were often made in terms of Aristotelian principles even though the conditions imagined were "contrary to fact" and impossible within Aristotelian natural philosophy. From such analyses intelligible and plausible alternatives to Aristotelian physics and cosmology emerged and demonstrated that things could be otherwise than was dreamt of in Aristotle's philosophy.

But if the Condemnation of 1277 beneficially stimulated speculation outside the bounds of Aristotelian natural philosophy, it may also have adversely affected scientific development. In emphasizing God's inscru-

---

27. The summary presented here of the relations between God and space is drawn from my book, *Much Ado About Nothing: Theories of Space and Vacuum from the Middle Ages to the Scientific Revolution* (Cambridge: Cambridge University Press, 1981), esp. 260–64.

28. See Grant, "Condemnation of 1277," 240–41.

table will and His absolute power to do as He pleased, the conservative theologians encouraged a philosophical trend in which confidence in demonstrative certainty, and ultimately confidence in the ability of science to acquire certain truth about the physical world, was weakened. The imaginary physical conditions that were frequently conjured up in the Middle Ages were usually contrary to the "common course of nature" *(communis cursus nature)*, which represented the operation of nature interpreted in accordance with Aristotelian natural philosophy. But, it was asked, if God could intervene at will in the causal order, how could scientific principles and laws be absolute, so as to guarantee a "common course of nature"? John Buridan (ca. 1295–ca. 1358), perhaps speaking for many arts masters who wished to defend Aristotelian science as the best means of understanding the physical world, conceded, as he had to, that God could interfere in natural events and alter their course at any time. To alleviate the effect of such uncertainty, however, Buridan urged natural philosophers to proceed as if nature *always* acted with regularity and followed its "common course."[29] On this assumption he believed that "for us the comprehension of truth with certitude is possible."[30] The scientific principles from which these certain truths are derivable are themselves indemonstrable, "but they are accepted because they have been observed to be true in many instances and false in none."[31] Since the ultimate principles depend on experience rather than strict logical demonstration or a priori grounds, any of Buridan's certain truths could be overturned by a single empirical counterexample. A degree of uncertainty thus lurked within Buridan's concept of certitude. On methodological grounds Buridan also found a place for the principle of Ockham's razor:

29. Ockham had also adopted this attitude (see Sylla, "Autonomous and Handmaiden Science," 359), as did others who sought to assign meaning and significance to the "common course of nature."

30. Buridan, *Questions on the Metaphysics*, bk. 2, question 1. The interpretations of Buridan and Oresme (below) are based on my article "Scientific Thought in Fourteenth-Century Paris: Jean Buridan and Nicole Oresme," in *Machaut's World: Science and Art in the Fourteenth Century*, edited by Madeleine Pelner Cosman and Bruce Chandler, Annals of the New York Academy of Sciences 314, 105–24, esp. 109 (New York: New York Academy of Sciences, 1978). On the quotation from Buridan, see also William A. Wallace, *Prelude to Galileo: Essays on Medieval and Sixteenth-Century Sources of Galileo's Thought* (Dordrecht and Boston: D. Reidel, 1981), 345.

31. From Buridan's *Questions on the Metaphysics*, bk. 2, question 2, as translated by Ernest A. Moody, "Buridan, Jean," *Dictionary of Scientific Biography*, edited by Charles C. Gillispie, 16 vols. (New York: Scribners, 1970–1980), 2:605.

that if more than one explanation could "save the phenomena," the simplest should be chosen.

But even if one accepted the simplest explanation as true, how could the best and simplest explanation be determined with certainty? Nicole Oresme (ca. 1320–1382), one of the most brilliant natural philosophers and theologians of the fourteenth century, found experience and human reason inadequate for the proper determination of physical truth. Only faith could furnish us with genuine truth. The fourteenth-century emphasis on God's free and unpredictable will, encapsulated in the concept of God's absolute power, had eroded confidence in human ability to arrive at demonstrated truth in both theology and natural philosophy. In the process of defending God's absolute power to act as He pleased, theologians not only showed the inconclusiveness of certain philosophical proofs traditionally employed to demonstrate what God could or could not do, or to prove His existence or attributes, but they also revealed the limitations of natural philosophy by demonstrating the radically contingent nature of the world. Led by William of Ockham (ca. 1285–1349), many theologians concluded that neither reason nor experience could provide certain knowledge of any necessary connection between causes and their alleged effects. Both reason and experience were consequently deemed inadequate to demonstrate fundamental truths about God and His physical creation, both of which were generally perceived as less knowable during the fourteenth century than in the thirteenth. Where demonstrative certainty about nature was the goal of most natural philosophers in the thirteenth century, probable knowledge was the most that was thought attainable by many in the fourteenth century. While the latter were hardly skeptics, their attitude toward nature, when compared with that of thirteenth-century Scholastics, appears to mark a loss of confidence in human ability to acquire certain knowledge—apart from faith and revelation—about the true nature of God and the world. It was within this intellectual environment that a new trend developed in which physical problems were couched in hypothetical form without existential implication. The phrase *secundum imaginationem*, "according to the imagination," was regularly employed to characterize the innumerable hypothetical possibilities that were formulated in both natural philosophy and theology without any regard for physical reality or application to the world. In marked contrast, the key figures in the later Scientific Revolution—Copernicus, Galileo, Kepler, Descartes, and Newton, to name only the greatest—were confident, perhaps naively so, that nature's essential structure and operation were knowable. They were thus encouraged to search after nature's true laws of the physical world. With them, hypothetical conditions were but heuristic devices to arrive at physical truth. Things were quite different in the fourteenth century.

The Impact of Science on Theology

With a diminished confidence in the certainty of theological and scientific claims, theologians of the fourteenth century turned their attention to hypothetical problems posed *secundum imaginationem.* The *Sentences* (*Sententiae,* or opinions) of Peter Lombard (d. ca. 1160), written around 1150, provided a major point of departure for consideration of these problems. Divided into four books devoted, respectively, to God, the Creation, the Incarnation, and the sacraments, the *Sentences* served for some four centuries as the standard text on which all theological students were required to lecture and comment. Although the second book, devoted to the six days of creation, afforded ample opportunity to consider specific scientific topics such as the nature of light, the four elements, the problem of the supracelestial waters, and the order and motion of the celestial spheres and planets, there was an even more direct impact of natural philosophy on theology involving the attempt to define the relationship of God to the world and His creatures. The injection of science, mathematics, and logic into commentaries on Peter Lombard's *Sentences* grew to such proportions that in 1366 the University of Paris decreed that except where necessary those who read the *Sentences* should avoid the introduction of logical or philosophical material into the treatment of the questions.[32] Despite such appeals, however, Scholastic commentators ap-

---

32. Denifle and Chatelain, *Chartularium Universitatis Parisiensis,* 3.144. The statute is cited and discussed by John E. Murdoch, "From Social into Intellectual Factors: An Aspect of the Unitary Character of Late Medieval Learning," in *Cultural Context,* ed. Murdoch and Sylla, 276. Some 160 years later, John Major (1469–1550), in the introduction to the second book of his *Sentence Commentary* (1528), declared that "for some two centuries now, theologians have not feared to work into their writings questions which are purely physical, metaphysical, and sometimes purely mathematical" (translated by Walter Ong, *Ramus, Method, and the Decay of Dialogue: From the Art of Discourse to the Art of Reason* [Cambridge, Mass.: Harvard University Press, 1958], 144).

parently found it "necessary" to introduce such matters frequently and extensively.

That science and mathematics were applied to the exegesis of the creation account in medieval commentaries on the second book of Peter Lombard's *Sentences* comes as no surprise. Since later antiquity, science and mathematics had been used extensively in hexaemeral commentaries on Genesis—for example, by Sts. Basil, Ambrose, and Augustine—a practice that continued throughout the Middle Ages (and well into the seventeenth century) and reached enormous proportions in the lengthy (over a million words), popular, encyclopedic commentary of Henry of Langenstein (composed between 1385 and 1393), which employed almost every scientific subject in its biblical exegesis and made apparent the ease with which science could be introduced into the analysis of creation.[33] During the late Middle Ages, however, science and mathematics were also applied extensively to theological problems that were largely or wholly unrelated to the creation account in Genesis. Themes, techniques, and ideas from natural philosophy and mathematics were frequently used in problems that concerned God's omnipresence, omnipotence, and infinity, as well as His relations to the beings of His own creation and to comparisons between created species. Mathematical concepts were regularly drawn from proportionality theory, the nature of the mathematical continuum, convergent and divergent infinite series, the infinitely large and small, potential and actual infinites, and limits, which included boundary conditions involving first and last instants or points.[34]

Not only were these concepts applied to theological problems, but the latter were frequently formulated in the language of mathematics and measurement. Such concepts were employed to describe the manner in which spiritual entities could vary in intensity and how such variations could best be represented mathematically by application of the peculiarly

33. Henry's *Lecturae super Genesim*, to use its Latin title, has been analyzed by Nicholas Steneck, *Science and Creation in the Middle Ages: Henry of Langenstein (d. 1397) on Genesis* (Notre Dame, Ind.: University of Notre Dame Press, 1976); see esp. 21.

34. The basic research on the application of concepts of mathematics and measurement to theology has been done by John E. Murdoch in at least two articles on which I have relied: "*Mathesis in Philosophiam Scholasticam Introducta:* The Rise and Development of the Application of Mathematics in Fourteenth Century Philosophy and Theology," in *Arts libéraux et philosophie au Moyen Âge: Actes du quatrième Congrès international de philosophie médiévale, Université de Montréal, 27 août–2 septembre 1967* (Montreal: Institut d'études médiévales; Paris: J. Vrin, 1969), 215–54; and "From Social into Intellectual Factors," 271–339.

medieval doctrine known as "the intension and remission of forms or qualities" (or occasionally as "the configuration of qualities"); they were also to determine the manner in which upper and lower limits, or first and last instants, could be assigned to various processes and events, as in problems concerning free will, merit, and sin. In the fourteenth century Robert Holkot conceived a dilemma requiring *either* that limits be placed on free will *or* that we concede that God might not always be able to reward a meritorious person and punish one who was sinful.[35] Thus he imagined a situation in which a man is alternately meritorious and sinful during the final hour of his life: he is meritorious in the first proportional part of that last hour and sinful in the second proportional part; he is again meritorious in the third proportional part and again sinful in the fourth proportional part, and so on through the infinite series of decreasing proportional parts up to the instant of his death. Since the instant of death cannot form part of the infinite series of decreasing proportional parts of the man's last hour of life, it follows that there is no last instant of his life, and therefore no last instant in which he could be either meritorious or sinful. As a result, God does not know whether to reward or punish him in the afterlife, which was an unacceptable consequence of the doctrine of free will.[36] One could only conclude that free will cannot be assumed to extend to every imaginable sequence and pattern of choices, a point that Holkot buttressed with eight more continuum arguments.

The mathematical concepts already mentioned, and others as well, were applied to many other problems, especially those concerned with infinites. In this category were included speculations about God's infinite attributes (namely, His power, presence, and essence); the kinds of infinites He could possibly create; the infinite distances that separated Him from His creatures, a problem related to the widely discussed concept

35. The following illustration appears in Holkot's (or Holcot's) *Sentence Commentary,* bk. 1, question 3, the Latin text of which is quoted by Murdoch ("From Social into Intellectual Factors," 327 n. 102) from the edition of Lyon, 1518. For the interpretation of this difficult argument I am indebted to my student Mr. Peter Lang.

36. Ockham argued that if God wished, He could save a man who died without grace (see Sylla, "Autonomous and Handmaiden Science," 358). With respect to Holkot's argument, Ockham might have replied that God could save the man regardless of his state of grace at the final moment of life and despite God's ignorance of that state. The startling aspect of Holkot's argument, however, is that God could be in ignorance about a person's state of grace or sin at the last moment of life.

of the perfection of species;[37] the possible eternity of the world; whether God could improve upon something He had already made, especially whether He could make endlessly better and better successive worlds or whether He could create an ultimate, best possible world.[38] A host of problems was concerned with the behavior of angels, namely, how, if at all, an angel could occupy a place; whether it could be in two places simultaneously; whether two or more angels could occupy one and the same place simultaneously; whether angels moved between two separate places with finite or instantaneous speed. In all these problems about angels, basic concepts that had been developed in discussions of the motion of material bodies were applied directly or used as the standard of comparison. The motion of angels was one of the most popular contexts for the intense medieval debate about the nature of the continuum: whether it consisted of parts that are infinitely divisible or was composed of indivisible, mathematical atoms that could be either finite or infinite in number.[39] In contemplating the range of theological topics to which mathematics and mathematical concepts were applied, one may reasonably conclude that in the fourteenth century theology had been quantified.

Further examples of the quantification of theology could easily be supplied, since the process was ubiquitous in *Sentence Commentaries*. But just as the influence of theology, with its emphasis on God's absolute power, had encouraged, and even facilitated, the formulation in natural philosophy of hypothetical speculations about natural impossibilities, so also did the importation into theology of concepts, ideas, and techniques from mathematics and natural philosophy influence and encourage theology to express many of its problems in a scientific and logicomathematical format that was essentially hypothetical and speculative, or, as would be said in the Middle Ages, *secundum imaginationem*. Why theological arguments should have been expressed hypothetically in a logicomathematical format is by no means obvious. The hypothetical character of the arguments is probably attributable to the Condemnation of 1277

---

37. On the application of scales and measurements to the perfection of species, see Murdoch, *"Mathesis in Philosophiam Scholasticam Introducta,"* 238–39.

38. See Steven J. Dick, *Plurality of Worlds: The Origins of the Extraterrestrial Life Debate from Democritus to Kant* (Cambridge: Cambridge University Press, 1982), 31–35; and Armand Maurer, "Ockham on the Possibility of a Better World," *Mediaeval Studies* 38 (1976): 291–312.

39. Murdoch, *"Mathesis in Philosophiam Scholasticam Introducta,"* 217 n. 4.

and its long aftermath. Either because it was the safer course to pursue or perhaps because of the widespread conviction among theologians that God's nature and the motives for His actions were not directly knowable by human reason and experience, it became rather standard procedure to couch theological problems in hypothetical form. That the format of the problems was frequently quantitative and logicomathematical and involved measurements and comparisons between all sorts of spiritual and incorporeal entities is perhaps also explicable by the educational background of theological students and masters. With their overwhelming emphasis on natural philosophy and logic, and training in geometry, they may have found it quite natural to formulate, and even recast, their hypothetical theological problems in the quantitative languages that had formed their common educational background and that had been fashioned by natural philosophers in the first thirty or forty years of the fourteenth century.

Whatever the reasons for the hypothetical and quantitative format, it is no exaggeration to detect in all of this a major change in the techniques of theology, the like of which had never been seen before. Under the seductive influence of science, mathematics, and logic, theology found major expression in a quantified format within which solutions to a host of hypothetical theological problems were sought by various kinds of measurements, especially in problems that involved relationships between God and His creatures.[40] Traditional theological questions were often recast in a quantitative mold that allowed the easy application of mathematical and logical analysis. Yet this massive influx of quantitative apparatus appears to have had little, if any, impact on the content of theology. But if content was unaffected, the traditional methodology of theology had been transformed by the emphasis on natural philosophy and mathematics. It is this transformation that marks the fourteenth century and the late Middle Ages as an extraordinary period in the history of the relations between science and theology in the Western world.

The impact of science on theology was not all of this kind, however. The application of science to the interpretation of the creation account in Genesis was quite traditional and generally lacked the quantitative and hypothetical, imaginary character that dominated other aspects

40. Murdoch, "From Social into Intellectual Factors," 292.

of theology. Basic procedures for the application of science to the creation account had been laid down by St. Augustine in his *Commentary on Genesis*[41] and were faithfully summarized by St. Thomas Aquinas (ca. 1225–1274) in the latter's own commentary on the six days of creation in the *Summa theologiae*. In considering whether the firmament was made on the second day, Aquinas observes that Augustine had insisted upon two hermeneutical points in the explication of scriptural texts:

First, the truth of Scripture must be held inviolable. Secondly, when there are different ways of explaining a Scriptural text, no particular explanation should be held so rigidly that, if convincing arguments show it to be false, anyone dare to insist that it still is the definitive sense of the text. Otherwise unbelievers will scorn Sacred Scripture, and the way to faith will be closed to them.[42]

These two vital points constituted the basic medieval guidelines for the application of a continually changing body of scientific theory and observational data to the interpretation of physical phenomena described in the Bible, especially the creation account. The scriptural text must be assumed true. When God "made a firmament, and divided the waters that were under the firmament, from those that were above the firmament,"[43] one could not doubt that waters of some kind must be above the firmament.[44] The nature of that firmament and of the waters above it were, however, inevitably dependent on interpretations that were usually derived from contemporary science. It is here that Augustine and Aquinas cautioned against a rigid adherence to any one interpretation lest it be shown subsequently untenable and thus prove detrimental to the faith.

In conformity with his own admonitions Aquinas adopted no single interpretation of either "firmament" or the "waters" above the firmament. Instead he enumerated different historical interpretations that were compatible with Scripture and patiently explained how the application of

41. *De Genesi ad litteram* 1.18, 19, and 21. These passages are partly translated and partly summarized by Stanley L. Jaki, *Science and Creation: From Eternal Cycles to an Oscillating Universe* (New York: Science History Publications, 1974), 182–83.

42. St. Thomas Aquinas, *Summa theologiae: Latin Text and English Translation, Introductions, Notes, Appendices and Glossaries,* vol. 10, *Cosmogony* (la65–74), translated by William A. Wallace, O.P., part 1, question 68, 1 (the second day), pp. 71–73 (New York and London: Blackfriars in conjunction with McGraw-Hill Book Co. and Eyre & Spottiswoode, 1967).

43. Genesis 1:7 (Douay-Rheims translation).

44. An opinion expressed by St. Augustine, *De Genesi ad litteram*, 2.5, and quoted approvingly by Aquinas, *Summa theologiae* (trans. Wallace), part 1, question 68, 2, p. 79.

different scientific theories implied different and sometimes conflicting consequences. The firmament created on the second day was susceptible of two interpretations: it could be the sphere of the fixed stars or part of the atmosphere where clouds condense.[45] The first of these opinions could be interpreted in a variety of ways, each dependent on the material nature assigned to the firmament, that is, whether it was compounded of the four elements (Empedocles), or of a single element such as fire (Plato), or indeed consisted of a fifth element wholly different from the other four (Aristotle). For each of these possibilities Aquinas explained in what sense it was or was not compatible with the creation of a firmament on the second day.

Aquinas approached the meaning of the "waters" above the firmament in a similar manner. Each of a variety of possible significations was made to depend on the material nature attributed to the firmament.[46] Thus if the sphere of the fixed stars is the firmament and is composed of the four elements, the waters above the firmament could then plausibly be interpreted as the ordinary element water, but if the firmament is not compounded of the four regular elements, the waters above the firmament must be something other than the regular element water. In the latter event, "water" may be interpreted in the Augustinian manner as the unformed matter of which bodies are made. Its designation as aqueous may even derive from its transparent nature rather than its fluidity. After all, those waters may be solid like ice, that is crystalline, as in the "crystalline heaven of some authors."[47] Should the firmament be construed as that part of the atmosphere where clouds are formed, the waters above the firmament would be identical with those that are evaporated below and rise up to fall as rain. Because of the solidity of the celestial spheres these evaporated waters could not rise beyond the moon and *a fortiori* would never rise above the celestial region itself. Indeed, they could not even survive the heat of the fiery region immediately below the moon and would never reach the celestial spheres. With the presentation of these differing opinions Aquinas felt he had accomplished his objective. Because they were all compatible with the scriptural text, he saw no need—and indeed no way—to choose among them.

45. *Summa theologiae* (trans. Wallace), part 1, question 68, 1, pp. 73–75.
46. Aquinas, *Summa*, part 1, question 68, 2, pp. 79–83.
47. Ibid., p. 81.

Occasionally the literal meaning of scriptural statements conflicted directly with universally accepted scientific theories and observations. In such instances the scriptural text had to be reinterpreted, as in the case of Psalms 103:2, where God is said to have stretched out the firmament like a tent. Because of the near unanimous opinion that the earth is spherical, it was necessary that the firmament also be spherical, a condition that a tent could not fulfill. Under these circumstances Augustine and medieval Scholastics generally agreed that it was the biblical exegete's duty to demonstrate that the description of the firmament as a tent was not contrary to the scientific truth of a spherical firmament.[48] Augustine admonished against the development of a special Christian science that would attempt to explain the literal meaning of difficult texts that conflicted with well-founded scientific truths. Such attempts would undermine the credibility of Christianity. Augustine's attitude was thus compatible with both literal and allegorical interpretations of Scripture. The literal meaning of a text was always preferable, even where multiple interpretations were unavoidable, as with the supracelestial waters. But wherever a scriptural passage conflicted with a scientifically demonstrated proposition—as happened in Psalms 103:2—the scientific interpretation must prevail to prevent any erosion of confidence in scriptural truth. Under such circumstances, an allegorical interpretation was required so that truth and Scripture would be in harmony.

During the late Middle Ages broad and liberal, rather than narrow and literal, interpretations were the rule in biblical exegesis involving physical phenomena. An important illustration of this tendency is the famous passage that describes God's miraculous intervention on behalf of the army of Joshua (Joshua 10:12–14). By commanding the sun to stand still over Gibeon, God lengthened the day and allowed Israel to triumph over the Amorites. Since it was the sun—not the earth—that was ordered to come to rest, it followed that night and day were the consequence of the sun's daily revolution around an immobile earth rather than a result of the earth's daily rotation around its own axis. Here the Bible was in

---

48. Jaki, *Science and Creation*, 182–83, provides the references to Augustine's *Commentary on Genesis*. Presumably, William of Conches thought he was following Augustine's advice when he insisted upon an allegorical interpretation of "firmament" as air, rather than taking it literally as anything celestial, beyond or above which it was impossible for water of any kind to exist (see Lemay, "Science and Theology at Chartres," 229–31).

conformity with the best of Greek and medieval astronomy. Yet Nicole Oresme challenged this seemingly routine interpretation. "When God performs a miracle," he explained,

we must assume and maintain that He does so without altering the common course of nature, in so far as possible. Therefore, if we can save appearances by taking for granted that God lengthened the day in Joshua's time by stopping the movement of the earth or merely that of the region here below—which is so very small and like a mere dot compared to the heavens—and by maintaining that nothing in the whole universe—and especially the huge heavenly bodies—except this little point was put off its ordinary course and regular schedule, then this would be a much more reasonable assumption.[49]

Despite the plain statement of Scripture that the sun stopped in its course, Oresme argued that the same effect could be produced more economically and with less interruption of the common and regular course of nature by the assumption of a *real* daily axial motion for the much smaller earth. The sun's cessation of motion could thus be construed as only apparent and not real, an appearance produced when God caused the real axial rotation of the earth to cease. On the assumption that God always acted in the simplest and least disruptive manner, He surely would have stopped the smaller earth and not the sun, from which it followed that the apparent daily motion of the sun results from a real rotation of the earth. But Scripture plainly states that the sun, not the earth, stood still. Oresme's assumption would conflict not only with this clear biblical statement but with many others that also speak of the sun's motion or the earth's immobility.[50] Such passages, Oresme countered, may not reflect literal truth but merely conform "to the customary usage of popular speech just as it [that is, Holy Scripture] does in many other places, for instance, in those where it is written that God repented, and He became angry and became pacified, and other such expressions which are not to be taken literally."[51]

However, despite persuasive arguments in favor of the earth's axial

49. Nicole Oresme, *Le livre du del et du monde*, edited by Albert D. Menut and Alexander J. Denomy, C.S.B., translated by Albert D. Menut, 537 (Madison: University of Wisconsin Press, 1968), 537. The passage is reprinted in Grant, *Source Book*, 509.

50. E.g., Genesis 15:12; Ecclesiastes 1:5; 2 Samuel 2:24; Psalms 92:1; Ephesians 4:26; and James 1:11.

51. *Le livre du del et du monde*, 531; Grant, *Source Book*, 507.

rotation, Oresme knew that it was beyond his powers to demonstrate it scientifically. In the end, faithful to the admonitions of Augustine and Aquinas, he retained the literal meaning of the Bible and rejected the earth's rotation. Although he adopted the traditional opinion, Oresme's interpretation of the Joshua passage was more daring than Galileo's in 1615. As a confirmed Copernican, Galileo interpreted the Joshua text literally. With the sun at the center of the planetary system, Galileo assumed that it controlled the motions of all the planets. Hence

> when God willed that at Joshua's command the whole system of the world should rest and should remain for many hours in the same state, it sufficed to make the sun stand still. Upon its stopping, all the other revolutions ceased. . . . And in this manner, by the stopping of the sun, without altering or in the least disturbing the other aspects and mutual positions of the stars, the day could be lengthened on earth—which agrees exquisitely with the literal sense of the sacred text.[52]

Oresme's interpretation was radically different and far more striking because it was contrary to the literal meaning of the text, which, in this instance, agreed fully with Aristotelian cosmology and Ptolemaic astronomy. Since Oresme's consideration of the earth's diurnal rotation was in the end merely hypothetical, it caused no apparent theological consternation. Whether the same indifference would have prevailed if Oresme had concluded in favor of the reality of the earth's daily axial rotation is simply unanswerable, as is the question whether he might have suffered a fate similar to that which befell Galileo some two hundred and fifty years later.

We may reasonably conclude that the application of science to medieval scriptural exegesis was effected without noticeable constraints or interference. Indeed, the text of Holy Scripture was more often compelled to conform to the established truths of science than vice versa. The application of science to Scripture is perhaps best characterized by flexibility. Though the literal meaning was preferred, provision was made for allegorical interpretations. Potential conflict lurked, however, in passages where the literal meaning contradicted what were thought to be scien-

---

52. "Letter to Madame Christina of Lorraine, Grand Duchess of Tuscany, Concerning the Use of Biblical Quotations in Matters of Science," in *Discoveries and Opinions of Galileo*, translated by Stillman Drake, 213–14 (Garden City, N.Y.: Doubleday, 1957).

tifically demonstrated truths. While theologians found it easy to place an allegorical interpretation on the passage in Psalms 103:2—no one believed that the firmament was shaped like a tent—they would eventually prove unyielding, as Galileo would learn to his sorrow, on the many passages that mentioned the sun's motion and the earth's immobility. Galileo's insistence on an allegorical interpretation of those passages, on the grounds that he could scientifically demonstrate the earth's motion, clashed with the interpretation of the theologians who rejected his demonstrations and insisted on the traditional, literal sense. Ironically, to legitimate their positions both sides quite properly appealed to Augustine's conception of scriptural interpretation. During the Middle Ages no similar conflict erupted, not even on the always vexing problem of the eternity of the world. The medieval theologian-natural philosopher was generally free to propose and adopt a single interpretation—though encouraged not to embrace it unreservedly if it were not scientifically demonstrated—or to enunciate multiple interpretations without firm commitment to any one of them.

Did Theology Inhibit Scientific Inquiry?

We must finally confront an unavoidable question on the relations of medieval science and theology: How, if at all, did the latter affect the freedom of inquiry of the former? The attempts to ban and expurgate the physical works of Aristotle during the first half of the thirteenth century bear witness to theological fears about the potential power of uncontrolled philosophical learning. The Condemnation of 1277 marked the culmination of theological efforts to contain and control natural philosophy. The bishop of Paris and his theological colleagues sought to restrict, under penalty of excommunication, categorical claims for a number of ideas in natural philosophy. It was now forbidden, for example, to deny creation and assert the eternity of the world, to deny the possibility of other worlds, and to deny that God could create an accident without a subject in which to inhere. Although these restrictions fell equally on masters of arts and theologians at the University of Paris, the arts masters were more seriously affected than their theological colleagues. Not only were they obliged to comply with the Condemnation of 1277, but, in the absence of professional credentials in theology, they had been required, since 1272, to swear an oath that they would avoid disputation of purely theological questions and were generally discouraged from introducing theological matters into natural philosophy.[53]

Despite such restrictions, however, arts masters were free to uphold almost all of Aristotle's scientific conclusions and principles, provided that they conceded to God the power to create events and phenomena that were contrary to those conclusions and principles and which were therefore naturally impossible in the Aristotelian system. They were thus free to support Aristotle and deny the existence of other worlds if only they would allow that God could create them if He wished. Even the eternity of the world, which was to the relations between science and religion in the Middle Ages what the Copernican theory was to the sixteenth and seventeenth centuries and what the Darwinian theory of evolution was to the nineteenth and twentieth centuries, could be proclaimed hypothetically when "speaking naturally" *(loquendo naturaliter)*, that is, when considering a question in natural philosophy. Indeed, on the assumption that there was a fixed quantity of matter in the world and that the world was eternal, Albert of Saxony concluded in the fourteenth century that over an infinite time this limited quantity of matter would, of necessity, furnish the bodies for an infinite number of human forms. It followed that on the day of resurrection, when every soul receives its material body, the same finite quantity of matter would be received by an infinite number of human souls, a clearly heretical consequence, since one and the same body would have to receive a plurality of souls. To this dilemma Albert's response was typical for natural philosophers who regularly contended with theological restrictions: "The natural philosopher is not much concerned with this argument because when he assumes the eternity of the world, he denies the resurrection of the dead."[54] By such appeals to the hypothetical, medieval natural philosophers could con-

---

53. For the statute, see Grant, *Source Book,* 44–45. On John Buridan's complaint against theological restrictions in his discussion of the vacuum, see 50–51.

54. For the text and discussion based on Albert's *Questions on Generation and Corruption,* see Anneliese Maier, *Metaphysische Hintergründe der spätscholastischen Naturphilosophie* (Rome: Edizioni di Storia e Letteratura, 1955), 39–40. Maier also notes (41) that, in coping with the same question, Marsilius of Inghen declared that, in truth, the world had a beginning and will come to an end. Whether, on the assumption of the eternity of the world, an infinity of souls would receive the same matter is a theological question and of no concern in a work on natural philosophy. Although he sought to avoid the question, Marsilius did allow that God could, if He wished, assign one matter to many men.

sider almost any condemned and controversial proposition. Nevertheless, they were not permitted to proclaim such beliefs categorically, and to the extent that their discussions touched theology or had theological implications, they were inhibited and frustrated, as when John Buridan complained that in his analysis of the vacuum, which touched upon faith and theology, he was reproached by the theological masters for intermingling theological matters.[55]

With the arts masters forbidden to apply their knowledge to theology, we are left with the theologians as the class of scholars who applied science to theology and theology to science during the Middle Ages. Not only were they thoroughly trained in natural philosophy and theology, but some were also significant contributors to science and mathematics, as the names of Albertus Magnus, John Pecham, Theodoric of Freiberg, Thomas Bradwardine, Nicole Oresme, and Henry of Langenstein testify. Because they were trained in both natural philosophy and theology, medieval theologians were able to interrelate science and theology with relative ease and confidence, whether this involved the application of science to scriptural exegesis, the application of God's absolute power to alternative possibilities in the natural world, or even the frequent invocation of scriptural texts in scientific treatises in support of scientific theories and ideas. Theologians had a remarkable degree of intellectual freedom[56] and, for the most part, did not allow their theology to hinder or obstruct inquiry into the structure and operation of the physical world. If there was any real temptation to produce a "Christian science," they successfully resisted it. Biblical texts were not employed to "demonstrate" scientific truths by blind appeal to divine authority. When Nicole Oresme inserted some fifty citations to twenty-three different books of the Bible in his scientific treatise *On the Configurations of Qualities and Motions,* he did so only by way of example or for additional support, but in no sense to demonstrate an argument.[57]

55. For the text of Buridan's complaint, see Grant, *Source Book,* 50–51.

56. For an elaborate defense of this claim, see Mary Martin McLaughlin, *Intellectual Freedom and Its Limitations in the University of Paris in the Thirteenth and Fourteenth Centuries* (New York: Arno Press, 1977; Ph.D. diss., Columbia University, 1952), chap. 4 ("The Freedom of the Theologian as Scholar and Teacher"), 170–237, and chap. 5 ("Intellectual Freedom and the Role of the Theologian in the Church and in Society"), 238.

57. See Marshall Clagett, ed. and trans., *Nicole Oresme and the Medieval Geometry of Qualities and Motions: A Treatise on the Uniformity and Difformity of Intensities Known as "Trac-*

Ironically, rather than inhibiting scientific discussion, theologians may have inadvertently produced the opposite effect, as suggested by the impact of the doctrine of God's absolute power described above. Theological restrictions embodied in the Condemnation of 1277 may have actually prompted consideration of plausible and implausible alternatives and possibilities far beyond what Aristotelian natural philosophers might otherwise have considered, if left to their own devices. While these speculations did not lead to the abandonment of the Aristotelian worldview, they generated some of the most daring and exciting scientific discussions of the Middle Ages.

That medieval theologians combined extensive and intensive training in both natural philosophy and theology, and possessed exclusive rights to interrelate the two, may provide a key to explain the absence of a science-theology conflict in the extensive medieval commentary literature on the *Sentences* and Scripture. For the host of issues they regularly confronted, the medieval theologian-natural philosophers knew how to subordinate the one discipline to the other and to avoid conflict and confrontation. Indeed, they were in an excellent position to harmonize the two disciplines while simultaneously pursuing all manner of hypothetical and contrary-to-fact conditions and possibilities. Compared to the situation in late antiquity, when Christianity was struggling for survival, and the difficult times that lay ahead, the late Middle Ages—except for the 1260s and 1270s—was a relatively tranquil period in the long interrelationship between science and theology.

---

tatus de configurationibus qualitatum et motuum" (Madison: University of Wisconsin Press, 1968), 134–35.

10 ∞ The Fate of Ancient Greek Natural Philosophy in the Middle Ages

Islam and Western Christianity

The enduring impact of ancient Greek science and natural philosophy on the civilizations of Islam and Latin Christianity is one of the great success stories in the history of the world. The successful transmission of Greek science into Arabic and then of Greek and Arabic science into Latin compels us to speak of "Greco-Islamic-Latin" science in the Middle Ages. It was Greco-Islamic-Latin science and natural philosophy that unquestionably set the stage for the Scientific Revolution of the seventeenth century, which would otherwise have been impossible. The transmittal of science and natural philosophy from Greek to Arabic and from Greek and Arabic to Latin was largely a one-way process, a one-way belt of transmission. There was little, if any, backward movement—that is, there were no meaningful translations from Arabic to Greek and from Latin to Arabic and Greek—and therefore no significant interactions between Western Christianity and Islam.

But if there were no mutual interactions in science and natural philosophy between Latin Christianity and Islam, the two religions on which I shall focus, there were important contrasts in the way each religious tradition responded to, and utilized, the scientific heritage it received. Perhaps the differences in their long-term responses to secular pagan philosophical and scientific learning were shaped to a lesser or greater extent by the culture and civilization in which each was born and the manner in which each came into being.

Major Differences That Transcend Science and Natural Philosophy

Christianity was born inside the Roman Empire and was spread slowly and quietly, but persistently. By comparison with Islam, Christianity was

253

disseminated at a snail's pace. Not until three hundred years after the birth of Christ was Christianity effectively represented throughout the Roman Empire. Only in 313, by the Edict of Milan, or Edict of Toleration, was Christianity given full equality with other religions in the empire. And it was not until 392—almost four centuries after the birth of Christ—that Christianity became the state religion, when Emperor Theodosius ordered the closing of pagan temples and forbade pagan worship.

In striking contrast, Islam was spread over an enormous geographical area in a remarkably short time. In less than one hundred years after the death of Muhammad in 632, Islam became the dominant religion in a vast area stretching from the Straits of Gibraltar in the West to India in the East. Such a rapid spread could only have occurred by conquest. Where Christianity spread slowly, by proselytizing, Islam came from outside the Roman world as an alien intruder, and although its converts were pagans and often former Christians, the mind-set of the invaders was one that viewed Greek learning as alien, as is illustrated by the fact that Muslims distinguished two kinds of sciences: the Islamic sciences, based on the Koran and Islamic law and traditions, and the foreign sciences, or "pre-Islamic" sciences, which encompassed Greek science and natural philosophy. We might say that the slow spread of Christianity provided Christians an opportunity to adjust to Greek secular learning, whereas Islam's rapid dissemination made its relations with Greek learning much more problematic.

Another dramatic difference concerns the relationship between church and state. From the outset, Christianity recognized the state as distinct from the church. The separation is encapsulated in these momentous words of Jesus: "Render therefore unto Caesar the things which are Caesar's; and unto God the things that are God's" (Matthew 22.21). Thus did Jesus acknowledge the state and implicitly urge his followers to be good citizens. Although church and state were contending powers throughout the Middle Ages, each acknowledged the independence of the other. They regarded themselves as two swords—although, all too often, they were pointed at each other. Even when the church asserted supremacy over the state, however, it never attempted to establish a theocracy by appointing bishops and priests who were also to function as secular rulers. The tradition of the Roman state within which Christianity developed and the absence of explicit biblical support for a theocratic state were powerful constraints on unbridled and grandiose papal ambitions and, above all, made the imposition of a theocratic state implausible.

In Islam church and state are one. Religion cannot be understood apart from politics, and vice versa.[1] "The function of the state was to guarantee the well-being of the Muslim religion, so that all who lived within the state could be good, practicing Muslims."[2] Where religion is strong, as it was in medieval Islam, it is likely to dominate secular activities, such as natural philosophy. To avoid this consequence, at least one of the following conditions would be essential: (1) regard natural philosophy as a discipline that is distinct and independent from theology; or (2) a secular state protects natural philosophy; or (3) religious authorities regard natural philosophy favorably. While we shall see that the first and third conditions were met in the Latin West, none of the three conditions was met in medieval Islam.

A third significant difference between Islam and the medieval Christian West is organizational and structural. Islam has no overriding, central authority to determine its orthodoxy, whereas the Latin West had the papacy to ensure adherence to the faith and to combat heresy. In brief, Islam is a kind of democratic religion that relies on consensus, whereas medieval Christendom was a centralized religion, headed by a single individual, the pope, who, in principle, had supreme authority to determine and shape religious opinion and belief. From such a major structural difference, one might suppose that papal-dominated, centralized Christianity would have been far more restrictive and oppressive toward secular Greek learning than consensus-seeking Islam. The record shows, however, that the Catholic Church was favorably disposed toward secular learning, especially Aristotelian natural philosophy. In Islam, however, Aristotelian natural philosophy and the philosophers who studied it were often treated with hostility. As we shall see, it was in Islam, rather than in the Latin West, that secular learning, philosophy, and, the "foreign sciences" in general were subject to significant constraints and confronted considerable obstacles and prejudice.

Although there are *hadiths,* or traditions, in Islam that praise the

1. See Reynold A. Nicholson, *A Literary History of the Arabs* (Cambridge: Cambridge University Press, 1953), 182.

2. Edward Grant, *The Foundations of Modern Science in the Middle Ages* (Cambridge: Cambridge University Press, 1996), 183.

unending quest for knowledge, "implying the endlessness of knowledge iself,"[3] there are others that see a world that is steadily deteriorating. At least two hadiths express this attitude. In the first, the Prophet proclaims that "time has come full circle back to where it was on the day when first the heavens and the earth were created," and in the second he declares that "[t]he best generation is my generation, then the ones who follow and then those who follow them."[4] Both hadiths were often cited and commentaries were made upon them. "They suggest a universe running down, an imminent end to man and all his works."[5] Such hadiths may have served as powerful elements in Islamic thought and custom. They may have encouraged those who sought to preserve the status quo, or who wished to turn the clock back as far as possible, to create a society as close as possible to that which existed in the days of the Prophet Muhammad. It is plausible to suppose that such hadiths exerted an influence on attitudes toward Greek science and natural philosophy, the foreign sciences, which did not exist within Islam when it began.

Indeed, the influence of such hadiths may have affected the way Islam responded to the invention of the printing press. Although it takes us beyond the Middle Ages, it is not irrelevant to mention the time and manner in which printing was introduced into Islam. Although the printing press had been in use in the West since around 1460, and its virtues were obvious, it was not introduced into Islam until 1727,[6] when Ibrahim Müteferrika, described as "a renegade from the Hungarian nobility," established the first press. Even when introduced, the forces that had always opposed it exacted a price: only secular works could be printed, not sacred texts, including the Koran. In *The Ottoman Centuries,* Lord Kinross declares that

With the aid of a committee of twenty-five translators, he [Müteferrika] published a flow of works revealing to his adopted compatriots the mysteries of

---

3. See Tarif Khalidi, "The Idea of Progress," *Journal of Near Eastern Studies* 40 (October 1981): 280.

4. Khalidi, "The Idea of Progress," 279. In note 6, Khalidi says: "See Wensinck, *Concordance et Indices de la Tradition Musulmane* (Leiden/New York: E. J. Brill, 1936–1988), s.v. 'Zaman,' 'Umma.'"

5. A. J. Wensinck, *A Handbook of Early Muhammadan Tradition, Alphabetically Arranged* (Leiden: E. J. Brill, 1960), s.v. 'Hour,' where *hadiths* about knowledge disappearing in the last days are cited. Cited in Khalidi, "The Idea of Progress," 279.

6. See Lord Kinross, *The Ottoman Centuries: The Rise and Fall of the Turkish Empire* (New York: Morrow Quill Paperbacks, 1977), 381.

such objects of study as geography and cartography, in which he himself specialized; physics and astronomy, including a translation of Aristotle with information for the first time on the telescope and microscope, on magnetism and the compass, on the theories of Galileo; on mathematics in its various branches, with the discussion of the ideas of Descartes; and finally on medicine.[7]

As an indication that the Ottoman government still did not realize the power of the printing press, Müteferrika's death in 1745 resulted in the cessation of printing until 1783, a hiatus of nearly forty years.[8]

There are undoubtedly other significant differences between Islam and Christianity, but we must now narrow our focus. Although Greek science and natural philosophy may have been regarded as foreign sciences in Islam, most of Greek science and natural philosophy were translated into Arabic and studied over the centuries. Significant contributions were made in science and natural philosophy by scholars in the Islamic world. Indeed, from around 1100 to 1500, sciences such as optics, astronomy, mechanics, mathematics, and medicine reached a higher state in Islam than in the medieval West. In what follows, however, I shall ignore the exact sciences, which posed no significant doctrinal problems for Islam or Christianity, and focus rather on natural philosophy, almost exclusively Aristotle's natural philosophy, which did indeed pose major problems for Islam and Christianity.

During the Middle Ages, Aristotelian natural philosophy was inherently more important than any single identifiable exact science. In the broadest sense, natural philosophy was the study of change and motion in the physical world. It was one of Aristotle's three subdivisions of theoretical knowledge, or knowledge for its own sake. As the very name suggests, the domain of natural philosophy was the whole of nature. It did not represent any single science, but could, and did, embrace bits and pieces of all sciences. In this sense, natural philosophy was "The Mother of All Sciences." But medieval natural philosophy was far more significant than is indicated by the mere fact that embedded within it were bits and pieces of different modern sciences. In a culture such as that of the Middle Ages, in which the tools for scientific research and inquiry were largely absent, how could nature be interpreted and analyzed in order to arrive at some understanding of a world that would otherwise be un-

7. Ibid., 382.   8. Ibid.

knowable and inexplicable? The most powerful weapon available was hu-man reason, employed in the manner that Aristotle had used it. The idea was to come to know what things seemed to be—and this could be done by empirical means—and then to determine what made them that way, a process that was largely guided by metaphysical and a priori consider-ations. In the ancient and medieval worlds, Aristotle's works represented the apotheosis of reason. Without reason, science cannot exist. It is the first indispensable element in the development of science and it was the characteristic feature of medieval natural philosophy. For these reasons, a comparison of the status of natural philosophy in medieval Islam and in the Latin West should tell us much about the potentiality for science within each civilization, and therefore provide some insight into a peren-nially perplexing question: Why did Islam, which reached a higher state of scientific development in the Middle Ages than the Latin West, fail to continue its development, while the West, which started much later, sur-passed Islam by 1600. One major result of any comparison between the relations of these two religions to Aristotelian natural philosophy will reveal that, in contrast with the West, Aristotelian natural philosophy in Islam had an uneasy and uneven existence. Let us see why.

Islam

Throughout the history of medieval Islam, the role of Greek philosophy was problematic. At any particular time, there were those who viewed it favorably, while others, undoubtedly a considerable majority, viewed it, at best, with indifference, and perhaps even with some degree of hostil-ity. Occasionally the attitude of this or that caliph was instrumental in altering attitudes toward natural philosophy, but more often attitudes to-ward natural philosophy and Greek thought were governed by Muslim religious leaders, who exercised great influence in particular regions or cities. Not only was Greek philosophy regarded as a foreign science, but the term *philosopher (faylasufs)* was often employed pejoratively.

In the intellectual hierarchy of medieval Islamic society, scholars dis-tinguish three levels.[9] Because Islam was a nomocracy, the first level was

9. Toby Huff, *The Rise of Early Modern Science* (Cambridge: Cambridge University Press, 1993), 69.

comprised of legal scholars. The religious law and traditions were valued above all else, and therefore valued even more than theology. Next in order came the mutakallimun, scholars who used Greek philosophy to interpret and defend the Muslim religion. The mutakallimun emphasized rational discourse, to which they added the authority of revelation. And, finally, at the bottom, were the *falasifa,* the Islamic philosophers, who followed rational Greek thought, especially the thought of Aristotle. Not surprisingly, the philosophers placed greatest reliance on reasoned argument while downplaying revelation. The philosophers sought to develop natural philosophy in an Islamic environment, and, as A. I. Sabra has put it, did so, "often in the face of suspicion and opposition from certain quarters in Islamic society."[10]

Of the three Islamic groups just distinguished, namely, legal scholars, who were almost always traditionalists, the mutakallimun, and philosophers, the traditionalists made no real use of Greek philosophy, largely because they found it a threat to revealed truth and the Islamic faith. In their bitter struggle with each other and with the traditionalists, the mutakallimun and the philosophers made much use of Greek philosophy. The mutakallimun were primarily concerned with the Kalam, which, according to A. I. Sabra, is "an inquiry into God, and into the world as God's creation, and into man as the special creature placed by God in the world under obligation to his creator."[11] Thus Kalam is a theology that used Greek philosophical ideas to explicate and defend the Islamic faith.

Two groups of mutakallimun have been identified: the Mu'tazilites, who were the more extreme, and the Ash'arites.[12] Both groups shared an attitude "against the passive acceptance of authority in matters of faith." It was their intention to replace the "passive acceptance of authority" with "a state of knowledge [*'ilm*] rooted in reason."[13] The Mu'tazilites were regarded as Islamic rationalists who equated the power of reason with that of revelation.[14] They are said to have "made an outstanding contribution to Islamic thought by the assimilation of a large number of Greek

10. A. I. Sabra, "Science and Philosophy in Medieval Islamic Theology," in *Zeitschrift für Geschichte der Arabisch-Islamischen Wissenschaften* 9 (1994): 3.

11. Ibid., 5.

12. See Arthur Hyman and James J. Walsh, eds., *Philosophy in the Middle Ages: The Christian, Islamic, and Jewish Traditions* (Indianapolis: Hackett Publishing Co., 1973), 205.

13. Sabra, "Science and Philosophy in Medieval Islamic Theology," 9.

14. Huff, *The Rise of Early Modern Science,* 111.

ideas and methods of argument."[15] These arguments and methods were not adopted for their own sake but rather for their utility in understanding the Islamic religion. In the ninth century, the Mu'tazilites gained the support of caliphs like al-Mamun and Mutassim, as well as influential intellectuals. The supportive caliphs persecuted those who opposed the Mu'tazilite belief that the Koran was created. They implemented a virtual inquisition. Because many thought their rationalism was extreme, the Mu'tazilites were regarded as heretics by many Sunni Muslims.[16] Their ascendancy ended with the rule of the Sunni caliph al-Mutawakkil, who destroyed their movement.[17]

The Asharites, who followed the teaching of al-Ash'ari (d. 935), are the second group of mutkallimun. They broke with Mu'tazilism and replaced it as the main representatives of Kalam. Ash'arism, however, was a complicated movement, with some of its followers emphasizing rationalism, while others argued in the traditionalist mode.[18] Although the mutakallimun, both Mutazilites and Asharites, were severe critics of the philosophers, they were, in turn, themselves regarded as too rational and were bitterly opposed by more conservative Muslims, both from the Sunni and the Shiite sides.

In treating of attitudes toward natural philosophy and science in medieval Islam, it is essential to have a good sense of the relationships between Muslim traditionalism and Muslim rationalism, which were engaged in an ongoing, and bitter, struggle about the role of Islam in intellectual life. George Makdisi provides a useful way to distinguish between Muslim traditionalism and Muslim rationalism by explaining that

[t]he traditionalists made use of reason in order to understand what they considered as the legitimate sources of theology: scripture and tradition. What they could not understand they left as it stood in the sources; they did not make use

15. W. Montgomery Watt, *Islamic Philosophy and Theology: An Extended Survey* (Edinburgh: Edinburgh University Press, 1985), 54.

16. Ibid., 55.

17. See Pervez Hoodbhoy, *Islam and Science: Religious Orthodoxy and the Battle for Rationality* (London: Zed Books, 1991), 99–100.

18. For a good account, see George Makdisi, "Ash'ari and the Ash'arites in Islamic Religious History," Parts 1 and 2, in *Studia Islamica* 17 (1962): 37–80 and 18 (1963): 19–39; reprinted in George Makdisi, *Religion, Law and Learning in Classical Islam*, Variorum Collected Studies Series (Hampshire, Great Britain, and Brookfield, Vermont: Gower Publishing Co., 1991), 1.

of reason to interpret the sources metaphorically. On the other hand, the rationalists advocated the use of reason on scripture and tradition; and all that they deemed to contradict the dictates of reason they interpreted metaphorically in order to bring it into harmony with reason.[19]

The antithetical approaches of the Muslim traditionalists and the Muslim rationalists can be illustrated directly from the mutakallimun themselves, namely, from the Mutazilites and Asharites. What was one to make of anthropomorphic statements in the Koran that speak of "the face of Allah, His eyes and hands, his sitting on His throne, and His being seen by the Faithful in Paradise."[20] The strong tendency in Islam was to take such statements literally. Thus al-Ash'ari himself, for whom reason in theology was still important, declared that:

We confess that God is firmly seated on His throne . . . We confess that God has two hands, without asking how . . . We confess that God has two eyes, without asking how. . . . We confess that God has a face. . . .[21]

Mutazilites, however, viewed these same statements metaphorically. God has no bodily parts; He has no parts or divisions; He is not finite. They also say that "He cannot be described by any description which can be applied to creatures, in so far as they are created. . . . The senses do not reach Him, nor can man describe Him by analogy. . . . Eyes do not see Him, sight does not reach Him, phantasy cannot conceive Him nor can He be heard by ears."[22] I am unaware of any analogous discussion in the Christian West during the Middle Ages. Medieval Latin theologians regarded anthropomorphic descriptions of God as metaphorical pronouncements.

The Philosophers

Of the three groups distinguished earlier, the least popular were the philosophers, whom the mutakallimun and conservative Muslims attacked because they used natural philosophy and logic to acquire truth for its

19. Makdisi, "Ash'ari and the Ash'arites in Islamic Religious History," Part 2, in *Studia Islamica* 18 (1963): 22; reprinted in Makdisi, *Religion, Law and Learning in Classical Islam*, 1, 22.

20. See A. J. Arberry, *Revelation and Reason in Islam: The Forwood Lectures for 1956 Delivered in the University of Liverpool* (London: George Allen & Unwin; New York: Macmillan, 1957), 22.

21. Ibid.   22. Ibid., 23.

own sake, which usually signified that they were ignoring religion. One of the most significant Ash'arite thinkers, the famous al-Ghazali (1058–1111), leveled a devastating attack against philosophy. He was fearful of the detrimental effects on the Islamic religion of subjects like natural philosophy, theology (actually metaphysics), logic, and mathematics. In his famous quasi-autobiographical treatise, *Deliverance from Error,* he explained that religion does not require the rejection of natural philosophy, but that there are serious objections to it because nature is completely subject to God, and no part of it can act from its own essence. The implication is obvious: Aristotelian natural philosophy is unacceptable because it assumes that natural objects can act by virtue of their own essences and natures. That is, Aristotle believed in secondary causation—that physical objects are capable of causing effects in other physical objects. Al-Ghazali found mathematics dangerous because it uses clear demonstrations, thus leading the innocent to think that all the philosophical sciences are equally lucid. A man will say to himself, al-Ghazali related, that "if religion were true, it would not have escaped the notice of these men [that is, the mathematicians] since they are so precise in this science."[23] Ghazali explains further that such a man will be so impressed with what he hears about the techniques and demonstrations of the mathematicians that "he draws the conclusion that the truth is the denial and rejection of religion. How many have I seen," al-Ghazali continues, "who err from the truth because of this high opinion of the philosophers and without any other basis."[24] Although al-Ghazali allowed that the subject matter of mathematics is not directly relevant to religion, he included the mathematical sciences within the class of philosophical sciences (these are mathematics, logic, natural science, theology or metaphysics, politics, and ethics) and concluded that a student who studied these sciences would be "infected with the evil and corruption of the philosophers. Few there are who devote themselves to this study without being stripped of religion and having the bridle of godly fear removed from their heads."[25]

In his great philosophical work, *The Incoherence of the Philosophers,*

---

23. Translated in W. Montgomery Watt, *The Faith and Practice of al-Ghazali* (London: George Allen & Unwin, 1953), 33.
24. Ibid.
25. Ibid., 34.

al-Ghazali attacked ancient philosophy, especially the views of Aristotle. He did this by describing and criticizing the ideas of al-Farabi and Avicenna, two of the most important Islamic philosophical commentators on Aristotle. After criticizing their opinions on twenty philosophical problems, including the eternality of the world; that God knows only universals and not particulars; and that bodies will not be resurrected after death, al-Ghazali declares:

All these three theories are in violent opposition to Islam. To believe in them is to accuse the prophets of falsehood, and to consider their teachings as a hypocritical misrepresentation designed to appeal to the masses. And this is blatant blasphemy to which no Muslim sect would subscribe.[26]

Al-Ghazali regarded theology and natural philosophy as dangerous to the faith. He had an abiding distrust of philosophers and praised the "unsophisticated masses of men," who "have an instinctive aversion to following the example of misguided genius." Indeed, "their simplicity is nearer to salvation than sterile genius can be."[27] As one of the greatest and most respected thinkers in the history of Islam, al-Ghazali's opinions were not taken lightly.

In light of al-Ghazali's attack on the philosophers, it is not surprising to learn that philosophers were often subject to persecution by religious leaders. Many religious scholars regarded philosophy, logic, and the foreign Greek sciences generally as useless, and even ungodly, because they were not directly useful to religion. Indeed, they might even make one disrespectful of religion.[28] In the thirteenth century, Ibn as-Salah ash-Shahrazuri (d. 1245), a religious leader in the field of tradition (hadith), declared in a *fatwa* that "[h]e who studies or teaches philosophy will be abandoned by God's favor, and Satan will overpower him. What field of learning could be more despicable than one that blinds those who cultivate it and darkens their hearts against the prophetic teaching of Muhammad...."[29] Logic was also targeted, because, as Ibn as-Salah, put it,

26. From *Al-Ghazali's Tahafut al-Falasifah* [*Incoherence of the Philosophers*], Pakistan Philosophical Congress Publication 3, translated into English by Sabih Ahmad Kamali (Lahore, Pakistan: Pakistan Philosophical Congress, 1963), 249.

27. Ibid., 3.

28. Huff, *The Rise of Early Modern Science,* 68.

29. Ignaz Goldziher, "The Attitude of Orthodox Islam toward the 'Ancient Sciences,'" in *Studies on Islam,* edited and translated by Merlin L. Swartz, 205 (New York/Oxford: Oxford University Press, 1981).

"it is a means of access to philosophy. Now the means of access to some-thing bad is also bad."[30] Ibn as-Salah was not content to confine his hos-tility to words alone. In a rather chilling passage, he urges vigorous ac-tion against students and teachers of philosophy and logic, because

[t]hose who think they can occupy themselves with philosophy and logic mere-ly out of personal interest or through belief in its usefulness are betrayed and duped by Satan. It is the duty of the civil authorities to protect Muslims against the evil that such people can cause. Persons of this sort must be removed from the schools and punished for their cultivation of these fields. All those who give evidence of pursuing the teachings of philosophy must be confronted with the following alternatives: either (execution) by the sword or (conversion to) Islam, so that the land may be protected and the traces of those people and their sci-ences may be eradicated. May God support and expedite it. However, the most important concern at the moment is to identify all of those who pursue philoso-phy, those who have written about it, have taught it, and to remove them from their positions insofar as they are employed as teachers in schools.[31]

Although numerous others shared the attitude of Ibn as-Salah, logic continued to be used as an ancillary subject in scholastic theology (Ka-lam) and in many orthodox religious schools. But there was enough hostil-ity toward philosophy and logic in Islam to prompt philosophers to keep a low profile. Those who taught did so privately to students who might have sought them out. Following the translations in the early centuries of Islam, Greek philosophy, primarily Aristotle's, received its strongest sup-port from a number of individuals scattered about the Islamic world. Numbered among the greatest of Islamic natural philosophers are al-Kindi (801–873), Al-Razi (ca. 854–925 or 935), Ibn Sina (Avicenna) (980–1037), and Ibn Rushd (Averroes) (1126–1198). All were persecuted to some extent.

Al-Kindi's case reveals important aspects of intellectual life in Islam. The first of the Islamic commentators on Aristotle, al-Kindi was at first favorably received by two caliphs (al-Mamun and al-Mutassim), but his luck ran out with al-Mutawakkil, the Sunni caliph mentioned earlier. According to Pervez Hoodbhoy, "It was not hard for the ulema to con-vince the ruler that the philosopher had very dangerous beliefs. Mutaw-

30. Ibid.
31. Ibid., 206.

wakil soon ordered the confiscation of the scholar's personal library. . . . But that was not enough. The sixty year old Muslim philosopher also received fifty lashes before a large crowd which had assembled. Observers who recorded the event say the crowd roared approval with each stroke."[32] The other four scholars were also subjected to some degree of persecution and a number of them had to flee for their safety.

Persecutions and harassment of those who advocated the use of reason to explicate revelation are unknown in the medieval Latin West after the mid-twelfth century, when Bernard of Clairvaux and other traditional theologians opposed the application of reason to theology. Bernard undoubtedly had much in common with Islamic traditionalist theologians. In his relentless assault on Peter Abelard, Bernard was convinced that Abelard's heresies, as he saw them, were the result of an excessive reliance on reason, as he makes clear in a letter to a cardinal of the church. "He has defiled the Church," Bernard declares,

he has infected with his own blight the minds of simple people. He tries to explore with his reason what the devout mind grasps at once with a vigorous faith. Faith believes, it does not dispute. But this man, apparently holding God suspect, will not believe anything until he has first examined it with his reason.[33]

Bernard's hostile attitude lingered on into the first forty years of the thirteenth century, but only at the University of Paris (though not at Oxford), where church authorities first banned the books of Aristotle from public or private use, then sought unsuccessfully to censor them. By the 1240s, however, Aristotle's books of natural philosophy were taught and read at the University of Paris. Indeed, they had become the core of the curriculum in the arts faculty of that great medieval university.[34] After the 1240s, and for the rest of the Middle Ages, attacks on reason would have been regarded as bizarre and unacceptable. Some theologians were opposed to certain of Aristotle's ideas, but, like St. Bonaventure, they used Aristotelian natural philosophy, and fully recognized that they could not do theology without it. Scholars were sometimes accused of heresy, and occasionally the church tried to curb the excessive use of logic and natu-

32. See Pervez Hoodbhoy, *Islam and Science*, 111.

33. *The Life and Letters of St. Bernard of Clairvaux*, translated by Bruno Scott James, letter 249, p. 328 (London: Burns Oates, 1953).

34. For a brief account of the reaction to Aristotle's works at the University of Paris, see Grant, *The Foundations of Modern Science in the Middle Ages*, 70–80.

ral philosophy in theological treatises, but I know of no instance where religious authorities sought to prevent the study of natural philosophy because it threatened religion. Indeed, as time passed, Aristotelian natural philosophy only became more entrenched in the medieval universities. By the time of the Galileo affair in the seventeenth century, the Catholic Church went to great lengths to defend and protect Aristotle's natural philosophy.

How different it was in Islam, if we judge by a question that Averroes posed in the twelfth century in his treatise *On the Harmony of Religion and Philosophy*. In this treatise, Averroes seeks to determine "whether the study of philosophy and logic is allowed by the [Islamic] Law, or prohibited, or commanded—either by way of recommendation or as obligatory."[35] In the thirteenth century, Ibn as-Salah ash-Shahrazuri, an expert on the tradition of Islam and whom we have already met, issued a written reply (*fatwa*) to a question which asked, in Ignaz Goldziher's words,

whether, from the point of view of religious law, it was permissible to study or teach philosophy and logic and further, whether it was permissible to employ the terminology of logic in the elaboration of religious law, and whether political authorities ought to move against a public teacher who used his position to discourse on philosophy and write about it.[36]

What is remarkable in all this is the fact that in the twelfth century, Averroes, and, in the thirteenth century, Ibn as-Salah, were grappling with the question whether, from the standpoint of the religious law, it was legitimate to study science, logic, and natural philosophy, even though these disciplines had been readily available in Islam since the ninth century. Averroes felt compelled to justify their study, while Ibn as-Salah, astonishingly, denied their legitimacy. I know of no analogous discussions in the late Latin Middle Ages in which any natural philosopher or theologian felt compelled to determine whether the Bible permitted the study of secular subjects. It was simply assumed that it did.

---

35. Averroes, *On the Harmony of Religion and Philosophy: A Translation with Introduction and Notes, of Ibn Rushd's Kitab fasl al-maqal, with Its Appendix (Damima) and an Extract from Kitab al-kashf 'an manahij al-adilla*, translated by George F. Hourani, 44 (London: Luzac, 1976).

36. Goldziher, "The Attitude of Orthodox Islam toward the 'Ancient Sciences,'" 205.

The Madrasas and the Universities

Despite the enormous obstacles faced by Islamic natural philosophers and scientists, either in the form of active harassment and even persecution, or simply as indifference, the remarkable feature about medieval Islamic science and natural philosophy was the high level to which they attained. As I have already mentioned, the level of achievement in the sciences between 1100 and 1500 was higher in Islam than in the Christian West. It is more difficult to compare natural philosophy, partly because the Latin West derived some of its ideas from Islamic treatises. In the exploration of Aristotle's works and in the departures they made from Aristotle's thought the West may have advanced beyond Muslim scholars. But in other ways, certain Muslim scholars went beyond anything contemplated in the West. This is especially true in the attitudes of some Islamic natural philosophers toward theologians and religion.

For example, al-Razi (ca. 854–925 or 935), known as Rhazes in the West, was a famous physician whose major medical work was translated into Latin, actually attacked religion, denying miracles attributed to the prophets of Islam, Judaism, and Christianity (he is said to have written a treatise titled *The Tricks of the Prophets*). He refused to accept authority in either religion or science and believed that the sciences continually progressed because scientists build upon the knowledge they inherit from their predecessors. He accepted an atomic theory of matter akin to that of Democritus. Because of his independent views, al-Razi was severely criticized by his successors and many of his works have disappeared.[37] Avicenna thought, in the words of Shlomo Pines, that al-Razi "should have confined himself to dealing with boils, urine, and excrement and should not have dabbled in matters beyond the range of his capacity."[38]

Averroes (Ibn Rushd) wrote famous commentaries on Aristotle, which are extant in the original Arabic, as well as in Latin and Hebrew translations. He is unusual because in his treatise *On the Harmony of Religion and Philosophy* he insisted that only the philosophers are competent to

---

37. My information about al-Razi is drawn from the article "Al-Razi, Abu Bakr Muhammad Ibn Zakariya" by Shlomo Pines in the *Dictionary of Scientific Biography*, 16 vols., 11:323–26 (New York: Charles Scribner's Sons, 1970–1980).

38. Ibid., 326.

judge Scripture, because they use demonstrative arguments. By contrast, the mutakallimun are incompetent to do so because they use dialectical reasoning based on popularly accepted premises.[39] The kind of hostility that al-Razi and Averroes showed to the theologians, as well as al-Razi's attacks on religion, have no counterparts in the medieval West.

By contrast, Ibn Khaldun (1332–1406) was more like al-Ghazali and defended religion against philosophy and logic. But his great fame does not derive from any defense of religion against the foreign sciences. Rather, it derives from his extraordinary treatise, known as the *Muqaddima* ("The Introduction"), which consisted of the introduction and first book of a lengthy world history. According to Franz Rosenthal, "The *Muqaddima* was indeed the first large-scale attempt to analyze the group relationships that govern human political and social organization on the basis of environmental and psychological factors."[40] Arnold Toynbee spoke in superlative terms about Ibn Khaldun, declaring that in his *Muqaddima* "he has conceived and formulated a philosophy of history which is undoubtedly the greatest work of its kind that has ever yet been created by any mind in any time or place."[41] George Sarton's assessment is more critical, but nonetheless highly laudatory. Sarton did not rate Ibn Khaldun as a great historian, but regarded him as

the greatest theorician of history, the greatest philosopher of man's experience, not only of the Middle Ages, but of the whole period extending from the time of the great classical historians down to that of Machiavelli (1532), Bodin (1576), and even Vico (1725). Badly composed as the Muqaddama is, with many repetitions, and poorly written sometimes to the point of obscurity, it remains one of the noblest and most impressive monuments of medieval thought. A comparison between Ibn Khaldun and Machiavelli is not to the disadvantage of the earlier writer.[42]

Although comparisons are difficult, there is no reason to believe that Islamic natural philosophers were inferior to those in the Latin West in

39. Averroes, *On the Harmony of Religion and Philosophy*, 24.
40. Franz Rosenthal, "Ibn Khaldun," in *Dictionary of Scientific Biography*, 7:321 (1973).
41. Toynbee's assessment of Ibn Khaldun is cited in *An Arab Philosophy of History: Selections from the Prolegomena of Ibn Khaldun of Tunis (1332–1406)*, translated and edited by Charles Issawi, ix (Princeton, N.J.: Darwin Press, 1987). Toynbee's remarks were taken from his *A Study of History*, vol. 3.
42. George Sarton, *Introduction to the History of Science*, 3 vols. in 5 parts, 3:1775 (Baltimore: Carnegie Institution of Washington, 1927–1948).

the late Middle Ages. But the fate of natural philosophy in Islam differed radically from that in the West. To gain a proper sense of the difference, we must compare the madrasas in Islam with the universities in the West. A madrasa was a charitable trust, which was established freely by an individual Muslim, known as a *waqif*, who endowed the trust with substantial funds to be used for a public purpose. The founder had great latitude in determining the conditions for the operation of the madrasa he had founded with his own property. "The legal status of the *madrasa* allowed the founder to retain complete control over the administrative and instructional staff of the institution."[43] But the founder of a madrasa had to accept one condition: the terms of the foundation could not violate the tenets of Islam.[44]

The madrasa was essentially a school for the study of the religious sciences and subordinate and related subjects. Excluded from its curriculum were the "foreign sciences," that is, the philosophical and natural sciences.[45] Those who wished to study natural philosophy or the sciences for their own sakes had to either teach themselves, or make arrangements for private instruction with someone knowledgeable in such matters.[46] Occasionally nonreligious courses were taught in the madrasas on an optional basis. In his splendid book, *The Mantle of the Prophet,* Roy Mottahedeh explains that "*Madreseh* learning had formerly been a conspectus of higher learning, with its optional courses in Ptolemaic astronomy, Avicennian medicine, and the algebra of Omar Khayyam. But ... even the mullahs recognized that their learning really was 'religious' learning, and only a few enthusiasts studied the traditional nonreligious sciences such as the old astronomy in private."[47] However, only those subjects were taught that illuminated the Koran or the religious law. One such subject was logic, which was found useful in semantics and in avoiding "simple errors of inference," although philosophical logic, popular in the West, was usually avoided.[48] The primary function of the madrasa, however, was "to

43. Article "Madrasa" in *Encyclopedia of Islam,* 2nd ed., 12 vols. (Leiden: Brill, 1960–2005), vol. 5 (1986), p. 1128, col. 2.

44. See George Makdisi, *The Rise of Colleges: Institutions of Learning in Islam and the West* (Edinburgh: Edinburgh University Press, 1981), 36.

45. Ibid., 77.

46. Ibid., 78.

47. See Roy Mottahedeh, *The Mantle of the Prophet* (New York: Pantheon Books, 1985), 237.

48. On the subject of logic in Islam, see John Walbridge's excellent article, "Logic in the Is-

preserve learning and defend orthodoxy."[49] In Iran, the madrasas existed into the twentieth century, limping on until the end of World War II.

The Medieval University in the Latin West

Apart from a few works of Aristotle's logic, the Christian West had virtually no knowledge of Aristotle's natural philosophy for approximately eleven hundred years after the birth of Christianity. It was not until the twelfth and thirteenth centuries that it made the bulk of Greek natural philosophy and science a part of its intellectual heritage. Islam began its serious appropriation of Greek science in less than two centuries after its founding and by 1000 had translated into Arabic virtually all that it would receive.

But if the West took approximately eleven hundred years to receive Aristotle's natural philosophy, where Islam acquired it in only two to three hundred years, the West wasted no time in making the most of what it received. By 1200, the new translations facilitated the transition from the cathedral school system to the university system, represented initially by the universities of Paris, Oxford, and Bologna. Despite some difficulties at the University of Paris, the new universities, and numerous others that would follow in the course of the next three centuries—approximately sixty-five existed by 1500—unhesitatingly chose Aristotle's logic and natural philosophy to form the curriculum of their arts faculties. In a complete university, we find four faculties: arts, theology, medicine, and law. All students were required to obtain a bachelor's degree in arts. If they wished to enter one of the three higher faculties of theology, medicine, and law, they were expected to obtain the master of arts degree. This meant that virtually all, if not all, theologians, physicians, and lawyers had been thoroughly trained in logic and natural philosophy, as were also those who were content to acquire only a master of arts degree. A university education in the Middle Ages was in no way intended to teach religion or theology. Theology was only taught in theology faculties to theology students. It was a jealously guarded intellectual preserve.[50]

---

lamic Intellectual Tradition: The Recent Centuries," in *Islamic Studies* 39, no. 1 (Spring 2000): 55–75. On attitudes toward philosophical logic, see 68.

49. Mottahedeh, *The Mantle of the Prophet*, 91.

50. For a brief description of the medieval university and its faculties and curriculum, see

Because all theologians were thoroughly trained in logic and Aristotle's natural philosophy, they used these subjects extensively in their theological commentaries, which all theological students were expected to produce. They posed questions that were answerable only by the application of logic and natural philosophy. One of the most powerful logical tools theologians used was the law of noncontradiction where it was assumed that not even God could perform a contradiction. Richard of Middleton, for example, asks "whether God could do contradictory things simultaneously,"[51] and concludes that He cannot. Theologians incessantly inquired whether God could perform this or that act. Their object was to determine whether God could or could not do something by applying the law of noncontradiction. If no contradiction was involved, God could perform the act; if there was a contradiction He could not. For example, Hugolin of Orvieto, in the fourteenth century, applied the law of noncontradiction to determine "Whether God could make the future not to be?"[52] and "Whether God could make a creature exist for only an instant?"[53] and Gregory of Rimini applied it to a question in which he inquired whether God could make someone sin.[54]

During the Latin Middle Ages theology became an analytical discipline with a heavy emphasis on logic and natural philosophy. Indeed, to the extent that medieval theologians increased the analytic content of their theological treatises, they seem simultaneously to have diminished their spiritual content. Medieval theological commentaries became exercises in natural philosophy and logic. From time to time church authorities sought to stem the tide by edicts that were intended to curtail, if not prevent, excessive reliance of theology on natural philosophy and logic.[55]

Grant, *The Foundations of Modern Science in the Middle Ages,* ch. 3 ("The Medieval University"), 33–53. For a lengthy, detailed account, see H. de Ridder-Symoens, ed., *A History of the University in Europe, Vol. 1: Universities in the Middle Ages* (Cambridge: Cambridge University Press, 1992).

51. "Quarto quaeritur utrum Deus possit simul contradictoria facere"; Richard of Middleton, *Super quatuor libros Sententiarum,* vol. 1, bk. 1, distinction 42, qu. 4, 374, col. 1–375, col. 2. Cited from facsimile reprint, Frankfurt: Minerva, 1963.

52. *Hugolini de Urbe Veteri OESA Commentarius in Quattuor Libros Sententiarum,* edited by Willigis Eckermann, O.S.A., 4 vols., vol. 2, book 1, distinction 40, qu. 3, art. 3, p. 341 (Würzburg: Augustinus Verlag, 1980–1988).

53. Hugolin of Orvieto, ibid., vol. 3, book 2, distinction 2, unique question, art. 3, pp. 97–99.

54. *Gregorii Arimensis OESA Lectura super Primum et Secundum Sententiarum,* 7 vols, vol. 3, bk. 1, distinctions 42–44, qu. 1, p. 359 (Berlin: Walter de Gruyter, 1979–1987).

55. For a summary account of the reaction of churchmen to the invasion of theology by

Their efforts were in vain. Theology had become too dependent on logic and natural philosophy.

During the history of medieval Islam, a continual struggle raged among theologians, philosophers, and religious teachers. Only religious subjects constituted the curriculum of the madrasas, while the foreign sciences—logic, natural philosophy, and the exact sciences—were either ignored or taught only as ancillary subjects to shed light on religion. The university system in the Christian West was radically different. Universities taught a nonreligious analytic curriculum based on logic, science, and natural philosophy. So great was the surge toward analyticity, that theology was transformed into a large collection of problems that could only be resolved by the use of logic and natural philosophy. This practice continued routinely for four centuries and laid the basis for a rationalistic society.

Islam and the West differed not because one civilization taught, studied, and wrote about analytic subjects that the other ignored. Both civilizations taught, studied, and wrote about logic, natural philosophy, and the sciences. But in contrast with Islam, the West taught, studied, and wrote about these disciplines in universities that fully supported them. This was possible because the university curriculum was enthusiastically approved by church and state. Anyone with a university education had studied, and perhaps even commented on, Aristotle's natural philosophy and done so for its own sake, not for the sake of better understanding or explicating Scripture. In Islam, the foreign sciences, which comprised the analytic subjects derived ultimately from the Greeks, were rarely taught in religious schools such as the madrasas, which formed the core of Islamic higher education. The analytic subjects were there, though they were marginal. But why were they marginal? Why did they not have equal status with religious and theological subjects? Why did they have to be taught as ancillary subjects, or taught privately and unobtrusively? We have now come full circle, since the answer to these questions requires reiteration of all the arguments and quotations that I have already presented. In light of the obstacles faced by natural philosophy in Islam, the high level of achievement that it attained is quite remarkable.

---

natural philosophy and logic, see Monika Asztalos, "The Faculty of Theology," in *A History of the University in Europe, Vol. 1: Universities in the Middle Ages*, edited by H. de Ridder-Symoens, 420–33 (Cambridge: Cambridge University Press, 1992).

But as long as religious traditionalists opposed or ignored analytic studies, they could not attain the required degree of acceptance to be a potent intellectual force in Islamic society.

I should like to conclude with a hypothetical scenario. Let us assume that a reassessment of the traditional interpretation of the madrasas should reveal that Islamic society did embrace analytic studies. What if it were shown that the madrasas laid heavy emphasis on rational subjects such as Aristotelian logic, natural philosophy, mathematics, and astronomy. And let us suppose that this rationalistic curriculum has been extremely stable since 1300, a period of seven hundred years! And like the West these rationalistic subjects had as one of their functions the explication of Islamic revelation. In effect, let us assume that Islam's educational system in the madrasas was as rationalistic as that which prevailed in the medieval Latin West.

If this should prove to be an accurate characterization of Islamic education since 1300, certain fundamental questions arise. Why did Islamic education remain so static for seven centuries, while in the West, the analogous curriculum, based on medieval Aristotelian learning, was largely abandoned in the seventeenth century, after approximately four centuries, to be replaced by a new approach to science that is associated with the Scientific Revolution? With the implementation of the new science in the West, why did Islamic scholars not appropriate what they could from the new science? Why did they continue on for centuries with an outmoded curriculum that had been abandoned in the West? Why did Islam not borrow the new science and learning from the West during the seventeenth to nineteenth centuries, just as the West had borrowed much of their science and natural philosophy from Islam in the twelfth and thirteenth centuries? Were Muslims too proud to borrow from the West? Were they fearful that Western ideas would endanger the faith? Were they simply uninterested? Or did they regard it as unimportant and perhaps even irrelevant? Or did they regard the new science as adding little or nothing to the science they already had in the madrasas and beyond, and perhaps even viewed Western science as a step backward from that standard?

This last possibility would make the hypothesis of a rationalistic curriculum in the madrasas seem far-fetched and implausible. And yet Professor Seyyed Hossein Nasr is convinced that Islamic science is so radically different from Western science that it could not have profited from it. As Professor Nasr sees it, Islamic science was as important for religious and spiritual life as it was for the acquisition of knowledge about the physical world. Islamic cosmological sciences not only provided "the necessary background and knowledge for particular disciplines of practical import such as medicine and agriculture," but they had "a direct practical effect upon the inner life of man," because "they are directly related to man's real existential problem which is to traverse the perilous caves and valleys of the 'mountains' of the physical and psychic worlds to reach safely the sky of the world of the Spirit."[56]

Continuing in the same vein, Professor Nasr explains that

[t]he traditional cosmological sciences ... concern man in an ultimate sense and on a level not to be compared with the modern sciences. The traditional cosmologies are related to man's inner perfection and to his ultimate end. They are inseparable from angelology and eschatology. They provide the background for that process of spiritual maturing which enables man to become God's vice-regent in actuality rather than only potentially. . . .[57]

For Seyyed Hossein Nasr science and religion merge, and even fuse, to form a vast spiritual enterprise. If his characterization of Islamic science is reasonably accurate, we might conclude that Muslims, satisfied with their own science, would have had no desire, and indeed, no need, to import Western science. Professor Nasr seems to regard Islamic science as the product of a more holistic approach, in contrast to the narrower, more focused science produced in the West.

A recent investigation into cultural differences may offer some support to those who would distinguish between Western and Islamic sci-

56. Seyyed Hossein Nasr, *Islamic Science: An Illustrated Study* (Westerham, Kent, Great Britain: World of Islam Festival Publishing Co., 1976), 236.

57. Ibid., 237. Professor Nasr also believes that modern science, that is, Western science, has led to the destruction of nature. By contrast, ". . . Islamic metaphysics and cosmology were able to create an extensive science of the physical and of the psychic worlds which far from destroying nature only accented the equilibrium that exists in the cosmic order and emphasized the harmony between man and his environment. While the Islamic sciences taught man a great deal about the world about him and enabled man to rule over this world, they also set limits to his power to destroy the earth and pointed in a thousand ways to the fact that man's end is to journey to a world beyond and not to be satisfied through pride or ignorance with imprisonment within the cosmic crypt which man's forgetfulness has made to appear as his natural state" (239).

ence along cultural lines. Dr. Richard Nesbitt, a social psychologist at the University of Michigan, and his colleagues challenge the widely held view among Western philosophers and psychologists that "the same basic processes underlie all human thought, whether in the mountains of Tibet or the grasslands of the Serengeti."[58] The basic processes that all humans followed were alleged to embrace "a devotion to logical reasoning, a penchant for categorization and an urge to understand situations and events in linear terms of cause and effect." However, in comparing East Asians and European Americans, Dr. Nesbitt and his colleagues arrived at a radically different assessment. They "found that people who grow up in different cultures do not just think about different things: they think differently." Easterners, they discovered, "appear to think more 'holistically,' paying greater attention to context and relationship, relying more on experience-based knowledge than abstract logic and showing more tolerance for contradiction. Westerners are more 'analytic' in their thinking, tending to detach objects from their context, to avoid contradictions and to rely more heavily on formal logic."

This is an intriguing analysis and is compatible with Professor Nasr's understanding of Islamic science and certainly fits what we know about Western analyticity. But it would require a great deal more investigation and discussion of medieval Islamic and Western natural philosophy and theology before we can assert with any confidence that the differences between them derive from cultural differences between East and West of the kind described by Dr. Nesbitt and his colleagues.

In light of all these uncertainties, it seems proper to conclude that we are as yet unable to answer the most vital questions about the course of science in Islam. Was it the kind of science Professor Nasr has described: as much concerned with the spiritual world as with the physical world; or was it more akin to medieval Western science and natural philosophy, and therefore incorporating a strong current of rationalistic thought? If the latter, why did Islam ignore Western science for so long? Answers to such questions would contribute mightily toward a proper understanding of Islamic attitudes toward science and natural philosophy from the Middle Ages to the present.

58. I rely here on the article "How Culture Molds Habits of Thought" by Erica Goode in the Science Times section of the *New York Times* for August 8, 2000. All the quotations in this paragraph are from Ms. Goode's article.

11 ∽ What Was Natural Philosophy in the Late Middle Ages?

If by natural philosophy we understand everything relevant to nature and natural phenomena, it seems plausible to infer that the subject matter of natural philosophy embraces all inquiries and questions about the physical world. The first humans must have been aware of nature, which was all around them and involved in everything they did. Nature was not invented. It was a given.[1] Long before the Greeks, the ancient civilizations of Egypt and Mesopotamia had already learned much about nature and its actions. But the ancient Greeks brought something new to the study of nature: they invented instructive ways of talking about it.

During the period 600 to 400 B.C., the foundations of Greek natural philosophy were laid by a group of thinkers known collectively as the pre-Socratics, who no longer explained natural phenomena, such as earthquakes, lightning, storms, and eclipses, as the actions of happy or angry gods, but as the actions of natural forces that regularly produced such effects. Not only did the pre-Socratics eliminate the gods as the causes of natural phenomena and replace them with natural causes, but they also devised a number of different approaches to explain the apparent diversity and change they observed in the world around them. In the process, they enunciated some of the most basic problems that would shape the discipline that would eventually be known as physics, or natural philosophy.

The problems that pre-Socratic philosophers identified, and with which they grappled, largely by abstract, rational arguments, ranged over the whole of what we might plausibly regard as natural philosophy. But there is no evidence that they reflected on the essential structure of natural phi-

1. See G. E. R. Lloyd, "The Invention of Nature," in *Methods and Problems in Greek Science,* edited by G. E. R. Lloyd, 41 (Cambridge, 1991).

losophy and how it fitted into the overall scheme of knowledge. These important tasks were left to Aristotle (384–322), who was the first to describe the role of natural philosophy in the overall scheme of natural knowledge. His analysis shaped that discipline for approximately two thousand years, from the fourth century B.C. to the seventeenth century A.D.

Aristotle

The concept of natural philosophy that reached the Latin Middle Ages in the twelfth and thirteenth centuries was embodied in the works of Aristotle, works that had been translated into Latin from Greek and Arabic. Aristotle distinguished three broad categories of knowledge that he regarded as scientific: the productive sciences, the practical sciences, and the theoretical sciences. Among the subdivisions of the theoretical sciences, Aristotle placed metaphysics first, then mathematics, and finally, physics, or natural philosophy. In contrast to metaphysics and mathematics, which were concerned with unchangeable entities, natural philosophy was concerned only with things that are changeable, exist separately, and also have within themselves an innate source of movement and rest.[2] From Aristotle's standpoint, natural philosophy embraces both animate and inanimate bodies and is applicable to the whole physical world, that is, to both the terrestrial and celestial regions. To derive knowledge about the ever-changing natural world, Aristotle believed that we must begin with sense perception, from which we rise to universal propositions. He emphasized the role of causes because nature operates by causes. An investigation of nature by means of physics, or natural philosophy, involved a study and analysis of those causes and the motions and changes they produce. Almost all the topics Aristotle seriously pursued in natural philosophy appear in a collection of his treatises that came to be known as the "natural books" *(libri naturales)*, which include *Physics, On the Heavens (De caelo), On the Soul (De anima), On Generation and Corruption (De generatione et corruptione), Meteorology,* and *The Short Physical Treatises (Parva naturalia),* which consists of a number of brief treatises,[3] and his biological works. Although, strictly speak-

---

2. For Aristotle's division of the sciences, see his *Metaphysics* bk. 6, ch. 1.

3. They are titled; *Sense and Sensibilia, On Memory, On Sleep, On Dreams, On Divination in Sleep, On Length and Shortness of Life, On Youth, Old Age, Life and Death,* and *Respiration.*

ing, Aristotle's *Metaphysics* and logical works were not classified as part of natural philosophy, they were always regarded as highly relevant to that discipline.

Just as important as his delineation of the scope of natural philosophy was Aristotle's style of doing natural philosophy. Above all else, Aristotle had the scientific temperament, constantly emphasizing the application of reason to the problems of natural philosophy. In the *Nicomachean Ethics* (10.7.1178a.5–8), Aristotle declares "that which is proper to each thing is by nature best and most pleasant for each thing; for man, therefore, the life according to intellect is best and pleasantest, since intellect more than anything else *is* man." Aristotle frequently emphasizes reasoned discourse and accords it the highest place.[4]

Characteristic Features of the New Europe that Emerged at the End of the Barbarian Invasions in the Eleventh Century

To appreciate the developments and influence of natural philosophy in the late Middle Ages, one must understand that it developed in a vibrant societal context that is contrary to common assumptions made about the Middle Ages, a period often described as the "Dark Ages" and usually regarded as a backwash of superstition and stupidity, or as a nineteenth-century historian put it: "a thousand years without a bath."[5] After the nadir of Western civilization was reached between 500 and 1000 A.D., a new people emerged who differed greatly from their predecessors in the Roman Empire period. As soon became apparent, the medley of peoples that had intermingled with the inhabitants of the Roman Empire formed a new and vibrant society that was unusually creative and inventive. One of the most distinctive features of this new society was its desire to utilize human reason for the proper understanding of the physical and spiritual worlds. This did not emerge full-blown, but it was there from the beginning in embryonic form until it reached virtual maturity by the end of the Middle Ages, around 1500.

These, and all the rest of Aristotle's treatises, are printed in *The Complete Works of Aristotle: The Revised Oxford Translation*, edited by Jonathan Barnes, 2 vols. (Princeton, 1984).

4. For instances of Aristotle's use of the term *reason*, see Troy W. Organ, *An Index to Aristotle in English Translation* (New York, 1966), 138.

5. Cited by C. Warren Hollister, *Medieval Europe: A Short History*, 7th ed. (New York, 1994), 1.

Significant advances were made in technology (the invention of eye-glasses, magnetic compass, mechanical clock, firearms and cannon, ship's rudder, cranks to convert continuous rotary motion to reciprocating motion); higher education (the invention of universities), and banking (bills of exchange, checks, marine insurance). Medieval medical schools were the first to dissect human cadavers for teaching purposes. In government, the Middle Ages can lay claim to the development of the nation-state; Magna Carta; and the English Parliament (the first representative government). Other momentous achievements include the invention of the basis of the modern corporation and polyphonic music; and, in law, the Middle Ages laid the foundation of the Western legal system. The Arabic number system was first introduced into the West during the Middle Ages. And, finally, Johannes Gutenberg's invention of the printing press in the 1450s may well represent the most revolutionary change made during the second millennium.

We cannot leave unmentioned the important fact that medieval explorers expanded the horizons of Europe as never before. The Vikings reached the shores of Newfoundland around 1000. Before 1500, European explorers (Bartholomew Diaz and Vasco da Gama) reached India by rounding the Cape of Good Hope followed, a few years later, by Christopher Columbus, who reached the New World and began the long period of European imperialism and colonization. These, and numerous other accomplishments, form the great beginnings of the uninterrupted development of Western Civilization, from around 1100 or 1200 to the present.

The University in the Middle Ages

One of the most important medieval achievements was the development, or even invention, of the university.[6] By 1200 at least three famous universities were in existence: the Universities of Paris, Bologna, and Oxford; by 1500, there were approximately seventy-five universities in Western Europe, all of which had at least one faculty, the faculty of arts. Some

6. For a standard, lengthy history of medieval universities, see Hastings Rashdall, *The Universities of Europe in the Middle Ages*, edited by F. M. Powicke and A. B. Emden, 3 vols. (Oxford, 1936; reprint, 1988); for a more recent study, see Hilde de Ridder-Symoens, ed., *A History of the University in Europe, Vol. 1: Universities in the Middle Ages* (Cambridge, 1992). For a brief, readable account, see Charles H. Haskins, *The Rise of Universities* (Ithaca, N.Y., 1957).

also had one or more of the higher faculties: law, medicine, and theology. The Universities of Paris and Oxford served as the basic models for most of the later universities, with Paris perhaps the most famous, largely because of its renowned school of theology.

The university as we know it today is an uninterrupted, evolutionary development from approximately 1200 to the present.[7] The legal and structural organization of the university derived from a unique Western conception: the corporation, or *universitas,* a fictional entity to which various legal rights were assigned. Members of various professions, crafts, or merchant guilds were eligible to form corporations, including masters and students at the recently formed universities. Each corporation had the legal right to elect its own officers and to run its own affairs, as long as its actions did not conflict directly with church or state. There were numerous corporations within a given university. For example, the students and masters in the arts faculty formed a corporation, as did the students and masters of each of the other faculties: law, theology, and medicine. Each faculty corporation had rights and privileges that enabled it to control its own internal affairs, and to preserve the integrity and relative freedom of its members. By the middle of the thirteenth century, the aggregate of all corporate entities within a given university was known as the *stadium generale.* As more universities came into being, the term *studium generale* was usually reserved for those that had at least three of the four basic faculties (arts, theology, law, and medicine). Thus masters and students at a *studium generale* were not only members of their respective corporations, but also members of the *studium generale.* Graduates of a recognized *studium generale* were automatically eligible— that is, were licensed—to teach at any other European university.

Most universities were urban institutions, and the greatest of them (Paris, Oxford, and Bologna) were international in scope. The students who were enrolled came from towns and cities scattered over all parts of Western Europe. They were noncitizens residing in foreign cities and needed some protection from municipal authorities and townspeople who might take advantage of them. It was the task of university corporations to protect, as much as possible, the well-being of their members. Although most students and masters were not members of the clergy,

7. I rely here on my earlier account in *The Foundations of Modern Science in the Middle Ages: Their Religious, Institutional, and Intellectual Contexts* (Cambridge, 1996), 34–51.

they were given clerical status while at their universities. This protected them while traveling and gave them an option of being tried in more lenient ecclesiastical courts than in the more uncertain civil courts.

There were no official requirements for admission to a university. Students who sought entry to universities were usually around fourteen or fifteen years of age. If the university rector approved, they were admitted, subject to the taking of an oath and the payment of a fee. Upon entry to the university, new students were required to attach themselves to a university master, who functioned as teacher and advisor, and undoubtedly played a major role in formulating a course of study.

Of the four faculties that comprised a full *studium generale,* the arts faculty was the filter through which all students passed prior to entering one of the three graduate programs in theology, law, or medicine. Before entering any of the specialized higher faculties, all students followed a common course of study in the arts faculty, which largely consisted of Aristotle's logic and natural philosophy. After fulfilling their course requirements and other obligations as undergraduates, students earned the bachelor of arts degree. If they wished to become teachers in an arts faculty, they were expected to continue studying for approximately two more years after which they were granted the master of arts degree. If a student then desired to become a theologian, lawyer, or physican, he had to seek admission to the relevant school and study for six or seven more years. Because the arts faculty provided a common, basic curriculum for all students who attained graduate degrees in theology, law, or medicine, it is apparent that theologians, lawyers, and physicians, who had all begun their studies in the arts faculty, were reasonably knowledgeable about logic and natural philosophy. Many utilized that knowledge in their written works. We may rightly conclude that during the period 1150 to around 1500, natural philosophy was institutionalized throughout Europe. Indeed, it was transformed in ways that Aristotle would probably have disapproved.

The Latin Translations of Greco-Arabic Natural Philosophy and Their Reception

The emergence of the university in the late twelfth century was accompanied by another major phenomenon that greatly facilitated its creation:

the influx of a large body of Greco-Arabic literature in science and natural philosophy that was previously unknown in the West and which was destined to transform its intellectual life. In the twelfth century, Western European scholars discovered the treasures of Greco-Arabic science and natural philosophy and immediately sought to translate them into Latin. They found Greek and Arabic treatises in the Arabic language in areas of Muslim Spain and Sicily that had been reconquered in the late eleventh century; and in Northern Italy, especially Venice, they discovered texts in science and natural philosophy in the Greek language by Greek authors ranging from the classical period to the end of antiquity. Most of what would be known of Greek and Arabic science and natural philosophy in the late Middle Ages was translated into Latin in these areas. For the history of natural philosophy, the translation of the works of Aristotle and his numerous commentators, both Greek and Arabic, was of fundamental significance. These works were new to the West and marked an explosive expansion of knowledge in that region. The most important of all commentators on Aristotle's natural philosophy in any language was Averroes (Ibn Rushd) (1126–1198), who lived in Muslim Spain and wrote in Arabic; among the Greek commentators of late antiquity, the most important were Alexander of Aphrodisias (fl. 198–209), Themistius (fl. late 340s–384/385), Simplicius (ca. 500–d. after 533), and John Philoponus (ca. 490–570s).

Translations were made from the mid-twelfth century to the end of the thirteenth century. The numerous translations into Latin of Aristotle's works on natural philosophy provided the newly awakened West with a corpus of treatises that met the most basic needs of the arts faculties in the new universities that had taken root by 1200. Aristotle's *natural books* came to serve as the basic curriculum for students studying natural philosophy in the arts faculties. But the introduction—and eventual acceptance—of Aristotle's works as the basis for the arts curriculum in the medieval university did not occur without a considerable degree of hostility from various theological authorities in the course of the thirteenth century. The reaction against Aristotle was concentrated at the University of Paris, where as early as 1210, an order was issued that forbade, under penalty of excommunication, the reading of Aristotle's natural books in public or in secret. The ban was repeated in 1215, but proved of no avail. In 1231, Pope Gregory IX established a three-man commission to "correct" Aristotle's texts—that is, make them compatible with

Christian teaching. This command was apparently never executed, because by 1255, the unexpurgated texts of Aristotle were being taught at the University of Paris.

By the 1270s, some theologians, including St. Bonaventure (John of Fidanza; 1221–1274), instituted a new tactic: they condemned ideas they thought dangerous to the faith. Conservative theologians prevailed upon the bishop of Paris, Etienne Tempier, to issue a condemnation of 13 articles in 1270; and in 1277, he cooperated with a condemnation of 219 articles. The 219 articles condemned in 1277 were repetitious and listed in no particular order. Many of them, however, were relevant to Aristotle's natural philosophy. A number of articles were directed against Aristotle's arguments for the eternity of the world, while others sought to condemn those ideas and arguments in which Aristotle claimed that some action, or other, was impossible, as, for example, the creation of the world, or the simultaneous existence of more than one world, and so on. These claims were condemned because, by calling such actions impossible, Aristotle was effectively saying that not even God could do those things. Hence those Aristotelian arguments were regarded as denials of God's absolute power to do whatever He wished, short of a logical contradiction. As we shall see, problems relevant to God's absolute power played a significant role in the substantive development of medieval natural philosophy.

By the end of the thirteenth century, however, despite perennial problems with Aristotle's ideas about the physical world that conflicted with the Christian religion, Aristotle's natural philosophy was fully accepted and integrated, not only into the curriculum of the arts faculties of European universities, but into the whole domain of European intellectual life. Indeed, it is ironic that in the sixteenth and seventeenth centuries, the Catholic Church and many of its theologians fought doggedly to save Aristotle's natural philosophy from the unrelenting assaults on it by numerous rival philosophies that had emerged after the Protestant Reformation. But what was the natural philosophy that dominated medieval thought and which endured for some four centuries?

The Subject Matter of Natural Philosophy

More to the point: what did university natural philosophers regard as the subject matter of their discipline? In the most general sense, Scholastic natural philosophers identified "mobile being" *(ens mobile)* as the basic subject matter of natural philosophy. Mobile being included not only bodies, but also the motion of immaterial substances, such as angels. All things in motion, not just bodies, are the province of natural philosophy, or, as an anonymous fourteenth-century author put it, "the whole of movable being is the proper subject of natural philosophy."[8] Although natural philosophy was primarily about the entire physical universe, it also included all immaterial substances, such as angels or intelligences, that were capable of motion.

The Literature of Natural Philosophy

Natural philosophy was the subject of university lectures which frequently came to be embodied in written texts. By the late thirteenth century, three major literary forms had evolved: (1) commentaries on the natural books of Aristotle, the most famous being the Aristotelian commentaries by Averroes and St. Thomas Aquinas; (2) treatises comprised solely of questions on Aristotle's natural books, say a *Questions on Aristotle's Physics,* or *Questions on Aristotle's On the Heavens,* or *Questions On Aristotle's On Generation and Corruption,* and so on; and (3) thematic treatises, or tractates, in each of which a particular theme or subject area is discussed systematically and at considerable length, as, for example, Nicole Oresme's *Treatise on the Uniformity and Difformity of Intensities* known as *Tractates de configurationibus qualitatum et motuum,* or Thomas Bradwardine's *Treatise on Proportions or Ratios.*[9]

Of these three categories, the most commonly used type was the questions format, which more than anything else shaped the medieval perception of the world. By virtue of this questioning approach to the world many interesting, and even strange, questions were formulated. Most of

---

8. The anonymous treatise appears in Bibliothèque Nationale 6752 and consists of 236 folios written in a clear hand. The treatise has never been edited or translated although it has been briefly discussed by Lynn Thorndike, "An Anonymous Treatise in Six Books on Metaphysics and Natural Philosophy," in *A History of Magic and Experimental Science,* 8 vols. (New York, 1923–1958), 3.568–84. On 761–66, Thorndike gives the Latin text of all chapter titles in BN 6752.

9. A detailed description of the three types of literature in natural philosophy appear in Edward Grant, *The Foundations of Modern Science in the Middle Ages,* 127–33, and *God and Reason in the Middle Ages* (Cambridge, 2001), 103–8.

them were about problems in Aristotle's works. Scholastic commentators inquired about all kinds of natural phenomena, prefacing their questions with the interrogative "whether," as, for example,

> whether the whole earth is habitable;
> whether spots appearing in the moon arise from differences in parts of the moon or from something external;
> whether the earth is spherical;
> whether a comet is of a celestial nature or [whether it is] of an elementary nature, say of a fiery exhalation;
> whether lightning is fire descending from a cloud;
> whether there are four elements, no more nor less;
> whether it is possible for an actual infinite magnitude to exist;
> whether the existence of a vacuum is possible.[10]

The structure of these questions, and all others in medieval natural philosophy, was remarkably constant. Every question began with an enunciation of the problem (step 1), usually asking whether (*utrum* was the Latin term) this or that is the case—for example, "whether there could be an infinite dimension" or "whether the earth always is at rest in the center of the universe."[11] As in a university disputation, which was the basis of the written medieval question, arguments were presented for or against the enunciated thesis (step 2). If the author offered a series of affirmative arguments, anywhere from one to ten, or even more, he would usually end up defending a version of the negative side. Or the reverse might obtain: the author presents a sequence of negative arguments, from which it could usually be inferred that he would ultimately defend the affirmative side. These initial arguments were called the "principal arguments" (*rationes principales*). They were followed by a statement of the opposite position (step 3), which might take the form "Aristotle says the opposite," or "Aristotle determines the opposite," or "The Commentator [Averroes] affirms the opposite," and so on. After presenting the opposite opinion,

10. The questions cited above were drawn from the fourteenth-century questions on Aristotle's works by John Buridan, Albert of Saxony, and Themon Judaeus. For these questions, and many more, see Edward Grant, *A Source Book in Medieval Science* (Cambridge, Mass., 1974), 199–210.

11. The first question is from Albert of Saxony's *Questions on the Physics*, bk. 3, qu. 11; the second is from John Buridan's *Questions on De caelo*, bk. 2, qu. 22. Translations of the enunciations of these questions appear in Grant, *A Source Book in Medieval Science*, 201, 205.

the author might then explain his understanding of the question, raise doubts about it, and even define ambiguous terms in the question (step 4). The author was now ready to express his own opinions, usually by way of distinct, numbered conclusions (step 5). When this task was completed, the author took the final step (step 6): a brief point-by-point response to each of the principal arguments enunciated at the outset of the question.[12]

This six-step format was used in the formulation of hundreds of questions during the course of the late Middle Ages. In every question, the objective was to present the affirmative and negative arguments and to choose, or "determine," the correct response. Aristotelian natural philosophy was comprised of hundreds of questions that were largely unrelated. Occasionally, an author referred from one question to another, and thereby linked one or more questions. Most questions, however, were left isolated and unconnected. As a result of this customary approach, medieval natural philosophers did not present an integrated, overall picture of the cosmos, but one that was highly fragmented.

In the late Middle Ages, natural philosophy is found not only in standard questions on this or that treatise of Aristotle's, but it turns up with great frequency in theological commentaries, most notably in commentaries on the four books of *Sentences* of Peter Lombard, a twelfth-century theological text on which, for more than four centuries, theological students were required to lecture and write commentaries. When natural philosophy entered the Christian West in the thirteenth century, it caused theologians some concern, but by the 1250s it had become an integral part of the university curriculum. Natural philosophy formed a link between the arts masters in the arts faculties and the theologians in the theology faculties. In order to enter a graduate school of theology, a potential student was expected to have the equivalent of a master of arts degree. To obtain a master of arts degree one had to be proficient in natural philosophy as taught in the arts faculty. Virtually every theologian was well versed in natural philosophy, a state of affairs that had monumental consequences for the development of that subject. A theologian could freely apply natural philosophy to theological problems, whereas an arts master

12. For an illustration of the formal structure of a medieval question, see Grant, *God and Reason in the Middle Ages*, 153–60. The question is by Nicole Oresme (in his *Questions on Aristotle's On the Heavens*) and asks "whether it is possible that other worlds exist."

could not. Theologians expanded the range of natural philosophy from questions on the physical world to questions on the nature of God, the Eucharist, and other articles of faith. But they did not stop there. In their theological commentaries on the *Sentences* of Peter Lombard, they also included straightforward questions on traditional themes in Aristotelian natural philosophy. Thus not only did theologians expand the horizons of natural philosophy, but, in applying natural philosophy to theology, theologians transformed theology into an analytical discipline, virtually devoid of religious content. Popes complained about the extensive use of natural philosophy in theology and tried to curtail it, but failed utterly, as we can see from a remark by John Major, an eminent theologian in the sixteenth century, who declared that "for some two centuries now, theologians have not feared to work into their writings questions which are purely physical, metaphysical, and sometimes purely mathematical."[13] As if to support John Major's opinion, John Murdoch declares that "genuine parts of fourteenth-century theological tracts ... successfully masqueraded as straightforward tracts in natural philosophy."[14] It is no exaggeration to claim that theologians contributed more to the development and advance of natural philosophy than did masters in the arts faculty.

The Substantive Nature of Medieval Natural Philosophy

Whether we study natural philosophy in the commentaries and questions treatises on Aristotle's books on natural philosophy, or whether we study natural philosophy as it was used and applied in theological commentaries, we will find that the substantive content of it is essentially the same.

In the arts faculty of universities, the masters, who were not theologians, sought to keep theology and natural philosophy distinct. Indeed, they had a special incentive to do this at the University of Paris, which, beginning in 1272, required arts masters to take an oath that they would not introduce theology into their questions, but if perchance they did,

13. Translated by Walter Ong, *Ramus, Method, and the Decay of Dialogue: From the Art of Discourse to the Art of Reason* (Cambridge, Mass., 1958), 144; also cited in Grant, *God and Reason in the Middle Ages*, 281–82.

14. John E. Murdoch, "From Social into Intellectual Factors: An Aspect of the Unitary Character of Late Medieval Learning," in *The Cultural Context of Medieval Learning*, edited by John E. Murdoch and Edith Dudley Sylla, 276 (Dordrecht, Holland, 1975).

they were required to resolve any issues in favor of the faith. But arts masters were likely to have excluded theology and matters of faith from natural philosophy by the very nature of things. They assiduously avoided the intrusion of theology into their arguments because they knew they were dealing with *natural* philosophy, not *supernatural* philosophy. To invoke religious or theological arguments to resolve problems in natural philosophy would be to invoke supernatural explanations rather than natural explanations. No Aristotelian natural philosopher could have accepted that, and none did in the Middle Ages.

To see how medieval natural philosophers viewed the world and its operations and avoided theological entanglements, one would do well to examine the attitudes of John Buridan, a fourteenth-century Scholastic, who was not a theologian, but was probably the greatest natural philosopher among medieval arts masters in the entire Middle Ages; and Nicole Oresme, who was a theologian-natural philosopher and Buridan's younger contemporary, perhaps even Buridan's student at the University of Paris.

Buridan

In his *Questions on Aristotle's De caelo,* Buridan asks whether something exists beyond the world and declares that: "Thirdly, I say that there is no body beyond the heaven or world, namely, beyond the outermost heaven. And Aristotle obviously assumes this. But what must be said about this according to the truth or constancy of faith, you ought to refer to the theologians."[15] Buridan obviously regarded theological pronouncements as irrelevant to the question, and chose to avoid them.

Elsewhere in his *Questions on De caelo,* Buridan presents an argument for believing that in an infinite future time, the world will have the potentiality for not-being. "As to this argument," he explains,

it must be noted that this argument is not about natural powers, but it is about supernatural power, because it was not by nature that the world was created, nor is it by nature that the world could be annihilated, but [rather] by supernatural power. And so what the argument concludes might well be conceded.

But now the question is restricted in the way we speak about what is sought: having assumed that the world is eternal and incorruptible in the way that Aristotle

---

15. My translation from *Ioannis Buridani Quaestiones super libris quattuor De caelo et mundo,* edited by Ernest A. Moody, bk. 1, qu. 20, 93 (Cambridge, Mass, 1942).

imagined it, and assuming that something could not be made from nothing, but that it is necessary that everything that exists be made from preassumed matter, just as is true [for] that which cannot be made otherwise than in a natural way.[16]

Buridan sought to defend Aristotelian natural philosophy as the best means of understanding the physical world. Readily conceding—as all medieval natural philosophers did—that God could interfere at any time and alter the natural course of events, or the "common course of nature" (communis cursus nature), as it was frequently expressed, Buridan nevertheless assumed that "in natural philosophy we ought to accept actions and dependencies as if they always proceed in a natural way."[17] Should a conflict arise between the Catholic faith and Aristotle's arguments, which, after all, are based only on sensation and experience, it is not necessary to believe Aristotle, as, for example, in the doctrine of the eternity of the world. And yet, if we wish to confine ourselves to a consideration of natural powers only, it is appropriate to accept Aristotle's opinion on the eternity of the world, as if it were true. Generally, Buridan was interested in arriving at truths about the regular operations of the physical world in the "common course of nature."

Buridan differed radically from Nicholas of Autrecourt (ca. 1300–d. after 1350), who was a skeptic arguing that scientific knowledge is impossible. He rejected Aristotle's concept of substance and believed that the concept of causality is fallacious. Buridan, however, argued that fundamental and indemonstrable principles of natural science need not be absolute, but can be derived by inductive generalization—that is, "they are accepted," he says, "because they have been observed to be true in many instances and to be false in none."[18]

Oresme

Nicole Oresme is a far more complex scholar than Buridan, in part, perhaps, because he was both a natural philosopher and a theologian. Oresme is well known as one who, in his writings, rejected the discipline of astrology and was very critical of magical claims and procedures. In contrast to

16. My translation from ibid., bk. 1, qu. 23, 112.
17. My translation from ibid., bk. 2, qu. 9, 164.
18. Translated by Ernest A. Moody in his article "Buridan, Jean," *Dictionary of Scientific Biography,* 16 vols. (1970–1980); Buridani, *Questions on the Metaphysics* (1518), bk. 2, qu. 2, fol. 9v, col. 2.

Buridan, however, Oresme was somewhat skeptical about the knowledge we could derive from natural philosophy. He was convinced on mathematical grounds that celestial motions were probably incommensurable and that precise data about them was unobtainable. Not only was nature's behavior necessarily approximate by virtue of the mathematics that described it, but Oresme was often enough dubious about human explanations of large cosmic problems. He was frequently content to propose alternative explanations for traditionally accepted Aristotelian conceptions of the cosmos, as he did for the problem of a plurality of worlds and the possible daily axial rotation of the earth. In these instances, Oresme was content to show that neither reason nor experience could demonstrate the truth. The arguments for a single world were no better than those for a plurality of worlds; and those in defense of the earth's immobility were no better—indeed in some ways they seemed less impressive—than those in favor of its axial rotation. Although Oresme eventually opted for the traditional opinions and therefore denied a plurality of worlds as well as the earth's axial rotation, he did so for theological, rather than scientific, reasons.

Oresme was convinced that human knowledge is uncertain. Only faith could furnish us with certainty. What is most noteworthy, however, is that although Oresme erodes confidence in human ability to determine natural causal truths with precision, he refrained from invoking God and theology to discredit arguments in natural philosophy. Rather, he used his profound knowledge of Aristotelian science and his considerable knowledge of mathematics to undermine the claims for certainty in natural philosophy. Thus he used reason to confound reason. For Oresme, theology could not decide an issue in natural philosophy; most theologians in the late Middle Ages would have agreed.

The Range of Medieval Natural Philosophy

The natural philosophy in a straightforward questions treatise on a book of Aristotle's natural philosophy was no different than the natural philosophy in a theological commentary. *But what was natural philosophy in the Middle Ages? Was it science?* Because the domain of natural philosophy was the whole of nature, it could not represent any single science, but it could, and did, embrace bits and pieces of all sciences. In this sense, natural philosophy was "The Mother of all Sciences." If you were to write the history of any particular science, and wished to cover its history from its earliest beginnings, you would have to range over many treatises in natural philosophy. For example, John Buridan offered an interesting and important explanation of mountain formation in his questions on Aristotle's *On the Heavens* and in his *Questions on the First Three Books of the Meteors.*[19] Anyone writing a history of geology would be obligated to include Buridan's opinions and assessments as part of the overall history of the subject. And yet there was no discipline of geology until the eighteenth or nineteenth century. Aristotle's *Meteorology*, for example, was a focal point for numerous scientific questions, such as possible motions of the earth, the ebb and flow of oceans, the nature of lightning, and others. These questions were discussed in natural philosophy long before any specific sciences emerged to claim one or another of these subjects.

To truly appreciate the richness and diversity of medieval natural philosophy, one must get a sense of the range of questions that were posed on themes in Aristotle's natural books, as well as the natural philosophy that was employed in questions embedded in theological commentaries. Certain categories of questions take us into subject areas whose very existence Aristotle denied as naturally impossible, and others that would have been utterly alien to him. Many of these questions derive from the medieval concept of God's absolute power to do anything short of a logical contradiction, a concept that emerged in the aftermath of the condemnation of 219 articles by the bishop of Paris in 1277, a condemnation that was primarily directed against the masters of arts in the University of Paris. The impact on natural philosophy was most pronounced with respect to the belief in the eternity of the world, against which the church authorities directed some 27 of the 219 condemned articles. Other significant themes that were affected by the Condemnation of 1277, and about which medieval natural philosophers posed questions, concerned the possibility of other worlds and the existence of void spaces.

---

19. For Buridan's question in his *Questions on the Four Books of Aristotle's On the Heavens,* see Grant, *A Source Book in Medieval Science,* 621–24. The question in which Buridan discusses earthquakes is titled "Whether the whole earth is habitable."

On the Eternity of the World

Numerous questions were proposed about the possible eternity of the world. It was a central theme in medieval natural philosophy because Aristotle had argued for the eternity of the world. He could not find any good reasons for believing that our world had come into being naturally from any previous material entities. As the centerpiece of Aristotle's natural philosophy, the eternity of the world posed a direct threat to the creation account in Genesis. The perceived threat to the Christian faith from Aristotle's belief in the eternity of the world is reflected in the 27 articles that condemned it. These vary greatly, attacking the doctrine from a variety of perspectives. The questions that were posed by natural philosophers on the possible eternity of the world exemplify this diversity, as can be seen from the following list of questions that were frequently posed by scholars, natural philosophers, and theologians in the late Middle Ages.

(1) Whether the universe could have existed from eternity.

(2) Whether there is eternal motion.

(3) Whether generations could have proceeded from eternity without a first generation.

(4) Whether the world will end at sometime.

(5) Whether the world is generable and corruptible or ungenerable and incorruptible.

(6) Whether the sky [or heaven] is generable and corruptible, augmentable and diminishable, and alterable.

(7) Whether God could create a motion anew before which there was neither a motion nor a mutation.

(8) Whether something created anew could be perpetuated; and whether something eternal could be corrupted.

(9) Whether, on the assumption of eternity, it could be demonstrated that every uncreated thing is incorruptible and that every incorruptible thing is ungenerated.[20]

---

20. For the names of those who discussed these questions, and the places where they discussed them, see Edward Grant, *Planets, Stars, and Orbs: The Medieval Cosmos, 1200–1687* (Cambridge, 1994), 682–86. Brief treatises were also written on the eternity of the world, among which those by St. Bonaventure and St. Thomas Aquinas were the most prominent.

Although some theologians—most notably St. Bonaventure—sought to demonstrate the absurdity of an eternal world, most were prepared to argue that neither the eternity nor the creation of the world were demonstrable, but that one had to accept the creation of the world as an article of faith. As Thomas Aquinas put it, "That the world had a beginning . . . is an object of faith, but not of demonstration or science."[21]

Are Other Worlds Possible?

If the creation account in Genesis strongly suggested a temporal beginning for the world, it also seemed to signify its uniqueness. Here, at least, Aristotle and Christianity seemed in agreement: there is only one world. This apparent unanimity of opinion was, however, deceptive. Although Aristotle's conclusion might be applauded, his derivation of it was offensive because he had argued that the existence of another world was impossible, or, as he put it, "there is not now a plurality of worlds, nor has there been, nor could there be."[22] To argue that creation of other worlds was impossible, even for God, was viewed as a restriction on God's absolute power to do as He pleased. Indeed, the response to Article 34, one of the 219 condemned in 1277, required natural philosophers to concede that God could create as many other worlds as He pleased. Despite a virtually unanimous conviction that God had not actually created other worlds, the condemnation of Article 34 in 1277 stimulated significant discussions in which Scholastic theologians and natural philosophers contemplated the consequences of a plurality of worlds for Aristotelian natural philosophy. To grapple with this problem, they asked questions such as:

(1) Whether there are, or could be, more worlds.
(2) Whether beyond this world, God could make another earth of the same species as this world.
(3) If there were several worlds, whether the earth of one would be moved naturally to the middle [or center] of another.

21. From Thomas Aquinas, *Summa Theologiae*, pt. 1, qu. 46, art. 1, in *St. Thomas Aquinas, Siger of Brabant, St. Bonaventure, On the Eternity of the World (De aeternitate mundi)*, translated from the Latin with an introduction by Cyril Vollert, Lottie H. Kendzierski, and Paul M. Byrne, 66 (Milwaukee, 1964).

22. Aristotle, *On the Heavens*, 1.9.279a.7–11; from *Aristotle On the Heavens* with an English translation by W. K. C. Guthrie (London, 1960), 91.

A general consideration involved the kind of plurality of worlds an author wished to discuss. Almost all who included a question on the plurality of worlds assumed a plurality of simultaneous worlds, which were but replicas of our own world. But they also recognized that the existence of other worlds might take other forms. The worlds might be successive, rather than simultaneous; or, they might be concentric to one another and lie within our world, or they might be concentric to our world and therefore encircle our world. Nicole Oresme considered all three kinds of worlds and concluded that "the contrary cannot be proved by reason nor by evidence from experience, but also I submit that there is no proof from reason or experience or otherwise that such worlds do exist. Therefore, we should not guess nor make a statement that something is thus and so for no reason or cause whatsoever against all appearances; nor should we support an opinion whose contrary is probable; however, it is good to have considered whether such an opinion is possible."[23]

Most Scholastic natural philosophers came to believe that that if God created other worlds, each of these worlds would be self-contained and operate independently of all other worlds. Thus, contrary to Aristotle's central argument that only one center and circumference could exist, and therefore only one world, Scholastics believed that it was at least possible that many worlds could coexist simultaneously, and consequently so also could many centers and circumferences. Various hypothetical situations were imagined in which many of Aristotle's cosmological and physical principles were subjected to analysis in other worlds, producing significant hypothetical departures from Aristotle, departures that were not used to reform the Aristotelian worldview.

Void Space Within and Beyond Our World

If our world is truly unique and created in the manner described in Genesis, Scholastics inquired whether our created cosmos occupied all the space in existence? Does any kind of space lie beyond our world? Secular natural philosophers in the arts faculties of the Universities of Paris and

23. Oresme discussed a plurality of worlds in his French translation and commentary on Aristotle's *On the Heavens*. See *Nicole Oresme: Le Livre du ciel et du monde*, edited by Albert D. Menut and Alexander J. Denomy; translated with an introduction by Albert D. Menut, bk. 1, ch. 24, 171 (Madison, Wis., 1968).

Oxford were reluctant to concede such a possibility that would have been so devastating in its implications for Aristotle's natural philosophy. Some theologians, however, assumed the existence of an infinite void space beyond our world. This assumption was not by way of a counterfactual, but was proposed as reality. Its most important proponent was Thomas Bradwardine, an eminent theologian who also attained fame as a mathematician and natural philosopher. Bradwardine became archbishop of Canterbury and died in 1349. In a treatise titled *In Defense of God against the Pelagians,* written around 1344, Bradwardine presented five corollaries, by means of which he depicts God as immutable and omnipresent. In the second corollary, Bradwardine declares that God is "also beyond the real world in a place, or in an imaginary infinite void"; and in the final corollary, he explains that "it also seems obvious that a void can exist without body, but in no manner can it exist without God."[24]

For Bradwardine, God's infinite omnipresence implies that He is infinitely immense. Does this mean that God is an extended being, spread out over an infinite extension? Since all extended things are divisible, it would follow that God would be a divisible being, a consequence that was unacceptable in the Middle Ages. Bradwardine resolves this dilemma by simply declaring that God "is infinitely extended without extension and dimension." Bradwardine's infinite void space was therefore a dimensionless space. Nicole Oresme also reveals his firm conviction that a space exists beyond our world. In rejecting Aristotle's argument that no place or void could exist beyond our world, Oresme proclaims that "the human mind consents naturally . . . to the idea that beyond the heavens and outside the world, which is not infinite, there exists some space whatever it may be, and we cannot easily conceive the contrary."[25] By identifying this space with God's real, infinite immensity, there is no doubt that Oresme regarded this space as an actually existent infinite void. By identifying God's infinite immensity with the infinite void, it is likely that Oresme agreed with Bradwardine that this infinite space is dimensionless. In the seventeenth century, Thomas Compton-Carleton (1599–1666), a Scholastic theologian, broke with his medieval predecessors and took the dramatic step of attributing dimensionality to infinite space, which he still

24. For the five corollaries, see Grant, *A Source Book in Medieval Science,* 556–57. For Bradwardine's discussion of the corollaries, see 557–60.
25. Oresme, *Le Livre du del et du monde,* 177.

regarded as God's infinite immensity, although we do not know whether he also assumed that God is a dimensional being.

If there is one area in which medieval ideas had a significant impact on seventeenth-century thought, it is in the realm of infinite void space. Medieval Scholastics were the first to divinize infinite void space. Henry More and Isaac Newton adopted that idea, but made God a three-dimensional extended being by assuming that an extended infinite void space was God's three-dimensional attribute.

Other medieval discussions about void space found a place in the natural philosophy of the seventeenth century. Scholastics conjectured what would happen if God decided to move the whole world with a rectilinear motion. They allowed that a void space would be left behind when the world moved out of its initial place. Nicole Oresme regarded such a motion as an absolute motion, since there would be no other body to which its motion could be compared.[26] Scholastic authors also assumed that God could annihilate matter within part, or all, of our world. He might, for example, destroy all matter below the convex surface of the moon; or all the matter in the entire world by annihilating everything within the concave surface of the outermost sphere. Once God had annihilated part, or all, of the matter of the world, while preventing the restraining shell-like spheres from collapsing to prevent formation of any vacuum, Scholastics then imagined a variety of scenarios in which bodies were assumed to move in various ways and under various conditions. For example, in a question inquiring "whether if a vacuum did exist, a heavy body could move in it," Albert of Saxony imagines that God creates a vacuum by annihilating all matter within the concave surface of the lunar sphere and then inquires how a body would fall through this vacuum: whether it would fall instantaneously with an infinite speed, because there is no medium to resist it, or whether it would fall in a finite time, however small.[27]

Much of this found more than an echo in the seventeenth century when natural philosophers assumed that God annihilated matter, or

26. For a brief discussion, see Grant, *The Foundations of Modern Science in the Middle Ages,* 125.

27. For a translation of Albert's question, see Grant, *A Source Book in Medieval Science,* 337–38. For a further discussion, see Edward Grant, *Much Ado About Nothing: Theories of Space and Vacuum from the Middle Ages to the Scientific Revolution* (Cambridge, 1981), 47–49.

substituted the human imagination for God and simply imagined that this or that part of the world is annihilated; or that the world moves with a rectilinear motion; or that other worlds exist. The invocation of God's absolute power to annihilate all matter below the moon, or anywhere in the world, proved to be a powerful methodological tool, as is evident by its adoption in the seventeenth century by non-Scholastics who undoubtedly derived it, without acknowledgment, from their Scholastic predecessors. Pierre Gassendi (1592–1655) appealed to repeated supernatural annihilations of parts of our world and of other imagined worlds, in order to demonstrate that an infinite three-dimensional space existed. Gassendi imagines that God destroys all matter and body below the lunar sphere but leaves the sphere itself intact. That God can do this "no one would deny, except a man who denies God's power." The annihilation of matter was a methodological tool Gassendi derived from the Middle Ages, as he illustrates when he declares that

there is nothing that prevents us from supposing that the entire region contained under the moon or between the heavens is a vacuum, and once this assumption is made, I do not believe that there is anyone who will not easily see things my way.[28]

Gassendi also argued that infinite space is immobile because if God were to move the world through that space, the space would remain motionless.

Thomas Hobbes (1588–1679), an admirer of Gassendi, also made the annihilation of matter a principle of analysis, although he did not invoke God as the annihilator, choosing to assume that matter was simply annihilated. But Hobbes, who loathed Scholastics, paid unwitting tribute to them when he declared that

[i]n the teaching of natural philosophy, I cannot begin better (as I have already shewn), than from *privation;* that is, from feigning the world to be annihilated.[29]

By means of this technique, Hobbes formulated his concepts of space and time.

---

28. See Craig B. Brush, ed. and trans., *The Selected Works of Pierre Gassendi* (New York, 1972), 386.

29. From Hobbes's *De corpore* (1655) as it appears *The English Works of Thomas Hobbes of Malmesbury*, edited by William Molesworth, 16 vols., i.91 (London, 1839–1845). For a discussion of Gassendi and Hobbes, see Grant, *Much Ado About Nothing*, 390 n. 169.

In his famous *Essay Concerning Human Understanding*, John Locke based his argument for the existence of a three-dimensional void space on the assumption that God could annihilate any part of matter. Should God do so, a vacuum would remain, "for it is evident," Locke explains, "that the Space, that was filled by the parts of the annihilated Body, will still remain, and be a Space without Body."[30]

We see that while Gassendi and Locke invoked God to annihilate the matter in question, Hobbes did not: he chose to "feign" it. It was easy to eliminate God and simply imagine hypothetical conditions for all "natural impossibilities," as did Walter Charleton, an English follower of Gassendi, when he summarized Gassendi's annihilation argument and explained that "nothing is more usual, nor laudable amongst the noblest order of *Philosophers*" than the assumption of "natural impossibilities."[31] But the inspiration "for the noblest order of Philosophers" to imagine all manner of "natural impossibilities" was clearly derived from the way medieval Scholastics had used God's absolute power to imagine various natural impossibilities in order to see how a world would, or could, function under such conditions. We should recognize, however, that medieval appeals to God's absolute power had little, if any, religious motivation or content. Wherever we find it used in Aristotelian treatises, it is almost never intended to make a religious point. It simply became a convenient vehicle for the introduction of highly imaginative questions, the responses to which compelled natural philosophers to apply Aristotelian natural philosophy to situations and conditions that were impossible in Aristotle's natural philosophy. In the process, some of Aristotle's fundamental principles were challenged. The invocation of God's absolute power made many aware that things might be quite otherwise than were dreamt of in Aristotle's philosophy. By the seventeenth century, it did not much matter whether God's absolute power was made the causal agent of some hypothetical condition, or whether it was the human imagination. The medieval emphasis on the analysis of imaginary conditions had been assimilated into mainstream seventeenth-century philosophy.

---

30. John Locke, *An Essay Concerning Human Understanding,* edited with an introduction, critical apparatus, and glossary by Peter H. Nidditch (Oxford, 1975), 177–78.

31. Walter Charleton, *Physiologia Epicuro-Gassendo-Charltoniana* (London, 1654), 63–64, art. 5.

Theology and Natural Philosophy

Natural philosophy was not just a discipline to be applied to natural phenomena. It was also regarded as an invaluable tool for explicating and analyzing theological questions. Considerable intellectual energy was expended on imaginary problems about God and God's powers. Indeed, questions about what God could or could not do preoccupied theological commentators in the late Middle Ages. The questions had little to do with religion and everything to do with logic and natural philosophy. Indeed such questions transformed medieval theology into an analytic discipline.

The Infinite and Infinity

Theologians had a special interest in the actual infinite, probably because God is conceived as an infinite being. Aristotle had regarded the existence of an actual infinite as impossible, though he assumed the possibility of a potential infinite based on the concept of infinite divisibility. Although most Scholastics agreed with Aristotle that a potential infinite was possible, they disagreed among themselves as to whether it is possible for an actual infinite to exist that is distinct from God, whose infinite omnipresent immensity was accepted by all. Scholastics were divided about this. Some were convinced that God could not create an actual infinite, because if He did, He would be unable to create anything larger, because there is nothing larger than an actual infinite. To assume that God could create something larger than an actual infinite magnitude would be a contradiction.[32] In arguing against the eternity of the world, St. Bonaventure had denied the possibility of eternity by arguing that one consequence of an eternal world would be unequal infinites, which he regarded as absurd.

Other Scholastic authors were called *infinitists* because they believed that God could indeed create an actual infinite. The infinites that Bonaventure regarded as unequal and absurd were deemed by others, for example, Robert Holkot, to be equal and by no means absurd. Gregory of

---

32. John Buridan argued this way, as did Richard of Middleton and Durandus de Sancto Porciano. See Grant, *God and Reason in the Middle Ages*, 232–33.

Rimini may have produced the most significant result among those who grappled with problems of the infinite. In inquiring "whether God could make some actual infinite," Gregory concluded that God could make three different kinds of actual infinite: infinite multitude, infinite magnitude, and an infinitely intense quality. In the course of his discussion, Gregory had occasion to discuss such terms as "part," "whole," "greater than," and "less than." He argued that these terms were also applicable to infinites in a special sense. Gregory had arrived at a momentous idea about the relationship between infinites, an idea that lies at the heart of the modern theory of infinite sets. He argues that one infinite can be part of another infinite, but that the infinite that is part is nevertheless equal to the infinite of which it is a part. Gregory concedes that "some infinite is less than some [other] infinite because the infinite which is the part does not contain all the things which the infinite that is the whole contains." Gregory provides no example, but setting the even numbers in one-to-one correspondence with the infinite set of natural numbers would illustrate his point, since there are as many even numbers as natural numbers. Thus the two infinite sets are equal, or, to use modern terminology, they have the same cardinality. Thus Gregory discovered the counterintuitive idea that in the domain of the infinite, a part can equal the whole.[33] Henry of Harclay, also in the fourteenth century, carried it a step further when, as John Murdoch reports, he "firmly believed that infinites can be, and often are, unequal."[34] Henry, however, did not develop the idea that one infinite can be greater than another. He left that to Georg Cantor in the nineteenth century.

Angels in Natural Philosophy

Because they were immaterial substances capable of motion, angels could be studied in the domain of natural philosophy. It fell to the theologians to perform this function. When angels were capable of performing an act that physical bodies could also perform, it was usually the case that angels did it differently. Theologians were expected to explain the differ-

---

33. For a full discussion, see Grant, ibid., 244–48.

34. John E. Murdoch, "*Mathesis in Philosophiam Scholasticam Introducta:* The Rise and Development of the Application of Mathematics in Fourteenth Century Philosophy and Theology," in *Arts Libéraux et Philosophie au Moyen Age* (Montreal, 1969), 223.

ences. They usually compared the behavior of angels and material bodies in questions specifically about angels. For example, all acknowledged that when a physical body moves from A to B, it passes through all the intermediate places. Is this true for immaterial angels? For the most part, it is true. An angel, like a body, may traverse a divisible distance between two places by passing through all the points continuously and successively. Thus Aquinas argued that angels do not traverse distances instantaneously. Most, though not all (for example, Gregory of Rimini), theologians agreed. Their arguments often lead into the domain of instants. Richard of Middleton, for example, argued that, since an instant is the smallest measure of time, it follows that if an angel moved through some medium in an instant, God could not move that angel through the same medium in any time less than an instant. But surely God, the strongest force of all, ought to be able to move an angel some distance in less time than an instant? But that is impossible, because there is no temporal measure smaller than an instant. It therefore followed that an angel could not move through a medium in an instant This argument is analogous to one in which God is said to be incapable of creating an infinite world because He would then be unable to create a greater world, since there can be nothing greater than an infinite. This is treated as a contradiction, from which it follows that God cannot create an infinite world.

Under the guise of considering the behavior of angels, theologians, like Gregory of Rimini, for example, included extensive discussions of mathematics, physics, and logic, often ignoring the angels, although the questions were ostensibly about angels. In a fifty-three-page discussion that was supposed to be about angels, Gregory discourses at great length on the nature of instants and the mathematical continuum. He cites Euclid's *Elements* a number of times and includes fourteen elaborate geometrical diagrams. In all the fifty-three pages, the word "angel" *(angelus)* occurs only once, on the very last page of Gregory's discussion.

Why Did Theologians Raise Unusual Questions?

Why did medieval theologians think it important to know "whether God could make some actual infinite" or how angels would move by comparison to physical bodies? What did they hope to achieve by inquiring "whether God could make a creature exist for only an instant"; or "whether angels could have foreknowledge of their fall in the first instant of their existence"; or "whether God could make the future not to be?" And why would they ask whether God can speak falsely, or whether God could erase the past, or whether God could make someone sin?[35]

Theologians were eager to raise such questions because the responses required them to use logicomathematical techniques they had learned in their university courses on natural philosophy. For example, Robert Holkot used the concept of infinite divisibility of a continuum and the doctrine of first and last instants to determine the limits of an imaginary theological problem: Can God always reward the meritorious and punish the unmeritorious? Holkot imagines a situation in which a man is alternately meritorious and sinful during the final hour of his life. Thus he is meritorious during the first proportional part of his last hour and sinful in the second proportional part; he is again meritorious in the third proportional part, and again sinful in the fourth proportional part; and so on through the infinite series of decreasing proportional parts up to the last instant, when death occurs. Because the instant of death does not form part of the infinite series of decreasing proportional parts of the man's final hour, it follows that there is no last instant of his life, and therefore no last instant in which he could be either meritorious or sinful. Since the man was neither meritorious nor sinful in his last instant of life, God cannot judge him.[36] By this example, Holkot shows that God could be in ignorance about a person's state of grace or sin in the last moment of life and thus indirectly sets limits on God's ability to make just rewards and punishments. Holkot follows this example with eight others. In all of them he uses the concept of first and last instants applied to the infinite divisibility of a continuum, as in the article just described.

Indeed, one wonders what theologians themselves thought about their efforts to do theology by the application of logic and natural philosophy to ostensible theological problems. Did they believe that they were contributing positively to knowledge and understanding about God and the faith? Did they regard the application of quantitative and analytic

35. For most of these questions, and others as well, see Grant, *God and Reason in the Middle Ages*, 251–52, 359.

36. Robert Holkot, *In quattuor libros Sententiarum quaestiones* (Lyon, 1518), bk. 1, qu. 3, fol. Biiiiv, col. 2. The Latin text is reproduced by Murdoch, "From Social into Intellectual Factors," 327 n. 101. For a summary of the argument, see Grant, *The Foundations of Modern Science in the Middle Ages*, 154.

methods to theological problems as, in some sense, enhancing their spiritual understanding of the faith? And did they regard it as important to determine what God could or could not do, or what He could or could not know? The theologians themselves fail to shed light on such questions. But somehow in addition to the personal pleasure they may have derived from the effort to resolve challenging, if bizarre, questions, by analytical means, we must, I believe, assume that medieval theologians regarded their efforts as in some sense advancing and buttressing their faith. To think otherwise would signify that they knowingly engaged in meaningless and empty puzzle-solving, analytic exercises that had no relevance to their faith. But in what sense they may have regarded their contributions as meaningful for the faith escapes my understanding. By the fourteenth century, medieval theologians were as much logicians and natural philosophers as they were theologians. They made theology a mixture of logic and natural philosophy. Consequently, the theology they produced was virtually unintelligible to those who lacked training in logic and natural philosophy. Nothing like the theology of the late Middle Ages had ever been seen before; and after its demise in the seventeenth century, nothing like it has been seen since.

Reason in Natural Philosophy

The most important aspect of medieval natural philosophy was its emphasis—perhaps even overemphasis—on reason. Aristotle's works were, of course, a great model for reasoned exposition. But reason had begun to challenge authority more than a century before the translation of Aristotle's works into Latin. The emphasis on reason was further reinforced when the questions format was used to organize medieval natural philosophy. In each question, as we saw, natural philosophers had to consider the pros and cons and subject both sides to careful scrutiny and analysis. In the Middle Ages, when such powerful tools for scientific research as systematic observation, controlled experiments, and the regular application of mathematics to physical phenomena were largely absent, how could nature be interpreted and analyzed so that scholars could arrive at some understanding of a world that would otherwise be unknowable and inexplicable? The most powerful available instrument was human reason, employed in the manner in which Aristotle had used it. The idea was to come to know what things seemed to be, which could be done by empirical means, and then to determine what made them that way, a process that was largely guided by metaphysical considerations. Although, as Jonathan Barnes has explained, "Aristotle was an indefatigable collector of facts—facts zoological, astronomical, meteorological, historical, sociological,"[37] he nevertheless relied essentially on a priori reasoning to form a picture of the structure and operation of the cosmos. Logic and reason were the basic means for understanding the way the world had to be in order to appear and function the way it does. This was the indispensable first major phase in the process that would eventually produce early modern and modern science.

The role of reason is dramatically highlighted when we examine its relationship to empiricism. Following Aristotle, numerous Scholastics emphasized the central importance of experience and observation. In his *Opus Majus* (part 6, ch. 1), Roger Bacon stressed the importance of experience when he declared the "[r]easoning draws a conclusion and makes us grant the conclusion, but does not make the conclusion certain, nor does it remove doubt so that the mind may rest on the intuition of truth, unless the mind discovers it by the path of experience." Bacon invokes fire as his basic example, asserting that "if a man who has never seen fire should prove by adequate reasoning that fire burns and injures things and destroys them, his mind would not be satisfied thereby, nor would he avoid fire, until he placed his hand or some combustible substance in the fire, so that he might prove by experience that which reasoning taught. But when he has had the actual experience of combustion his mind is made certain and rests in the full light of truth. Therefore reasoning does not suffice, but experience does."[38]

Despite these sentiments, which were also held by other medieval natural philosophers, and despite Aristotle's emphasis on observation and experience, it was relatively rare that observation or experience determined the resolution of any physical question. Aristotle himself provides the reason for this when he declares that our senses "give the most authoritative knowledge of particulars. But they do not tell us the 'why' of anything—e.g. why fire is hot; they only say that it is hot" (Aristo-

37. Jonathan Barnes, *Aristotle* (Oxford, 1982), 17.
38. See *The Opus Majus of Roger Bacon*, translated by Robert Belle Burke, 2 vols. (New York, 1962), 1.583.

tle, *Metaphysics* 1.1.981b.10–11). Aristotle and his medieval followers were primarily interested in the "why" of things, but paid lip service to the observational basis of knowledge about the physical world. In the core treatises of his natural philosophy, Aristotle constructed a picture of the cosmos that was far removed from its alleged observational foundation. Aristotle's world was one that conformed to his preconceived ideas of what the universe had to be like in order to function in a manner worthy of a divine cosmos. His medieval followers did likewise, but they added dimensions that Aristotle could not have foreseen and they came to regard various aspects of Aristotle's physics and cosmology as unsound. They admired Aristotle but they were not his slavish followers, as is evident from their numerous and significant departures from Aristotle's explanations of various phenomena.

Departures from Aristotle's Physics and Cosmology

Aristotle's explanations of natural and violent motion were either abandoned or considerably modified. His explanation of the natural up-and-down motion of a material body required that a moment of rest occur at the precise instant of transition between its upward and downward movements. Without the moment of rest, the two contrary motions would be one continuous motion, which Aristotle regarded as absurd. Scholastics used a crucial thought-experiment, which was of Arabic origin, to reject Aristotle's view. They imagined a bean thrown upward while a millstone descended. When the millstone struck the bean, there could be no moment of rest before the bean reversed its direction and descended with the millstone.[39]

Aristotle's explanation of projectile motion was also rejected. Aristotle held that when a body lost contact with its initial mover, the air, or the medium, caused its motion to continue. Most Scholastic natural philosophers disagreed, arguing that the motive force transmitted an impressed force, or impetus *(impetus)* into the moving body, or projectile, thereby causing its continuous motion. The motion would continue until the moving body met another force that either prevented its motion, or, as in some explanations, until the impressed force expended itself, at

39. For a discussion of the "moment of rest," see Grant, *God and Reason in the Middle Ages*, 170–72.

which point the body would fall with a natural motion. Those who used the latter explanation opted for a self-expending impressed force rather than a permanent impetus that could only be dissipated by forces external to the body.[40]

Aristotle had located the earth at the center of the world where it lay immobile. But Nicole Oresme argued that the alternative—the daily axial rotation of the earth—was as plausible as the earth's immobility. Indeed, his arguments for the earth's daily axial rotation seem more powerful than those in favor of the earth's total immobility. Oresme believed there were no good evidential arguments for choosing either alternative. In the absence of compelling evidence for either explanation, Oresme opted for a motionless earth because it was consonant with biblical texts.[41]

The major departures from Aristotle's views about the infinite, the possible existence of other worlds, extracosmic void space, and motion in void spaces within our cosmos, were discussed earlier. All of the departures from Aristotle mentioned here were significant and show that medieval natural philosophers took seriously the remarks about Aristotle by Albertus Magnus, who declared: "if . . . one believes him to be but a man, then without doubt he could err just as we can too."[42]

Conclusion

What was the legacy of medieval natural philosophy to the modem world? Before 1500, the exact sciences in Islam had reached lofty heights, greater than they achieved in medieval Western Europe, but they did so without a vibrant natural philosophy. By contrast, in Western Europe natural philosophy was highly developed, whereas the exact sciences were merely absorbed (from the body of Greco-Arabic scientific literature) and maintained at a modest level. After 1500, Islamic science effectively ceased to advance, but Western science entered upon a revolution that would culminate in the seventeenth century. What can we learn from this state of affairs?

40. Impetus theory is discussed in Grant, *The Foundations of Modern Science in the Middle Ages*, 93–98.

41. For medieval discussions of the earth's possible axial rotation, see Grant, ibid., 112–16.

42. Translated from Albertus's *Commentary on the Physics*, bk. 8, tract 1, ch. 14, by Edward A. Synan, "Albertus Magnus and the Sciences," *Albertus Magnus and the Sciences: Commemorative Essays 1980*, edited by James A. Weisheipl, 11 (Toronto, 1980).

Let me propose the following: that the exact sciences are unlikely to flourish in isolation from a well-developed natural philosophy, whereas natural philosophy is apparently sustainable at a high level even in the absence of significant achievements in the exact sciences. One or more of the exact sciences, especially mathematics, was practiced in a number of societies that never had a fully developed, broadly disseminated natural philosophy. In none of these societies had scientists attained as high a level of competence and achievement as they had in Islam. Was the subsequent decline of science in Islam perhaps connected with the relatively diminished role of natural philosophy in that society and to the fact that it was never institutionalized in higher education? This is a distinct possibility. In Islamic society, where religion was so fundamental, the absence of support for natural philosophy from theologians, and, more often, their open hostility toward that discipline, might have proved fatal to it and, eventually, to the exact sciences as well.

In retrospect, what legacy, if any, did medieval natural philosophers pass on to their non-Aristotelian, and largely anti-Aristotelian, successors in the early modern period? The answer, I am convinced, lies in a pervasive and deepseated spirit of inquiry that was a natural consequence of the widespread and intensive emphasis on reason that began in the Middle Ages. With the exception of revealed truths, reason was the ultimate arbiter for most intellectual arguments and controversies in medieval universities. It was quite natural for scholars immersed in a university environment to employ reason to probe into subject areas that had not been explored before, as well as to discuss possibilities that had not previously been seriously entertained. Reason and the spirit of inquiry appear to be natural companions. The spirit of inquiry that took hold in the Middle Ages may be aptly described as the spirit of "probing and poking around," a spirit that manifests itself through an urge to apply reason to almost every kind of question and problem that confront scholars of any particular period. Indeed, a vital aspect of "probing and poking around" involves an irresistible urge to raise new questions, which eventually give rise to even more questions. The spirit of "probing and poking around" may be appropriately characterized as nothing less than the spirit of scientific inquiry.

In the Middle Ages, reason was joined to an analytic questioning technique that was ubiquitous in university education and therefore widespread among the literate class. Questions were posed in natural philosophy that asked about the structure and operation of the physical world that Aristotle had described. Questions were also posed in theology about every aspect of faith and revelation. But the probing character of medieval questions went far beyond the straightforward and routine. Scholastic natural philosophers and theologians asked questions not only about what is, but also about what could be, but probably wasn't. Theologians exercised their logical talents by inquiring about what God could and could not do, or what He could and could not know. The criterion for judging God's infinite power was simple: if the claim or action led to a contradiction, God could not do it; if no contradiction was involved, God could do it Every question in the Scholastic arsenal produced pro and contra arguments that were intended to include all plausible and feasible positions.

What makes the "probing and poking around" approach so important is the fact that it was institutionalized in the medieval universities where it was the modus operandi for more than four centuries. Thus a spirit of inquiry took deep and extensive root in Western Europe. The myriad questions that were raised reflected the desires of an intellectual class that sought to know as much as it could by reason alone. The structural form of the question as it was used in the medieval universities was meant to provide a definitive answer to each question raised, although scholars might arrive at different, and conflicting, answers. Even if modern critics judge the questions and their responses to be trivial or of little utility, those who posed the questions and answered them regarded their efforts as of great importance. They were, after all, solving questions that ostensibly informed their contemporaries about the inner and outer workings of the world, as these were understood at the time. Not only did they provide their audience with answers to such questions, but they also included refutations of the arguments they found wanting.

And yet, despite the "probing and poking around" that produced numerous departures from Aristotle's natural philosophy, the intense questioning and probing did not transform medieval Aristotelian natural philosophy into a new way of doing science. The thought-experiments, the hypothetical questions, and the questions about what God could or could not do, or what He knows or does not know, which were so characteristic of the Middle Ages, were largely abandoned by the natural philosophers who produced the Scientific Revolution. The numerous departures from Aristotle's physics and cosmology by medieval natural philosophers were never incorporated into Aristotle's natural philosophy. No serious effort was ever made to transform and update the Aristotelian worldview. The numerous medieval departures and innovations were left as part of an unwieldy mass of unintegrated and conflicting ideas. The hundreds of medieval questions on the works of Aristotle were left as a mass of independent, but unrelated conclusions. If progress was to be made, the Aristotelian worldview had to be abandoned, as it was in the seventeenth century.

But if they abandoned Aristotle's explanations of cosmic operations, non-Scholastic natural philosophers also proceeded by way of questions. But the questions were now often only in their minds to guide them in their research and inquiries. The literary tradition of explicating a text by questions came to an end. The results non-Scholastic researchers published might not explicitly include the questions that guided the researcher and led to those results. Moreover, the questions they posed to themselves and to others were rarely about hypothetical, or imaginary, conditions, or about God's power to do or not to do some particular act, but were about the real world. Also noteworthy is the fact that natural philosophers in the seventeenth century answered the questions they posed to nature by appeals to observation, or by means of experiments, or by the application of mathematics. This became the way scientists would proceed to the present day. Non-Scholastic natural philosophers and scientists of the sixteenth and seventeenth centuries devised superior methods and techniques for resolving the problems that their Scholastic predecessors and contemporaries had grappled with.

Although scientists in the various sciences have evolved different techniques and procedures for answering the neverending parade of questions they generate, and without which modern science could not exist, the spirit of inquiry remains essentially what it was in the Middle Ages: an effort to advance a subject by "probing and poking around" with one or more questions to which answers are sought, after which more questions are posed, in a process that never ends. We are a questioning society that constantly seeks answers to queries about virtually everything, especially about nature, religion, government, and society.

The questioning method is the driving force in science, social science, and technology. Ironically, it is absent from modern theology, which no longer raises the kinds of questions that theologians in the Middle Ages characteristically posed. It would be difficult to imagine modern theologians asking about the limits of God's power and determining those limits by application of the law of noncontradiction. Not only did the scholars in the Middle Ages lay the basis for our probing society by means of an unending stream of questions, but they used reason as the fundamental criterion for arriving at their answers. By the seventeenth century natural philosophers saw that "pure" reason alone was often inadequate and they devised the experimental method to furnish evidence that reason alone could not provide. It was in this spirit that Isaac Newton began his work on the *Opticks* by proclaiming to his readers "My Design in this Book is not to explain the Properties of Light by Hypotheses, but to propose and prove them by Reason and Experiments."[43]

If modern science has progressed unrecognizably beyond anything known or contemplated in the natural philosophy and science of the Middle Ages, modern scientists are, nonetheless, heirs to the remarkable achievements of their medieval predecessors. The idea and the habit of applying reason to resolve the innumerable questions about our world, and of always raising new questions, did not come to modern science from out of the void. Nor did it originate with the great scientific minds of the sixteenth and seventeenth centuries, from the likes of Copernicus, Galileo, Kepler, Descartes, and Newton. It came out of the Middle Ages from many faceless Scholastic logicians, natural philosophers, and theologians, in the manner I have described. If you are skeptical about the medieval role in the advent of early modern science, I ask you to consider this question: Could a scientific revolution have occurred in the seventeenth century if the level of science and natural philosophy in Western Europe had remained what it was in the first half of the twelfth century? That is, could the dramatic changes in science and natural philosophy have occurred in the seventeenth century if medieval natural philosophers had not absorbed and developed the new Greco-Arabic science and natural philosophy that had been translated into Latin in the twelfth and thirteenth centuries? The response is obvious: no, it could not have occurred. We ought, therefore, to conclude that something important oc-

---

43. Quoted by Westfall, *Never at Rest: A Biography of Isaac Newton* (Cambridge, 1980), 642.

curred between approximately 1200 and 1600 that proved conducive to the emergence of a scientific revolution. Without the level that medieval natural philosophy attained, with its overwhelming emphasis on reason and analysis, and without the important questions that were first raised in the Middle Ages about other worlds, space, motion, the infinite, and without the kinds of answers they gave, we might, today, still be waiting for Galileo and Newton.

12 ∽ Aristotelianism and the Longevity of the Medieval Worldview

As the dominant intellectual system for the interpretation of the physical world, Aristotelianism endured for some four hundred and fifty years from the time of its reception in the Latin West at the end of the twelfth century to its general abandonment between 1600 and 1650. Why and how did it survive for so long? What was there about medieval Aristotelian Scholasticism that won it the allegiance of so many generations of students and scholars? At first glance, it would appear that historians of medieval science, and of medieval thought in general, would have placed the survival of Aristotelianism in the forefront of their speculations and analyses. In truth, the longevity of the Aristotelian worldview is not exactly a medieval problem. Since it continued as the dominant conception of the cosmos well beyond the Middle Ages and its death occurred in the seventeenth century, it is hardly surprising that medievalists have ignored the questions posed above. And yet the problem of the longevity of medieval Aristotelianism ought to form part of the legitimate concerns of the historian of medieval science, not only because the basic character of Latin Aristotelianism was formed in the late Middle Ages, between 1250 and 1400, but even more so because the factors that would make for its longevity were inherent in the very process that shaped it.

Before any reasons for the longevity of Aristotelianism are suggested, it will be well to explain briefly the two basic concepts of vital concern in this essay, namely, "Aristotelianism" and "medieval worldview." In the context of medieval natural philosophy, the fundamental core of Aristotelianism was composed of the physical, logical, and biological works of Aristotle, along with the late Greek and Arabic commentaries thereon. Taken as a whole, these works provided the framework and much of the detail of the medieval worldview, especially in physics and cosmology.

Aristotelianism in the narrow sense, then, comprised not only the core works mentioned above, but the innumerable commentaries and *questiones* on those works composed by medieval Latin Scholastics. Scholastic Aristotelianism, however, was much broader than the works of Aristotle and the Greek, Arabic, and Latin commentaries they generated. Already in the thirteenth century, much Aristotelian natural philosophy and metaphysics had been imported into theology, especially in the commentaries on the *Sentences* of Peter Lombard, that monumental twelfth-century theological treatise on which all bachelors in theology had to comment. Conversely, and almost inevitably, Aristotelian thought was, in turn, influenced by the demands and requirements of theology.

In this way, Aristotelianism extended much beyond the works of Aristotle and became the dominant, and, for some centuries, the sole intellectual system in Western Europe. It was, as we all know, the basis of the curriculum of the medieval university, where it remained entrenched for centuries. From the time the works of Aristotle entered Western Europe in the late twelfth century until perhaps 1600, or 1650, Aristotelianism provided not only the mechanisms of explanation for natural phenomena, but served as a gigantic filter through which the world was viewed and pictured.

As with all "worldviews," the medieval version had two fundamental but interrelated aspects. The first, often equated with the medieval worldview to the exclusion of the second, concerns the overall structural framework of the world as it was popularly conceived in the late Middle Ages. Largely drawn from the physical works of Aristotle—that is, from the Aristotelianism we have just described—but infiltrated at certain points with Christian ideas of the deity, angels, and soul, the structural frame of the world was, on the whole, remarkably simple. The cosmos was an enormous, finite, unique material sphere filled everywhere with matter. It was divided into two basic parts, celestial and terrestrial. Beginning with the lunar sphere and extending all the way to the sphere of the fixed stars, and even beyond to the empyrean sphere, the celestial region was conceived as filled with a perfect, incorruptible ether that moved with a perfect, uniform circular motion and from which the celestial spheres were formed. In contrast with the heavens, where the only activity was the uniform, circular motion of the spheres, the terrestrial region, lying below the concavity of the lunar sphere and descending to the geometric center of the universe, was characterized by incessant change as the bodies within it came into being and passed away. These terrestrial bodies were compounded of four elements, earth, water, air, and fire, each of which had its own natural place and the innate capacity for natural motion toward that place. The dominant element in any body determined the direction of its natural motion, which was always toward the natural place of the dominant element. When unimpeded, earthy bodies always fell naturally toward the center of the universe, and fiery bodies rose toward the lunar concavity. Watery bodies would rise in the natural place of earth and fall in the natural place of fire, while airy bodies rose in the natural places of earth and water and fell when located in the region of fire. Since the celestial region was judged to be more noble than the terrestrial, the former regularly influenced the behavior of organic and inorganic bodies in the latter. Despite the contact of the convex surface of the sphere of fire, which was the outermost surface of the terrestrial region, with the concave surface of the lunar sphere, which was the innermost surface of the celestial region, the influences were all unidirectional, from the celestial to the terrestrial.

The basic, skeletal frame described here was probably instrumental in the longevity of the Aristotelian worldview. In the judgment of C. S. Lewis, "The human imagination has seldom had before it an object so sublimely ordered as the medieval cosmos."[1] By the magnificent simplicity of its fundamental structure, it satisfied the European mind, psychologically and intellectually, for some four hundred and fifty years. It was this physical frame on which, and in which, the Christian God of the Middle Ages had exercised His wisdom and distributed angels and powers. Although additions to, and alterations of, the basic structure had occasionally been proposed and adopted in the course of the Middle Ages,[2]

1. *The Discarded Image: An Introduction to Medieval and Renaissance Literature* (Cambridge, 1964), 121.

2. As illustrations, we might mention that a few Scholastics in the fourteenth century (Thomas Bradwardine, Nicole Oresme, and perhaps Jean de Ripa) assumed the actual existence of an infinite, extracosmic void (see my article, "Place and Space in Medieval Physical Thought," in *Motion and Time, Space and Matter: Interrelations in the History of Philosophy and Science*, edited by Peter K. Machamer and Robert G. Turnbull, 137–67 (Columbus, Ohio, 1976); that there were those who insisted that the matter of the celestial and terrestrial regions was identical (for example, William Ockham, *Commentary on the Sentences*, bk 2, question 22, in *Guilelmus de Occam, O.FM., Opera plurima* (Lyon, 1494–1496; reprint Gregg Press, London, 1962), vol. iv: *Super 4 libros Sententiarum*, bk 2, question 22 ("Utrum in celo sit materia eius-

they posed no serious challenge to the worldview we have just described. And while many hypothetical suggestions had been made as to how God might have structured the world differently, and even made other worlds,[3] the passing centuries had seen the Aristotelian cosmos become ever more entrenched so that it seemed unthinkable, and even impious, to believe that He had actually made the basic frame of the world other than as it had been traditionally described. As Copernicus knew, and his followers would learn, Aristotelian cosmologists would not suffer rivals gladly.

But if Western Europe was largely agreed on the fundamental structure of the world as just described, it was by no means agreed on the second significant aspect of a worldview, namely, the details of cosmic operations. Aristotelian Scholastics, who were the principal architects of the medieval worldview, had no commonly shared conception of the manner in which the interrelationships between the basic components of the world were effected, and little consensus on the causes of a host of specific operations and activities that were deemed essential to cosmic efficacy and harmony. The operational aspect of the medieval worldview was thus characterized by diversity of opinion and lack of agreement. If the fundamental structure of the medieval cosmos was psychologically and emotionally satisfying, and therefore instrumental in perpetuating the system for centuries, it will be the argument of this essay that the secondary aspect of a worldview, namely, the details of cosmic operations,

---

dem rationis cum materia istorum inferiorum"), sig. Hiii, recto–Hiiii, verso (no foliation)); and that at least one Scholastic, Nicole Oresme, proposed a doctrine of place that clashed with Aristotle's (see *Nicole Oresme: Le livre du del et du monde*, edited by Albert D. Menut and Alexander J. Denomy; translated with an introduction by Albert D. Menut (Madison, Wis., 1968), bk 1, ch. 24, p. 173). Although other changes could be cited, these suffice to convey something of the nature of the alterations that were suggested.

3. Many of these suggestions followed as a consequence of the Condemnation of 1277, issued by Etienne Tempier, the bishop of Paris, and the general interpretation of God's absolute power in the fourteenth century. On the impact of the Condemnation of 1277, see Pierre Duhem, *Le système du monde: Histoire des doctrines cosmologiques de Platon à Copernic* (10 vols., Paris, 1913–1959), Quatrième Partie, "Le Reflux de l'Aristotelisme: Les condemnations de 1277," vol. 6, and Edward Grant, *Physical Science in the Middle Ages* (New York, 1971), 24–36. In altering situations within and without our world, God was frequently imagined to annihilate or create bodies. The possible consequences of such actions were then discussed. On the possibility of a plurality of worlds, see Duhem, *Le système du monde*, vol. 9, ch. 20, 363–430, and his *Etudes sur Léonard de Vinci, ceux qu'il a lus et ceux qui l'ont lu* (3 vols., Paris, 1906–1913; reprint 1955), vol. 2, 57–96, 408–23; for a recent summary of medieval views, see Steven J. Dick, "Plurality of Worlds and Natural Philosophy: An Historical Study of the Origins of Belief in Other Worlds and Extraterrestrial Life" (Ph.D. diss., Indiana University, 1977), 71–108.

also played a significant role in the long life of the Aristotelian cosmos. The diverse, and often conflicting, operational details of the medieval worldview were not, however, the cause of its longevity, but are the *explicanda* for which a cause or causes must be assigned. With the cause, or causes, identified, we must then describe how it, or they, served to prolong the life of the Aristotelian cosmos. Before all this, however, it is essential to convey a sense of the diversity of operational details, the causes of which will then be suggested.

For convenience, let us begin with the celestial region and proceed toward the earth at the center of the universe.[4] We have seen that all were agreed that the celestial region, composed of a near-perfect fifth element, or ether, was conceived as a region of incorruptibility and the ultimate source of all physical influence on that part of the world lying below the moon. It was the locale of the planets and fixed stars moving around the earth as center. But what was that celestial region really like? Was it, as St. Bonaventure argued, a fluid mass, or was it subdivided into a series of solid, and perhaps hollow, spheres, as Themon Judaeus would have it? Those who decided on spheres had then to determine their number. Based on a variety of circumstances and requirements, estimates varied from eight to eleven, with some accepting an outermost empyrean sphere, and others denying its existence. And what of the relationship between these orbs? Were they contiguous—that is, distinct and separate, as indicated by their diverse and contrary motions—as Michael Scot and Albert of Saxony believed; or did they form a continuous whole, sharing common surfaces by virtue of their identical, homogeneous composition, as Thomas Aquinas and others believed? What, or who, could be identified as the movers of celestial spheres? Angels, intelligences, souls, natural inclinations, and impressed forces were all suggested and partisans for each could be found. And what about relationships between celestial motions? Were they commensurable or incommensurable?[5] Although all were agreed that no material body existed beyond the last mobile sphere

4. The illustrations below are drawn largely from my article "Cosmology," in *Science in the Middle Ages*, edited by David C. Lindberg (Chicago: University of Chicago Press, 1978), 265–302.

5. On the problem of celestial commensurability or incommensurability, see *Nicole Oresme and the Kinematics of Circular Motion; Tractatus de commensurabilitate vel incommensurabilitate motuum celi*, edited with an introduction, English translation, and commentary by Edward Grant (Madison, Wis., 1971). Oresme argues that each of these alternatives determines a radically different world order. For the consequences of each, and Oresme's position, see 67–77.

to serve as its physical container or place, the question of the place of the last sphere was a persistent one. In his discussion of the problem, Averroes included five separate solutions of which he was aware. Four of them found supporters in the Latin Middle Ages, to which one must add a fifth developed in the sixteenth century.

Multiple solutions were also proposed for a wide range of problems concerned with the terrestrial region of perpetual generation and corruption. For example, Scholastics could not agree on the cause by which an element moved to its natural place;[6] nor could they agree whether the cause of violent motion was external or internal,[7] or whether a resistant medium was required for finite, temporal motion.[8] They were in disagreement as to whether an element in a compound retained its elemental form.[9] Some were of the opinion that, as geological changes caused the earth's center of gravity to shift, the entire earth moved as its new center of gravity sought to coincide with the geometric center of the universe.[10]

In fact, many, if not most, of the questions or problems that became part of the Scholastic *questiones* literature on Aristotle's physical treatises had a few major solutions which formed the basis of dispute. While in some instances a strong consensus for a particular opinion emerged, in many other problems, as, for example, those mentioned earlier, two or more interpretations were in serious contention. No resolution of most of these problems was really possible. How, for example, could one determine, with reasonable conclusiveness, whether the celestial region was a fluid mass or a system of hard spheres? Or what really moved the spheres? Or how many spheres really existed?

To convey a sense of the enormous range of physical problems on which serious disagreements probably occurred, we need only realize that in the fourteenth century Albert of Saxony included 107 questions in his *Questions on the eight books of Aristotle's Physics* and 35 in his *Questions on the two books of On generation and corruption;* that John Buridan considered 59 questions in his *Questions on De caelo* and Themon Judaeus attended to 65 in his *Questions on the four books of Aristotle's Meteorologica.* Excluding Aristotle's *Metaphysics*, which traditionally in-

---

6. See Edward Grant, *A Source Book in Medieval Science* (Cambridge, Mass., 1974), 263–64.

7. Ibid., 275–80.   8. Ibid., 253–62.

9. Ibid., 603–14.   10. Ibid., 621–24.

cluded a number of important physical questions, and the *Parva natura-lia,* or the *Small Physical Treatises,* the authors of the four physical treatises just mentioned considered a total of 266 questions.[11] If even half of these problems produced at least two serious solutions—and half is not an unreasonable estimate, and may even prove conservative—it is evident that whatever the unanimity on the macrostructure of the Aristotelian cosmos, it did not extend to its operational details.

What produced such a proliferation of theories and opinions about the details of cosmic operation? At least three reasons seem relevant and significant. First, there were Aristotle's own obscurities and ambiguities, which, in both large and small aspects of his thought, no amount of interpretation could resolve successfully with any large degree of unanimity. As with most cosmic system builders, there was often a maddening lack of detail in Aristotle's descriptions and arguments. In supplying those details, Scholastic commentators, with varying degrees of subtlety, often altered Aristotle's arguments and apparent intent, thereby generating new opinions and interpretations. The multiplication of opinions was aided and abetted in no small measure by the Greek and Arabic commentators whose works accompanied the introduction of Aristotle into the West. Major commentators, such as Simplicius, Averroes, and Avicenna, frequently furnished a variety of interpretations for this or that concept, principle, or argument. Scholastics would opt for one or another of them, or fashion new ones to compete with the old.

Opinions and theories were also easily multiplied in Aristotelian natural philosophy because "Aristotle's was the most capacious of philosophies" because "in principle it explained everything."[12] Aristotelian physical principles, such as potentiality-actuality, the four causes, matter and form, the constitution of the four elements, the doctrine of natural place, and others, were so broad and comprehensive that they were easily applied to competing theories and arguments. Not only were these basic principles never seriously challenged, but they found a range of application that would have surprised, if not shocked, Aristotle himself.

But even more significant than these in the multiplication of opinions, though largely ignored until now, is a third major reason, which will

---

11. For the enunciations of all 266 questions, see ibid., 199–210.

12. Charles Coulston Gillispie, *The Edge of Objectivity: An Essay in the History of Scientific Ideas* (Princeton, N.J., 1969), 11.

be central in the discussion to follow. Let us recall that the most common mode of expression in medieval natural philosophy was by means of a commentary on a traditionally recognized authoritative text. These commentaries often took the form of a series of *questiones,* or specific problems, which followed the order of the commented text and developed from it; or they could take the form of a straightforward commentary in which the commented text was discussed systematically section by section. In the *questiones,* which furnished most of the interesting cosmological discussion, each *questio* was subjected to a reasonably thorough analysis by means of a series of pros and cons, followed by the commentator's solution. By its very nature, the *questio* form encouraged differences of opinion. It was a vehicle par excellence for dispute and argumentation. Scholastic ingenuity was displayed by introducing new subtle distinctions, which, upon further development, would yield new opinions on a given question. It is thus hardly surprising that centuries of disputation within the *questiones* format should have produced a variety of opinions on a very large number of questions ranging over the full scope of Aristotelian physics and cosmology.

The ultimate consequence of this process must be viewed as of direct relevance to the longevity of the Aristotelian medieval worldview. For what emerged was a series of distinct and often intensively considered problems that remained isolated from, and independent of, other related *questiones,* to which allusions and references were minimal. As the major form of Scholastic literature in natural philosophy, the *questiones* produced an atomization of Aristotle's physical treatises into sequences of particular questions and problems that focused attention on the independent question and thus severed its connections and associations with other related issues treated in the same treatise or elsewhere in the Aristotelian corpus. Not only were related topics left unintegrated, but even single topics as, for example, the doctrine of place, were left in the form of a series of specific questions that were never organized into a larger, coherent whole, which might have drawn attention to glaring inconsistencies and weaknesses. It was the independent question that became the focal point of contention and with respect to which differing opinions were formulated.[13]

13. A significant aspect of the *questiones* format, and the commentary form generally, is that it tended to discourage the introduction of topics and ideas that had no counterpart in

But how did all this contribute to the longevity of the medieval Aristotelian worldview? Primacy of the independent question in medieval physical thought prevented any larger synthesis that might have forced a major overhaul or reconstitution of Aristotelian cosmology. It served to protect the satisfying macrostructure from any truly penetrating, critical inspection. The atomization of Aristotle's physical treatises resulted in an intellectual flotsam and jetsam of unrelated questions which actually concealed grave inconsistencies and discrepancies. Serious attempts to reconcile these might have encouraged efforts at a new synthesis, or perhaps riveted attention on the inadequate operational substructure that underlay the well-ordered and generally accepted macrostructure. Instead, the extreme atomization of physical thought in the *questiones* literature prevented medieval Scholastics from producing, or even attempting to produce, any comprehensive and systematic treatises on the scope and scale of a Cartesian or Newtonian *Principia*. No genuine effort was made to formulate a coherent and reasonably consistent cosmology within which the disparate elements scattered throughout the *questiones* could be brought together, evaluated, and assessed as part of a larger whole.

The closest medieval Scholasticism came to attempts at cosmological or physical syntheses was an occasional *Summa* in natural philosophy. During the first quarter of the fifteenth century, Paul of Venice (ca 1370–1429) composed a *Summa philosophie naturalis*, or *Summa naturalium*.[14] Here Paul subdivided natural philosophy into six parts corresponding to Aristotle's *Physics, De caelo, De generatione et corruptione, Meteorologica, De anima*, and *Metaphysics*. The order of the first four treatises was undoubtedly derived from Aristotle's opening remarks in the *Meteorologica*, where he explains that he had "already dealt with the first causes of nature and with all natural motion" *(Physics);* "with the ordered move-

---

the Aristotelian texts. Thus while a host of specific Aristotelian topics and themes were subjected to minute analysis, with a consequent multiplication of interpretations and opinions, subjects that were not considered at all by Aristotle could not be readily fitted into the traditional framework of questions. Thus it was *the independent question based on a problem specifically raised by Aristotle* that constituted the basis of medieval Scholastic literature. Despite this seemingly severe restriction, however, new ideas and concepts could be introduced as extensions, or implications, of traditional problems.

14. *Summa philosophie naturalis Magistri Pauli Veneti noviter recognita et a vitiis purgata ac pristine integritati restituta* (Venice, 1503). In the first edition published at Venice in 1476, the title given in the colophon is *Summa naturalium*.

ments of the stars in the heavens" (*De caelo,* bks 1 and 2); and "with the number, kinds, and mutual transformations of the four elements, and growth and decay in general" *(De caelo,* bks 3 and 4; *De generatione et corruptione).* It remains, then, to consider what is commonly called *Meteorology,* which is concerned with phenomena bordering "most nearly on the movement of the stars," that is, in the region immediately below the lunar sphere.[15] Faithful Aristotelian that he was, Paul of Venice not only followed the master's order of discussion, but considered the problems of each treatise in isolation. Under these circumstances, it is hardly surprising that Paul's *Summa* of natural philosophy is little more than a collection of six distinct Aristotelian treatises each with a set of its own *questiones.*[16] No more integration and synthesis was achieved than if the *questiones* on each treatise had been published separately.

By the fifteenth century, then, the Aristotelian *questiones* tradition had become so inflexible that not even a *Summa* could produce a higher synthesis or generate any significant rearrangement of problems. The individual treatise, with its rigidly compartmentalized, and largely unrelated, questions, reigned supreme. The *Summa* of natural philosophy thus represented little more than a convenient order in which to consider the different subject areas of that broad discipline. It was but an aspect of

15. *Meteorologica* 1.1.338a.20–338b.22, as translated by H. D. P. Lee in the Loeb Classical Library (Cambridge, Mass., 1962; London, 1962). Although Aristotle's remarks might have served as a point of departure for cosmic reflections, their only apparent effect was to provide an *order of discussion* for the subject matter of the four treatises mentioned. While Thomas Aquinas offers a brief commentary on Aristotle's introductory passage (see Aquinas, *In Aristotelis libros De caelo et mundo; De generatione et corruptione; Meteorologicorum expositio* [Turin/Rome, 1952], p. 392, col. 1), others, such as Themon Judaeus, Nicole Oresme, and the Coimbra Jesuit commentators of the late sixteenth century, chose to ignore it in the commentaries and *questiones* on the *Meteorologica.*

16. Even the order of discussion is puzzling, since Paul places the *Metaphysics* last, rather than first, which seems a priori more logical. John Dumbleton's fourteenth-century *Summa logicae et philosophiae naturalis* exhibits a similar tendency. "Parts ii-x [Part 1 is on logic] of Dumbleton's *Summa,* the only one produced by the early Mertonians on natural philosophy, is really a collection of certain *dubia* 'magnorum naturalium quinque'" (James A. Weisheipl, O.P., "Ockham and Some Mertonians," *Mediaeval Studies* 30 (1968): 200–201 (the bracketed phrase is mine). The "five great natural books" from which the *dubia* were drawn are Aristotle's *Physics, De caelo, Meteorologica, De generatione et corruptione,* and *De anima.* During the 1550s Petrus Fonseca conceived the idea of a course on Aristotelian philosophy for Jesuit schools (see Charles H. Lohr, "Renaissance Latin Aristotle Commentaries: Authors C," *Renaissance Quarterly* 28 [1975]: 717). To achieve this, he simply ordered commentaries on the separate works of Aristotle. Construction of an integrated worldview based on Aristotle, but not slavishly harnessed to the separate works of the corpus, probably never occurred to him.

the medieval and Renaissance penchant for displaying the organization of knowledge, a penchant nowhere better illustrated than in the *Margarita philosophica* of Gregor Reisch, first published near the close of the fifteenth century. Under Reisch's elaborate subdivision of philosophy,[17] natural philosophy, or physics, within which medicine is also included, is a theoretical, or speculative, discipline concerned with reality (as opposed to theoretical philosophy concerned with the purely rational subjects of the *trivium*, namely, grammar, rhetoric, and logic). The subjects of this division are drawn largely from pseudo- and genuine Aristotelian physical treatises, the first four of which are, not surprisingly, the *Physics, De caelo, De generatione et corruptione*, and *Meteorologica*, the order of treatment described in the last mentioned work. To these are added treatises on minerals, the elements, the soul (basically concerned with perception), animals and plants, the senses, memory, youth and old age, respiration, nourishment, health and sickness, the motion of the heart, life and death, and many others.

Although the organization of knowledge described here may perhaps reflect some deeper cosmic view, it is more likely that medieval and Renaissance divisions of knowledge were little more than traditional representations of the classification of the sciences formulated by Aristotle himself and elaborated subsequently by Augustine, Boethius, Cassiodorus, Hugh of St Victor, Domingo Gundisalvo, and others.[18] While such trees of knowledge were useful pedagogical devices, they were also a false façade. For then and now, they led many to believe that the ideas and explanations in the treatises sequenced and ordered in the various schema of knowledge were as tidy and harmonious as the outlines in which they were located.

In the absence of any genuine rival system, the Aristotelian worldview, with its well-ordered macrostructure and its richly diverse, but bewildering, inconsistent, and largely unexamined operational substructure, reigned unchallenged. By the time rival interpretations of any

17. Gregor Reisch, *Margarita philosophica*, mit einem Vorwort, einer Einleitung und einem neuen Inhaltsverzeichnis von Lutz Geldsetzer, *Instrumenta philosophica, Series thesauri*, 1 (Düsseldorf, 1973; reprint of the 4th ed., Basel, 1517), p.v., where Reisch furnishes a *partitio philosophie*.

18. See my introduction to the "Classification of the Sciences," in E. Grant, ed., *A Source Book in Medieval Science* (n. 6), 53–54.

consequence appeared, as happened in the sixteenth century, Aristotelianism, despite its numerous inconsistencies and multiplicity of opinions on almost every major issue, had acquired a degree of acceptance approaching that of Euclidean geometry before Bolyai, Lobachewsky, and Riemann.

Despite its sheltered and protected status in the conservative university environment, Aristotelianism was eventually faced with rival systems and modes of thought. The humanism that had generated a new interest in Greek antiquity and the influx, beginning in the fifteenth century, of Byzantine Greeks into the Latin West touched off a new wave of translation, now directly from Greek manuscripts. Old works were retranslated and new ones not previously known to the Latins were made available. It was in this new wave of translation, the likes of which had not been seen in Europe since the twelfth and early thirteenth centuries, that new ways of looking at the world became familiar in the West. With translations of the works of Plato, Proclus, Hero of Alexandria, and the Hermetic corpus, atomism, stoicism, Platonism, Neoplatonism, and Hermeticism emerged as flesh-and-blood doctrines. Lucretius's *De rerum natura,* the most complete account of atomism known, reappeared after centuries of obscurity to compete as a major cosmic system. The hostile view of atomism that Aristotle had presented could now be countered in detail.

But if by the sixteenth century Aristotelianism had not lost its intellectual dominance and appeal, it seemed to have lost its vitality. By the end of the fifteenth century, it had become uncreative and ossified. The responses and arguments formulated in the disputes of the thirteenth and fourteenth centuries were repeated in the fifteenth. As new universities were founded in eastern Europe in the late fourteenth and fifteenth centuries, and as the predominant Parisian and Oxford interpretations came to dominate there and in the established Italian universities, the commentators in those places made selections from among already formulated interpretations. The opinions they presented were a mere repetition, with occasional deviations, of well-established arguments and positions.

At the dawn of the sixteenth century, entrenched though it was, Aristotelianism had declined in vigor. At that point, one might well have pondered whether it could survive for long the influx of new ideas and philosophies that had already begun to enter Europe in the fifteenth century.

The new non-Aristotelian intellectual options available to scholars of the sixteenth century caused some to abandon Aristotelianism and to attack it. One need only mention Petrus Ramus, Francesco Patrizzi, and Giordano Bruno to realize that times had changed. Aristotelianism was under attack in a way that it had never been in the Middle Ages. Medieval disagreements with Aristotle, numerous though they were, were never regarded as a means of destroying the system, as was the case in the sixteenth century.

But the system was not destroyed. Paradoxically, the very influx of Greek texts and new translations that threatened the existence of Aristotelianism also served to impart new strength to it. The Aristotelian corpus was not only retranslated from the Greek, but the Greek texts were made available in printed editions. From the fifteenth century onward, the humanistic revival had encouraged the teaching of Greek, a trend which gained strength through the sixteenth century. And, as if to accentuate the new interest in Aristotle, the Greek texts and Latin translations of Aristotle's Greek commentators, such as Alexander, Philoponus, and Simplicius, and Themistius, which accompanied the new Aristotle, were read with as much interest as was Aristotle himself. Their interpretations, especially of the *Physics,* contained some new arguments and insights that were of fundamental importance. Thus a whole new dimension was added to Aristotelianism, which served to revive and refresh it. Sixteenth-century natural philosophers were now face to face with the real Aristotle and the more pristine interpretations of his thought. No longer were they dependent on translations from the Arabic. If it pleased them, they could now even abandon their old Arab guide, Averroes, for the Greek commentators. And, finally, they could also ignore the medieval Aristotelian tradition that was built primarily on translations from the Arabic and overreliance on Averroes and Avicenna. In the end, however, they followed many paths. Charles Schmitt[19] has aptly explained this Renaissance phase of Aristotelianism:

Rather than a singly close knit group of philosophers, scholastic Aristotelianism turns out to be a series of many different sects agreeing only on the most

---

19. Charles Schmitt, *A Critical Survey and Bibliography of Studies on Renaissance Aristotelianism, 1958–1969,* Saggi e testi 11 (Padua, 1971), 17–18.

fundamental issues. Some thinkers attempted to go back to the Greek text of Aristotle to obtain the truth, others followed Alexander, Themistius, Philoponus, or another ancient commentator, still others found the truth in Averroes or in Latin "Averroists" such as John of Jandun or Siger of Brabant, and yet others tended to see philosophy through the eyes of Thomas, Albert, Ockham, or Scotus. All things considered, there was quite a range of interpretation and differences more pronounced than one might think. Moreover, each of the individual thinkers underwent influences from sources other than that of his principal allegiance. For example, we know that Nifo was strongly influenced by Plato and Plotinus through Ficino; Pomponazzi by stoicism; and the Italian Aristotelian writers on logic by the medieval tradition stemming principally from Galen. What is perhaps more unusual is that at least some Aristotelians were influenced in one way or another by atomism through Lucretius and by popular traditions of craftmanship and technology. We know that even the most anti-Aristotelian thinkers of the period were significantly influenced by the Peripatetic tradition *malgré eux.*

If the new Greek texts and Latin translations had merely generated an interest in comprehending and establishing the meanings of the pristine Aristotle purged of medieval accretion and distortion, the new Aristotelianism would have qualified as the beginnings of the history of Aristotelian scholarship, but would have been an intellectual dead end. The reinvigoration of Aristotelianism after the bleak period of the fifteenth century derived not from a narrow philological approach in quest of the real Aristotle, but rather from its continued capacity to absorb the "new" into the old, where "new" is understood in terms of the recently introduced Greek authors and commentators whose works and ideas had not been part of medieval Aristotelianism. The disparities and disharmonies of the Middle Ages, which we emphasized earlier, were thus merely expanded and multiplied, as the new opinions, from whatever source, were assigned appropriate places in the traditional division of Aristotelian problems. The revitalized Aristotelianism was now so truly capacious that there was something for everybody and it managed to sustain itself as long as efforts to synthesize it into a coherent whole were avoided. From this standpoint, Aristotelianism acquired new strength and was able to perpetuate itself as much, if not more, on the basis of intellectual vigor than from its entrenched and traditionally privileged position.

The Aristotelian system was never reformed from within. It was destroyed from without on the basis of ideas developed by Copernicus, who attacked the macrostructure, and by Galileo, who not only upheld Copernicus, but also destroyed fundamental operational principles in the Aristotelian substructure. In challenging Aristotle and his followers, Galileo left an almost indelible historical impression that his Aristotelian opponents were inflexible, slavish partisans incapable of adopting, or even considering, new ideas. By "new," Galileo, of course, understood the Copernican heliocentric system and such of his own ideas as involved the abandonment of the concept of absolute heaviness and lightness. From this standpoint, he is undoubtedly correct, since these concepts were totally incompatible with the Aristotelian worldview. But if we count as "new" ideas and concepts developed in medieval Scholasticism as well as those introduced by the Greek authors and commentators mentioned above by Schmitt and made available in the late fifteenth and sixteenth centuries, then the problem of Aristotelianism is not inflexibility, but rather too much flexibility, too great a readiness to accept ideas and concepts that did not fit well, if at all, into Aristotle's natural philosophy. In the process of multiplying and absorbing new ideas from whatever sources, Aristotelians failed to notice the growing incoherence of the substructure. The capacity of Aristotelianism to absorb so much that was incompatible was possible only because of an absence of critical integration of the many disparate, conflicting, and unreconciled explanations, which formed its complicated operational substructure. Produced primarily by the atomization of Aristotelian Scholastic literature, that fragmented and confused operational substructure served inadvertently to protect the well-ordered macrostructure from critical scrutiny and enabled the medieval cosmos to retain its firm hold on the European mind. Thus did Aristotelianism live on until it fell under the onslaught that began with Copernicus and Galileo, who together provided not only the beginnings of a new cosmic macrostructure, but also laid the solid foundation of a new operational substructure on which the whole could appropriately rest.

Bibliography

Albert of Saxony. *Questiones et decisiones physicales insignium virorum. Alberti de Saxonia in octo libros Physicorum; tres libros De celo et mundo; duos libros De generatione et corruptione; Thimonis in quatuor libros Meteorum; Buridani in tres libros De anima; librum De sensu et sensato; librum De memoria et reminiscentia; librum De somno et vigilia; librum De longitudine et brevitate vite; librum De juventute et senectute Aristotelis. Recognitae rursus et emendatae summa accuratione et judicio Magistri Georgii Lokert Scotia quo sunt tractatus proportionum additis* [Lokert]. Paris, 1518.

Albertus Magnus (Albert the Great). *Alberti Magni Ordinis Fratrum Praedicatorum Opera Omnia,* ed. B. Geyer.

———. *Monasterii Westfalorum in aedibus Aschendorff,* Vol. 4, *Physica,* ed. P. Hossfeld, part 1 (bks. 1–4), 1987; part 2 (bks. 5–8), 1993. Vol. 5, part 1, *De caelo et mundo,* ed. P. Hossfeld, 1971.

Al-Ghazali. *Al-Ghazali's Tahafut al-Falasifah* [*Incoherence of the Philosophers*]. Translated into English by Sabih Ahmad Kamali. Pakistan Philosophical Congress Publication No. 3. Lahore, Pakistan: Pakistan Philosophical Congress, 1963.

Arberry, A. J. *Revelation and Reason in Islam: The Forwood Lectures for 1956 Delivered in the University of Liverpool.* London: George Allen & Unwin; New York: Macmillan, 1957.

Aristotle. *The Complete Works of Aristotle: The Revised Oxford Translation.* 2 vols. Edited by Jonathan Barnes. Princeton, N.J.: Princeton University Press, 1984.

———. *On the Heavens.* Translated by W. K. C. Guthrie. Cambridge, Mass.: Harvard University Press; London: Heinemann, 1960.

Asztalos, Monika. "The Faculty of Theology," in *A History of the University in Europe, Vol. 1: Universities in the Middle Ages,* edited by H. de Ridder-Symoens, 420–33. Cambridge: Cambridge University Press, 1992.

Averroes. *Commentary on Aristotle's Physics,* in *Aristotelis omnia quae extant Opera,* 9 vols. and 3 supplements. Venice, 1562–1574; reprint Frankfurt, 1962, as *Aristotelis Opera cum Averrois commentariis.*

———. *On the Harmony of Religion and Philosophy: A Translation with Introduction and Notes, of Ibn Rushd's Kitab fasl al-maqal, with Its Appendix (Damima) and an Extract from Kitab al-kashf 'an manahij al-adilla.* Edited and translated by George F. Hourani. London: Luzac, 1976.

Bacon, Roger. *Opera hactenus inedita Rogeri Baconi,* 16 fascicules, edited by
R. Steele and F. M. Delorme. Oxford, 1905–1940.

————. *The "Opus Majus" of Roger Bacon.* Translated by Robert Belle Burke.
2 vols. New York: Russell & Russell, 1962

————. *Roger Bacon and the Origins of "Perspectiva" in the Middle Ages: A
Critical Edition and English Translation of Bacon's "Perspectiva" with Introduction and Notes.* Edited and translated by D. C. Lindberg. Oxford, 1996.

————. *Roger Bacon's Philosophy of Nature: A Critical Edition, with English
Translation, Introduction, and Notes, of "De multiplicatione specierum" and "De
speculis comburentibus."* Edited and translated by D. C. Lindberg. Oxford, 1983.

Baconthorp, John. *Super quatuor sententiarum libros.* Venice, 1526.

Barnes, Jonathan. *Aristotle.* Oxford: Oxford University Press, 1982.

Baur, L., ed. *Die philosophischen Werke des Robert Grosseteste, Bischofs von Lincoln.* Munster: Achendorff, 1912.

Beaujouan, Guy. "L'enseignement de l'arithmetique elementaire a l'universite de
Paris aux xiiie et xive siécles," in *Homenaje a Millas-Vallicrosa,* 2 vols., 1.93–
124. Barcelona, 1954, 1956.

————. "Motives and Opportunities for Science in the Medieval Universities,"
in *Scientific Change: Historical Studies in the Intellectual, Social and Technical
Conditions for Scientific Discovery and Technical Invention, from Antiquity to
the Present: Symposium on the History of Science, University of Oxford, 9–15
July 1961,* edited by A. C. Crombie. New York, 1963.

Benjamin, F. S., Jr., and G. J. Toomer, eds. *Campanus of Novara and Medieval
Planetary Theory: "Theorica Planetarum."* Madison: University of Wisconsin
Press, 1971.

Bernard of Clairvaux, St. *The Life and Letters of St. Bernard of Clairvaux.* Translated by Bruno Scott James. London: Burns Oates, 1953.

Bonaventure, St. *S. Bonaventurae Opera Omnia, Vol. 1: Commentaria in primum librum Sententiarum.* Quaracchi, 1882.

————. *Saint Bonaventure's "De reductione artium ad theologiam": A Commentary with an Introduction and Translation.* Edited and translated by Sister
Emma Therese Healy. Saint Bonaventure, N.Y.: Franciscan Institute, 1955.

Bradwardine, Thomas. *De causa Dei contra Pelagium et De virtute causarum
. . . .* London, 1618.

Brush, Craig B., ed. and trans. *The Selected Works of Pierre Gassendi.* New York,
1972.

Bullough, Vern L. *The Development of Medicine as a Profession.* New York, 1966.

Buridan, John. *Iohannis Buridani Quaestiones super libris quattuor De caelo et
mundo.* Edited by E. A. Moody. Cambridge, Mass., 1942.

————. *In Metaphysicen Aristotelis; Questiones argutissime Magistri Ioannis
Buridani in ultima praelectione ab ipso recognitae et emissae ac ad archetypon
diligenter repositae cum duplice indicio materiarum videlicet in fronte quaestionum in operis calce.* Paris, 1518.

————. *Questiones super octo Phisicorum libros Aristotelis diligenter recognite et
revise magistro Johanne Dullaert de Gandavo.* Paris, 1509; reprint in facsimile

under the title *Johannes Buridanus, Kommentar zur Aristotelischen Physik,* Frankfurt, 1964.

Burke, R. B., trans. *The "Opus Majus" of Roger Bacon.* 2 vols. Philadelphia, 1928.

Burley, Walter. *Super octo libros Phisicorum.* Venice, 1501; reprint in facsimile with the title *In Physicam Aristotelis Expositio et quaestiones,* Hildesheim, 1972.

Campanella, Thomas. "The Defense of Galileo of Thomas Campanella . . . ." Edited and translated by Grant McColley. *Smith College Studies in History* 22, nos. 3–4 (April–July 1937).

Cantin, André. *Les sciences seculieres et la foi: Les deux voies de la science au jugement de S. Pierre Damien (1007–1072).* Spoleto: Centro Italiano di Studi sull'Alto Medioevo, 1975.

Čapek, Milie, ed. *The Concepts of Space and Time: Their Structure and Their Development.* Dordrecht, 1976.

Carmody, F. J., ed. *Al-Bitrûji. De motibus celorum: Critical Edition of the Latin Translation of Michael Scot.* Berkeley and Los Angeles: University of California Press, 1952.

Charleton, Walter. *Physiologia Epicuro-Gassendo-Charltoniana.* London, 1654.

Chauliac, Guy de. *La grande chirurgie de Guy de Chauliac.* Edited by E. Nicaise. Paris, 1890.

Chenu, M. D., O.P. *Nature, Man, and Society in the Twelfth Century: Essays on New Theological Perspectives in the Latin West.* Preface by Etienne Gilson. Selected, edited, and translated by Jerome Taylor and Lester K. Little. Chicago: University of Chicago Press, 1968; original French version published 1957.

Clagett, Marshall. *The Science of Mechanics in the Middle Ages.* Madison: University of Wisconsin Press, 1959.

————, ed. and trans. *Nicole Oresme and the Medieval Geometry of Qualities and Motions: A Treatise on the Uniformity and Difformity of Intensities Known as "Tractatus de configurationibus qualitatum et motuum."* Madison: University of Wisconsin Press, 1968.

Cohen, H. Floris. *The Scientific Revolution: A Historiographical Inquiry.* Chicago: University of Chicago Press, 1994.

Copleston, Frederick, S.J. *A History of Philosophy.* 9 vols. Westminster, Md.: Newman Press, 1946–1975.

Courtenay, William J. "Nominalism and Late Medieval Religion," in *The Pursuit of Holiness in Late Medieval and Renaissance Religion,* edited by Charles Trinkaus and Heiko A. Oberman. Leiden, 1974.

Crombie, A. C. *Medieval and Early Modern Science.* 2 vols. Garden City, N.Y.: Doubleday Anchor Books, 1959.

Cunningham, A. "How the *Principia* Got Its Name; or, Taking Natural Philosophy Seriously." *History of Science* 29 (1991): 377–92.

Darlington, Oscar G. "Gerbert, the Teacher." *American Historical Review* 52 (1946–1947): 467–70.

Day, C. "Jean Buridan and the Classification of the Sciences." Ph.D. diss., Indiana University, 1986.

Deferrari, Roy J., Sister M. Inviolata Barry, and Ignatius McGuiness. *A Lexicon of St. Thomas Aquinas Based on the "Summa Theologica" and Selected Passages of His Other Works.* Baltimore: The Catholic University of America Press, 1948.

Delhaye, Philippe. "La place des arts libéraux dans les programmes scolaires du xiiie siecle," in *Arts libéraux et philosophie au moyen âge: Actes du quatriéme congrés international de philosophie médiévale, Université de Montréal, Canada, 27 août–2 septembre 1967.* Montreal: Institut d'études mediévalés; Paris, Librairie philosophique J. Vrin, 1969.

Denifle, Heinrich, and Emile Chatelain, eds. *Chartularium Universitatis Parisiensis.* 4 vols. Paris: Ex typis Fratrum Delalain, 1889–1897.

Dick, Steven J. "Plurality of Worlds and Natural Philosophy: An Historical Study of the Origins of Belief in Other Worlds and Extraterrestrial Life." Ph.D. diss., Indiana University, 1977.

———. *Plurality of Worlds: The Origins of the Extraterrestrial Life Debate from Democritus to Kant.* Cambridge: Cambridge University Press, 1982.

Domingo Gundisalvo. *On the Division of Philosophy,* partially translated by M. Clagett and E. Grant, in *A Source Book in Medieval Science,* edited by Edward Grant, 62–65. Cambridge, Mass.: Harvard University Press, 1974.

Duhem, Pierre. *Etudes sur Léonard de Vinci ceux qu'il a lus et ceux qui l'ont lu.* 3 vols. Paris, 1906–1913.

———. *Les origines de la statique.* 2 vols. Paris, 1905–1906.

———. *Le systéme du monde: Histoire des doctrines cosmologiques de Platon à Copernic.* 10 vols. Paris, 1913–1959.

———. "Le temps et le mouvement selon les scholastiques." *Revue de philosophie* 23 (1913): 459–60.

———. *Un fragment inédit de l' "Opus tertium" de Roger Bacon. Précedé d'une étude sur ce fragment.* Quaracchi: Ex typographia Collegii S. Bonaventurae, 1909.

Duns Scotus, John. *Quaestiones in lib. II sententiarum,* in *Opera omnia.* Lyons, 1639; reprint Hildesheim, 1968.

Einstein, A. *Ideas and Opinions. Based on Mein Weltbild.* 3rd ed. Edited by C. Seelig and others, with new translations and revisions by S. Bargmann. New York, 1982

———. "On the Electrodynamics of Moving Bodies." Translated from "Zur Elektrodynamik bewegter Körper," *Annalen der Physik* 17 (1905), in *The Principle of Relativity: A Collection of Original Memoirs on the Special and General Theory of Relativity,* by H. A. Lorentz, A. Einstein, H. Minkowski, and H. Weyl, with notes by A. Sommerfeld; translated by W. Perrett and G. B. Jeffrey. New York, 1952; first published 1923.

Elders, Leo, S.V.D. *Faith and Science: An Introduction to St. Thomas' "Expositio in Boethii De Trinitate."* Rome: Herder, 1974.

Emden, A. B. *A Biographical Register of the University of Oxford to A.D. 1500.* 3 vols. Oxford, 1957–1959.

Evans, Gillian R. "The Rithmomachia: A Mediaeval Mathematical Teaching

Aid?" *Janus: Revue international de l'histoire des sciences* 63 (1976): 257–73.

Fisher, N. W., and Sabetai Unguru. "Experimental Science and Mathematics in Roger Bacon's Thought." *Traditio* 27 (1971).

French, Roger, and Andrew Cunningham. *Before Science: The Invention of the Friars' Natural Philosophy.* Aldershot, U.K.: Scolar Press, 1996.

Friedlein, G., ed. *Boetii De institutione arithmetica libri duo; De institutione musica libri quinque.* Leipzig, 1867.

Funkenstein, Amos. *Theology and the Scientific Imagination from the Middle Ages to the Seventeenth Century.* Princeton, N.J.: Princeton University Press, 1986.

Gaietanus de Thienis. *Recollecte . . . super octo libros Physicorum cum annotationibus textuum.* Venice, 1496.

Galileo Galilei. "Letter to Madame Christina of Lorraine, Grand Duchess of Tuscany, Concerning the Use of Biblical Quotations in Matters of Science," in *Discoveries and Opinions of Galileo,* translated by Stillman Drake, 213–14. Garden City, N.Y.: Doubleday, 1957.

————. "The Second Day," in *Galileo Galilei Dialogue Concerning the Two Chief World Systems—Ptolemaic and Coperican,* translated by Stillman Drake. Berkeley and Los Angeles: University of California Press, 1962.

————. *Two New Sciences.* Translated by Stillman Drake. Madison: University of Wisconsin Press, 1974.

Gassendi, Pierre. *The Selected Works of Pierre Gassendi.* Edited and translated by Craig B. Brush. New York, 1972.

George, W. Corner, trans. *Anatomical Texts of the Earlier Middle Ages.* Washington, D.C., 1927.

Gilson, Étienne. *History of Christian Philosophy in the Middle Ages.* London, 1955.

Glorieux, P. *La littérature quodlibétique.* 2 vols. Belgium, 1925, vol. 1; Paris, 1935, vol. 2.

Godfrey of Fontaine. *Les Quodlibets onze-quatorze de Godefroid de Fontaines (Texte inèdit),* Les Philosophes belges, Textes et études 5.1–2, edited by J. Hoffmans. Louvain, 1932.

Goldstein, R. B., ed. and trans. *The Arabic Version of Ptolemy's Planetary Hypotheses.* Philadelphia: American Philosophical Society, 1967.

Goldziher, Ignaz. "The Attitude of Orthodox Islam toward the 'Ancient Sciences,'" in *Studies on Islam,* edited and translated by Merlin L. Swartz. New York: Oxford University Press, 1981.

Goode, Erica. "How Culture Molds Habits of Thought." *New York Times,* August 8, 2000.

Grant, Edward. "Aristotelianism and the Longevity of the Medieval World View." *History of Science* 16 (1978): 93–106.

————. "Celestial Matter: A Medieval and Galilean Cosmological Problem." *Journal of Medieval and Renaissance Studies* 13 (1983): 165–71.

————. "Celestial Perfection from the Middle Ages to the Late Seventeenth Century," in *Religion, Science, and Worldview: Essays in Honor of Richard S.*

Westfall, edited by M. J. Osler and P. L. Farber, 149–62. New York: Cambridge University Press, 1985.

———. "The Condemnation of 1277, God's Absolute Power, and Physical Thought in the Late Middle Ages." *Viator* 10 (1979): 211–44.

———. "Cosmology," in *Science in the Middle Ages,* edited by David C. Lindberg, 265–302. Chicago, 1978.

———. "Eccentrics and Epicycles in Medieval Cosmology," in *Mathematics and Its Applications to Science and Natural Philosophy in the Middle Ages: Essays in Honor of Marshall Clagett,* edited by Edward Grant and J. E. Murdoch, 189–214. New York: Cambridge University Press, 1987.

———. *The Foundations of Modern Science in the Middle Ages: Their Religious, Institutional, and Intellectual Contexts.* Cambridge: Cambridge University Press, 1996.

———. *God and Reason in the Middle Ages.* Cambridge: Cambridge University Press, 2001.

———. *In Defense of the Earth's Centrality and Immobility: The Scholastic Reaction to Copernicanism in the Seventeenth Century.* Transactions of the American Philosophical Society, vol. 74, pt. 4. Philadelphia, 1984

———. "Jean Buridan and Nicole Oresme on Natural Knowledge." *Vivarium* 31, no. 1 (1993): 84–105.

———. "Late Medieval Thought, Copernicus, and the Scientific Revolution." *Journal of the History of Ideas* 23 (1962): 197–220.

———. "Medieval and Renaissance Scholastic Conceptions of the Influence of the Celestial Region on the Terrestrial." *Journal of Medieval and Renaissance Studies* 17 (1987): 1–23.

———. "Medieval and Seventeenth-Century Conceptions of an Infinite Void Space beyond the Cosmos." *Isis* 60 (1969): 39–60.

———. "Medieval Natural Philosophy: Empiricism without Observation," in *The Dynamics of Aristotelian Natural Philosophy from Antiquity to the Seventeenth Century,* edited by Cees Leijenhorst, Christopher Lüthy, and Johannes M. M. H. Thijssen, 141–68. Leiden: Brill, 2002.

———. *Much Ado About Nothing: Theories of Space and Vacuum from the Middle Ages to the Scientific Revolution.* Cambridge: Cambridge University Press, 1981.

———. "A New Look at Medieval Cosmology, 1260–1687." *Proceedings of the American Philosophical Society* 129 (1985): 417–32.

———, ed. and trans. *Nicole Oresme and the Kinematics of Circular Motion: "Tractatus de commensurabilitate vel incommensurabilitate motuum celi."* Madison: University of Wisconsin Press, 1971.

———, ed. and trans. *Nicole Oresme: "De proportionibus proportionum" and "Ad pauca respicientes."* Madison: University of Wisconsin Press, 1966.

———. *Physical Science in the Middle Ages.* New York, 1971; reprint Cambridge: Cambridge University Press, 1977.

———. "Place and Space in Medieval Physical Thought," in *Motion and Time, Space and Matter: Interrelations in the History of Philosophy and Science,* edited by Peter K. Machamer and Robert G. Turnbull, 137–67. Columbus, Ohio, 1976.

———. *Planets, Stars, and Orbs: The Medieval Cosmos, 1200–1687.* Cambridge: Cambridge University Press, 1994.

———. "Scientific Thought in Fourteenth-Century Paris: Jean Buridan and Nicole Oresme," in *Machaut's World: Science and Art in the Fourteenth Century,* Annals of the New York Academy of Sciences 314, edited by Madeleine Pelner Cosman and Bruce Chandler, 105–24. New York: New York Academy of Science, 1978.

———. *A Source Book in Medieval Science.* Cambridge, Mass.: Harvard University Press, 1974.

———. "Ways to Interpret the Terms 'Aristotelian' and 'Aristotelianism' in Medieval and Renaissance Natural Philosophy." *History of Science* 25 (1987): 335–58.

Gregory of Rimini. *Gregorii Arimensis OESA Lectura super Primum et Secundum Sententiarum.* 7 vols. Berlin: Walter de Gruyter, 1979–1987.

———. *Super primum et secundum sententiarum.* Franciscan Institute Publications, Text Series 7. Edited by E. M. Buytaert. Reprint St. Bonaventure, N.Y.: Franciscan Institute, 1955.

Grosseteste, Robert. *De artibus liberalibus,* in *Die philosophischen Werke des Robert Grosseteste, Bischofs von Lincoln, Beiträge zur Geschichte der Philosophie des Mittelalters,* edited by L. Baur, 9: 4–7. Munster, 1912.

Günther, Siegmund. "Geschichte des mathematischen Unterrichts im deutschen Mittelalter bis zum Jahre 1525, " in *Monumenta Germaniae Paedogogica,* vol. 3. Berlin: A. Hofmann & Co., 1887.

Hackett, Jeremiah. "Roger Bacon on *Scientia experimentalis,*" in *Roger Bacon and the Sciences: Commemorative Essays,* 277–315. Leiden: Brill, 1997.

Haskins, Charles H. *The Rise of Universities.* Ithaca, N.Y.,1957.

Healy, Sister Emma Therese, ed. and trans. *Saint Bonaventure's "De reductione artium ad theologiam": A Commentary with an Introduction and Translation.* Saint Bonaventure, N.Y.: Franciscan Institute, 1955.

Heath, T. *Aristarchus of Samos. The Ancient Copernicus. A History of Greek Astronomy to Aristarchus Together with Aristarchus' Treatise on the Sizes and Distances of the Sun and Moon.* Oxford: Clarendon Press, 1913.

Heiberg, J. L., ed. *Claudii Ptolemaei, opera quae exstant omnia, Vol. 2: Opera astronomica minora.* Leipzig: Tuebner, 1907.

Hillgarth, J. N. *Ramon Lull and Lullism in Fourteenth-Century France.* Oxford, 1971.

Hisette, Roland. *Enquête sur les 219 articles condamnés à Paris le 7 mars 1277.* Philosophes médiévaux 22. Louvain: Publications Universitaires; Paris: Vander-Oyez, 1977.

Hobbes, Thomas. *The English Works of Thomas Hobbes of Malmesbury.* 16 vols. Edited by William Molesworth. London, 1839–1845.

Holkot, Robert. *In quatuor libros sententiarum Quaestiones.* Lyon, 1518; reprint in facsimile Frankfurt, 1967.

Hollister, C. Warren. *Medieval Europe: A Short History.* 7th ed. New York, 1994.

Hoodbhoy, Pervez. *Islam and Science: Religious Orthodoxy and the Battle for Rationality.* London: Zed Books, 1991.

Horowitz, Tamara, and Gerald J. Massey, eds. *Thought Experiments in Science and Philosophy.* Savage, Md.: Rowman & Littlefield, 1991.

Huff, Toby. *The Rise of Early Modern Science.* Cambridge: Cambridge University Press, 1993.

Hugolin of Orvieto. *Hugolini de Urbe Veteri OESA Commentarius in Quattuor Libros Sententiarum.* 4 vols. Edited by Willigis Eckermann, O.S.A. Würzburg: Augustinus Verlag, 1980–1988.

Hugonnard-Roche, Henri. "L'hypothétique et la nature dans la physique parisienne du XIVᵉ siècle," in *La nouvelle physique du XIVe siècle,* edited by Stefano Caroti and Pierre Souffrin. Florence: Leo S. Olschki, 1997.

Hyman, Arthur, and James J. Walsh, eds. *Philosophy in the Middle Ages: The Christian, Islamic, and Jewish Traditions.* Indianapolis: Hackett, 1973.

Jaki, Stanley L. *Science and Creation: From Eternal Cycles to an Oscillating Universe.* New York: Science History Publications, 1974.

Jammer, Max. *Concepts of Space: The History of Theories of Space in Physics.* 2nd ed. Cambridge, Mass., 1969.

Jean de Ripa. "Jean de Ripa I Sent. dist. XXXVII: De modo inexistendi divine essentie in omnibus creaturis." Edited by André Combes and Francis Ruello, with an introduction by Paul Vignaux. *Traditio* 23 (1967): 191–267.

John of Damascus, St. *On the Orthodox Faith (De fide orthodoxa).* Translated by F. H. Chase Jr. New York: Fathers of the Church, 1958.

Katz, J., and R. Weingartner, eds. *Philosophy in the West: Readings in Ancient and Medieval Philosophy.* New York, 1965.

Keicher, Otto, ed. *Raymundus Lullus und seine Stellung zur arabischen Philosophie, mit einem Anhang, enthaltend die zum ersten Male veröffentlichte "Declaratio Raymundi per modum dialogi edita."* Beiträge zur Geschichte der Philosophie des Mittelalters 7.4–5. Minister, 1909.

Khalidi, Tarif. "The Idea of Progress in Classical Islam." *Journal of Near Eastern Studies* 40 (October 1981): 277–89.

Kibre, Pearl. "The *Quadrivium* in the Thirteenth Century Universities [with Special Reference to Paris]," in *Arts libéreaux et philosophie au moyen âge: Actes du quatriéme congrés international de philosophie médiévale, Université de Montréal, Canada, 27 août–2 septembre 1967.* Montreal: Institut d'études médiévalés; Paris: Librairie philosophique J. Vrin, 1969.

Kilwardby, Robert, O.P. *De Ortu Scientiarum.* Edited by Albert G. Judy, O.P. Toronto: British Academy and the Pontifical Institute of Mediaeval Studies, 1976.

King, Peter. "Mediaeval Thought-Experiments: The Metamethodology of Mediaeval Science," in *Thought Experiments in Science and Philosophy,* edited by Tamara Horowitz and Gerald J. Massey, 43–64. Savage, Md.: Rowman & Littlefield, 1991.

Kinross, Lord. *The Ottoman Centuries: The Rise and Fall of the Turkish Empire.* New York: Morrow Quill Paperbacks, 1977.

Kirschner, Stefan., ed. *Nicolaus Oresme Kommentar zur Physik des Aristoteles: Kommentar mit edition de Quaestionen zu Buch 3 unde 4 der Aristotelischen Physik sowie von vier Quaestionen zu Buch 5.* Stuttgart: Franz Steiner Verlag, 1997.

Koyré, Alexandre. "Le vide et l'espace infini au XIVe siècle." *Archives d'histoire doctrinale et littéraire du moyen âge* 24 (1949): 45–91.

Krafft, Fritz. "Guericke (Gericke), Otto Von," in *Dictionary of Scientific Biography,* 16 vols., edited by Charles C. Gillispie, 5.574–76. New York: Scribners, 1970–1980.

Kren, Claudia, ed. and trans. *The "Questiones super De celo" of Nicole Oresme.* Ph.D. diss., University of Wisconsin.

Leff, Gordon. *The Dissolution of the Medieval Outlook: An Essay on Intellectual and Spiritual Change in the Fourteenth Century.* New York, 1976.

———. *Paris and Oxford Universities in the Thirteenth and Fourteenth Centuries.* New York, 1968.

Laird, W. R. "The Scientiae mediae in Medieval Commentaries on Aristotle's *Posterior Analytics.*" Ph.D. diss., University of Toronto, 1983.

Larson, E. J., and L. Witham. "Scientists Are Still Keeping the Faith." *Nature* 386 (3 April 1997): 435–36.

Lemay, Helen R. "Science and Theology at Chartres: The Case of the Supracelestial Waters." *British Journal for the History of Science* 10 (1977): 226–36.

Lemay, Richard. "Gerard of Cremona," in *Dictionary of Scientific Biography,* 16 vols, edited by Charles C. Gillispie, 15.173–92. New York: Scribners, 1970–1980.

———. "The Teaching of Astronomy in Medieval Universities, Principally at Paris in the Fourteenth Century." *Manuscripta* 20, no. 3 (1976): 197–217.

Lerner, Ralph, and Muhsin Mahdi, eds. *Medieval Political Philosophy: A Sourcebook.* New York, 1963.

Lewis, C. S. *The Discarded Image: An Introduction to Medieval and Renaissance Literature.* Cambridge, 1964.

Lindberg, David C. *A Catalogue of Medieval and Renaissance Optical Manuscripts.* Subsidia Mediaevalia 4. Toronto, 1975.

———, ed. and trans. *John Pecham and the Science of Optics: "Perspectiva Communis."* Madison, Wis., 1970.

———. "On the Applicability of Mathematics to Nature: Roger Bacon and His Predecessors." *British Journal for the History of Science* 15 (1982): 3–26.

———. "The Transmission of Greek and Arabic Learning to the West," in *Science in the Middle Ages,* edited by David C. Lindberg, 52–90. Chicago, 1978.

———. *Roger Bacon's Philosophy of Nature: A Critical Edition, with English Translation, Introduction, and Notes of "De multiplicatione specierum" and "De speculis comburentibus."* Edited and translated by D. C. Lindberg. Oxford: Clarendon Press, 1983.

Little, A. G., ed. *Part of the "Opus tertium" of Roger Bacon.* Aberdeen, 1912.

Livesey, S. J. *Theology and Science in the Fourteenth Century: Three Questions on the Unity and Subalternation of the Sciences from John of Reading's Commentary on the Sentences.* Leiden, 1989.

Lloyd, G. E. R. *Aristotle: The Growth and Structure of His Thought.* Cambridge: Cambridge University Press, 1968.

———. "The Invention of Nature," in *Methods and Problems in Greek Science,* edited by G. E. R. Lloyd. Cambridge, 1991.

Locke, John. *An Essay Concerning Human Understanding.* Edited by Peter H. Nidditch. Oxford, 1975.

Lohr, Charles H. "Medieval Latin Aristotle Commentaries." *Traditio* 23 (1967): 33–413; 24 (1968): 149–245; 26 (1970): 135–216; 27 (1971): 251–351; 28 (1972): 281–396; 29 (1973): 93–197; "Supplementary Authors," 30 (1974): 119–44.

———. *Latin Aristotle Commentaries, II: Renaissance Authors.* Florence, 1988.

Lombard, Peter. *Magistri Petri Lombardi Parisiensis Episcopi Sententiae in IV Libris Distinctae.* 3rd ed. Grottaferrata [Rome]: Editiones Collegii S. Bonaventurae Ad Claras Aquas, 1971.

Maier, Anneliese. *Metaphysische Hintergründe der spätscholastischen Naturphilosophie.* Rome: Edizioni di Storia e Letteratura, 1955.

———. *Zwischen Philosophic und Mechanik: Studien zur Naturphilosophie der Spätscholastik.* Rome, 1958.

Makdisi, George. "Ash'ari and the Ash'arites in Islamic Religious History," Parts 1 and 2. *Studia Islamica* 17 (1962): 37–80 and 18 (1963): 19–39; reprinted in George Makdisi, *Religion, Law and Learning in Classical Islam,* Variorum Collected Studies Series. Hampshire, Great Britain, and Brookfield, Vermont: Gower Publishing Co., 1991.

———. *The Rise of Colleges: Institutions of Learning in Islam and the West.* Edinburgh: Edinburgh University Press, 1981.

Mandonnet, Pierre F., O.P. *Siger de Brabant et l'Averroisme latin au XIIIme siecle, IIme partie: Textes inedits.* 2d ed. Louvain: Institut supérieur de philosophic de l'Université, 1908.

Marsilius of Inghen(?). *Questiones subtilissime Johannis Marcilii Inguen super octo libros Physicorum secundum nominalium viam.* . . . Lyon, 1518.

Maurer, Armand. "Ockham on the Possibility of a Better World." *Mediaeval Studies* 38 (1976): 291–312.

McLaughlin, Mary Martin. *Intellectual Freedom and Its Limitations in the University of Paris in the Thirteenth and Fourteenth Centuries.* New York: Arno Press, 1977.

Miller, Donald G. "Pierre Duhem," in *Dictionary of Scientific Biography,* 16 vols., edited by Charles C. Gillispie, 4.225–33. New York, 1970–1980.

Minio-Paluello, L. "Aristotle: Tradition and Influence," in *Dictionary of Scientific Biography,* 16 vols., edited by Charles C. Gillispie, 1.267–81. New York: Scribners, 1970.

Molland, George. "Medieval Ideas of Scientific Progress." *Journal of the History of Ideas* 59 (1978): 561–77.

———. "Mathematics in the Thought of Albertus Magnus," in *Albertus Magnus and the Sciences: Commemorative Essays, 1980,* edited by J. A. Weishipl. Toronto: Pontifical Institute of Mediaeval Studies, 1980.

———. "Nichole Oresme and Scientific Progress," in *Miscellanea Mediaevalia,* veröffentlichungen des Thomas-Instituts der Universität zu Köln, Band 9: *Antiqui und Moderni,* 206–20. Berlin and New York, 1974.

Moody, E. A. "Buridan, Jean," in *Dictionary of Scientific Biography,* 16 vols., edited by Charles C. Gillispie, 2.605. New York: Scribners, 1970–1980.

————, ed. *Johannis Buridani, Quaestiones super libris quattuor De caelo et mundo.* Cambridge, Mass.: Mediaeval Academy of America, 1942.

Moody, E. A., and Marshall Clagett, eds. and trans. *The Medieval Science of Weights (Scientia de ponderibus): Treatises Ascribed to Euclid, Archimedes, Thabit ibn Qurra, Jordanus de Nemore, and Blasius of Parma,* with English introductions, English translations, and notes. Madison, Wis., 1952.

Mottahedeh, Roy. *The Mantle of the Prophet.* New York: Pantheon Books, 1985.

Murdoch, John E. "The Analytic Character of Late Medieval Learning: Natural Philosophy without Nature," in *Approaches to Nature in the Middle Ages,* edited by Lawrence D. Roberts, 171–213. Binghamton, N.Y.: Center for Medieval and Early Renaissance Studies, 1982.

————. "Bradwardine, Thomas," in *Dictionary of Scientific Biography,* 16 vols., edited by Charles C. Gillispie, 2.395. New York: Scribners, 1970–1980.

————. "Euclid: The Transmission of the Elements," in *Dictionary of Scientific Biography,* 16 vols., edited by Charles C. Gillispie, 4.443–48. New York: Scribners, 1970–1980.

Murdoch, J. E. "From Social into Intellectual Factors: An Aspect of the Unitary Character of Late Medieval Learning," in *The Cultural Context of Medieval Learning: Proceedings of the First International Colloquim on Philosophy, Science, and Theology in the Middle Ages, September 1973,* edited by John E. Murdoch and Edith D. Sylla, 272–348. Dordrecht and Boston: D. Reidel, 1975.

————. "*Mathesis in philosophiam scholasticam introducta:* The Rise and Development of the Application of Mathematics in Fourteenth Century Philosophy and Theology," in *Arts libéraux et philosophie au moyen âge: Actes du quatriéme congrés international de philosophie médiévale, Université de Montréal, Canada, 27 août–2 septembre 1967,* 215–54. Montreal: Institut d'études mediévalés; Paris, Librairie philosophique J. Vrin, 1969.

————. "Music and Natural Philosophy: Hitherto Unnoticed *Questiones* by Blasius of Parma (?)." *Manuscripta* 20, no. 2 (1976): 119–36.

Murray, Alexander. *Reason and Society in the Middle Ages.* Oxford: Clarendon Press, 1978.

Nasr, Seyyed Hossein. *Islamic Science: An Illustrated Study.* Westerham, Kent, Great Britain: World of Islam Festival Publishing Co., 1976.

Nicholson, Reynold A. *A Literary History of the Arabs.* Cambridge: Cambridge University Press, 1953.

Nichomachus of Gerasa. *Nichomachus of Gerasa: Introduction to Arithmetic.* Translated by Martin Luther D'Ooge, with studies in Greek arithmetic by Frank E. Robbins and Louis C. Karpinski. New York, 1926.

Oberman, Heiko. *The Harvest of Medieval Theology: Gabriel Biel and Late Medieval Nominalism.* Grand Rapids, Mich., 1967.

Ockham, William. *Opera plurima.* 4 vols. Lyons, 1494–1496; reprint in facsimile London, 1962.

———. *Quotlibeta septem; Tractatus de Sacramento altaris.* Strasbourg, 1491; reprint in facsimile Louvain, 1962.

———. *Tractatus de successivis Attributed to William Ockham,* Franciscan Institute Publications 1, 46, edited by Philotheus Boehner, O.F.M. St. Bonaventure, N.Y.: Franciscan Institute, 1944.

Ong, Walter J. *Ramus: Method and the Decay of Dialogue from the Art of Discourse to the Art of Reason.* Cambridge, Mass., 1958.

Oresme, Nicole. *De proportionibus proportionum,* in *Nicole Oresme "De proportionibus proportionum" and "Ad pauca respicientes."* Edited and translated by Edward Grant. Madison: University of Wisconsin Press, 1966.

———. *Nicole Oresme and the Kinematics of Circular Motion: Tractatus de commensurabilitate vel incommensurabilitate motuum celi.* Edited, translated, and with an introduction by Edward Grant. Madison: University of Wisconsin Press, 1971.

———. *Nicole Oresme: Le Livre du ciel et du monde.* Edited by Albert D. Menut and Alexander J. Denomy, translated with an introduction by Albert D. Menut. Madison: University of Wisconsin Press, 1968.

———. *Nicole Oresme Quaestiones super De generatione et corruptione.* Edited by Stefano Caroti. Munich: Verlag der Bayerischen Akademie der Wissenschaften, 1996.

———. "Nicholas Oresme's *Questiones super libros Aristotelis De anima:* A Critical Edition with Introduction and Commentary." Edited by Peter Marshall. Ph.D. diss., Cornell University, 1980.

Organ, Troy W. *An Index to Aristotle in English Translation.* New York, 1966.

Orme, Nicholas. *English Schools in the Middle Ages.* London, 1973.

Pecham, J. *John Pecham and the Science of Optics: Perspectiva communis.* Edited and translated by D. C. Lindberg. Madison, Wis., 1970.

———. *John Pecham Tractatus de perspectiva.* Edited by D. C. Lindberg. St. Bonaventure, N.Y., 1972.

Pedersen, Olaf. *A Survey of the Almagest.* Odense: Odense University Press, 1974.

———. "The Theorica Planetarum—Literature of the Middle Ages." *Classica et Mediaevalia: Revue Danoise de Philologie et d'Histoire* 23 (1962).

Peter Damian, St. *De divina omnipotentia,* in *Medieval Philosophy: From St. Augustine to Nicholas of Cusa,* edited by John F. Wippel and Allan Wolter, O.F.M., 143–52. New York: Free Press, 1969.

Petrus Aureoli. *Commentariorum in secundum librum sententiarum tomus secundus.* Rome, 1605.

Pines, Shlomo. "Al-Razi, Abu Bakr Muhammad Ibn Zakariya," in *Dictionary of Scientific Biography,* 16 vols., edited by Charles C. Gillispie, 11.323–26. New York: Scribners, 1970–1980.

Poulle, Emmanuel. "John of Ligneres," in *Dictionary of Scientific Biography,* 16 vols., edited by Charles C. Gillispie, 7.122–28. New York: Scribners, 1970–1980.

————. "John of Murs," in *Dictionary of Scientific Biography,* 16 vols., edited by Charles C. Gillispie, 7.128. New York: Scribners, 1970–1980.

Rashdall, Hastings. *The Universities of Europe in the Middle Ages.* 3 vols. Edited by F. M. Powicke and A. B. Emden. Oxford, 1936; reprint 1988.

Reisch, Gregor. *Margarita philosophica.* Basel: Michael Furterius, 1517; reprint Dusseldorf: Stern-Verlag Janssen & Co., 1973.

Richard of Middleton. *Super quatuor libros Senteniarum Petri Lombardi Quaestiones subtilissimae.* 4 vols. Brescia, 1591; reprint Frankfurt: Minerva, 1963.

Ridder-Symoens, H. de, ed. *A History of the University in Europe, Vol. 1: Universities in the Middle Ages.* Cambridge: Cambridge University Press, 1992.

Rosenthal, Franz. "Ibn Khaldun," in *Dictionary of Scientific Biography,* 16 vols., edited by Charles C. Gillispie, 7.321. New York: Scribners, 1970–1980.

Sabra, A. I. "Science and Philosophy in Medieval Islamic Theology." *Zeitschrift für Geschichte der Arabisch-Islamischen Wissenschaften* 9 (1994): 1–42.

Sarton, George. *Introduction to the History of Science.* 3 vols. in 5 parts. Baltimore: Carnegie Institution of Washington, 1927–1948.

Shapiro, Herman. "Motion, Time, and Place According to William Ockham." *Franciscan Studies* 16, no. 3 (1956).

Siger de Brabant. *Questions sur la Physique d'Aristote (texte inédit),* in *Les Philosophe Belges, Textes et Etudes,* vol. 15, translated by Philippe Delhaye. Louvain: Editions de l'Institut Supérieur de Philosophie, 1941.

Simplicus. *Commentaria in quatuor libros De caelo Aristotelis, Guillermo Morbeto interprete.* Venice: Hieronimus Scotus, 1540.

Siraisi, Nancy G. "The *libri morales* in the Faculty of Arts and Medicine at Bologna: Bartolomeo de Varignana and the Pseudo-Aristotelian Economics." *Manuscripta* 20 (1976): 105–18.

Sorenson, Roy A. *Thought Experiments.* New York and Oxford: Oxford University Press, 1992.

Steifel, Tina. "The Heresy of Science: A Twelfth-Century Conceptual Revolution." *Isis* 68 (1977): 347–62.

————. "Science, Reason, and Faith in the Twelfth Century: The Cosmologists' Attack on Tradition." *Journal of European Studies* 6 (1976): 1–16

Steneck, Nicholas. *Science and Creation in the Middle Ages: Henry of Langenstein (d. 1397) on Genesis.* Notre Dame, Ind.: University of Notre Dame Press, 1976.

Stock, Brian. *Myth and Science in the Twelfth Century: A Study of Bernard Silvester.* Princeton, N.J., 1972.

Suarez, Francisco. *Disputationes metaphysicae.* 2 vols. Paris 1866; reprint in facsimile Hildesheim, 1965; first printed 1597.

Sweeney, Leo, S.J., "Divine Infinity: 1150–1250." *The Modern Schoolman* 35 (1957).

Sylla, Edith. "Autonomous and Handmaiden Science: St. Thomas Aquinas and William of Ockham on the Physics of the Eucharist," in *The Cultural Context of Medieval Learning,* edited by John E. Murdoch and Edith D. Sylla. Dordrecht and Boston: D. Reidel, 1979.

————. "Medieval Quantification of Qualities: The 'Merton School.'" *Archive for History of Exact Sciences* 8, nos. 1–2 (1971).

Synan, Edward A. "Introduction: Albertus Magnus and the Sciences," in *Albertus Magnus and the Sciences: Commemorative Essays, 1980,* edited by James A. Weisheipl, O.P. Toronto, 1980.

Talbot, Charles H. "Medicine," in *Science in the Middle Ages,* edited by David C. Lindberg, 391–428. Chicago, 1978.

Themon Judaeus. *Questions on the Meteorology.* See Albert of Saxony.

Thomas Aquinas, St. *Cosmogony* (1a65–74), vol. 10 of *Summa theologiae: Latin Text and English Translation, Introductions, Notes, Appendices and Glossaries.* Translated by William A. Wallace. New York and London: Blackfriars Press/McGraw-Hill Book Co./Eyre & Spottiswoode, 1967.

————. *Opera Omnia. Parma.* 1852–1873; reprint New York, 1948–1950.

————. *S. Thomae Aquinatis In Aristotelis libros De caelo et mundo; De generatione et corruptione; Meteorologicorum Expositio cum textus ex recensione leonina,* ed. R. M. Spiazzi, Turin 1952.

————. *S. Thomae Aquinatis In octo libros De physico auditu sive Physicorum Aristotelis commentaria.* Edited by P. Fr. Angeli-M. Pirotta, O.P. Naples, 1953.

————. *St Thomas Aquinas Commentary on Aristotle's "Physics."* Translated by R. J. Blackwell, R. J. Spath, and W. E. Thirlkel; introduction by V. J. Bourke. New Haven, Conn., 1963.

————. *St. Thomas Aquinas, Siger of Brabant, St. Bonaventure, On the Eternity of the World (De aeternitate mundi),* translated from the Latin with an introduction by Cyril Vollert, Lottie H. Kendzierski, and Paul M. Byrne. Milwaukee, 1964.

————. *Summa Theologia, Vol. 8: Creation, Variety and Evil.* Latin text, English translation, introduction, and notes by Thomas Gilby, O.P. New York: Blackfriars Press, 1967.

Thomas of Strasbourg. . . . *Commentaria in IIII libros sententiarum.* Venice, 1564; reprint Ridgewood, N.J.: Gregg Press, 1965.

Thorndike, Lynn. *A History of Magic and Experimental Science.* 8 vols. New York: Columbia University Press, 1923–1958

————. "An Anonymous Treatise in Six Books on Metaphysics and Natural Philosophy," in *A History of Magic and Experimental Science,* 8 vols., *3.568–84.* New York: Columbia University Press, 1923–1958.

————, ed. and trans. *The "Sphere" of Sacrobosco and Its Commentators.* Chicago, 1949.

————. *University Records and Life in the Middle Ages.* New York, 1944.

Van Egmond, Warren. *The Commercial Revolution and the Beginnings of Western Mathematics in Renaissance Florence, 1300–1500.* Ph.D. diss., Indiana University, 1976.

Van Steenberghen, Fernand. *Siger de Brabant d'après ses oeuvres inédites.* 2 vols. Louvain, 1931–1942.

Vogel, Kurt. *Mohammed ibn Musa Alchwarizmi's Algorismus das früheste Lehrbuch zum Rechnen mit indischen Ziffern.* Nach der einzigen (lateinischen) Handscrift (Cambridge Un. Lib. Ms. Ii. 6.5) in Faksimile mit Transkription und Kommentar herausgegeben. Aalen, 1963.

Walbridge, John. "Logic in the Islamic Intellectual Tradition: The Recent Centuries." *Islamic Studies* 39, no. 1 (Spring 2000): 55–75.

Wallace, William A., trans. *Galileo's Early Notebooks: The Physical Questions, A Translation from the Latin, with Historical and Paleographical Commentary.* Notre Dame, Ind.: University of Notre Dame Press, 1977.

———. *Prelude to Galileo: Essays on Medieval and Sixteenth-Century Sources of Galileo's Thought.* Dordrecht and Boston: D. Reidel, 1981.

Watt, W. Montgomery. *The Faith and Practice of al-Ghazali.* London: George Allen & Unwin, 1953.

———. *Islamic Philosophy and Theology: An Extended Survey.* Edinburgh: Edinburgh University Press, 1985.

Weisheipl, James A., O.P. "Curriculum of the Faculty of Arts at Oxford in the Early Fourteenth Century." *Mediaeval Studies* 26 (1964): 143–85.

———. *Friar Thomas d'Aquino, His Life, Thought, and Work.* Garden City, 1974.

William of Auvergne. *Opera omnia.* 2 vols. Paris, 1674; reprint in facsimile Frankfurt, 1963.

Wippel, John F. "The Condemnations of 1270 and 1277 at Paris." *Journal of Medieval and Renaissance Studies* 7 (1977): 169–201.

Yates, Frances A. *Giordano Bruno and the Hermetic Tradition.* Chicago: University of Chicago Press, 1964.

Index

Abelard, Peter: attack against, 265, 151, 229

Abu'l Barakat, 205n26

Aegidius Romanus, 127–28

Alan of Lille, 31

Albert of Saxony, 35n58, 60n31, 97, 108, 129, 156, 199, 316, 317; assumed rotation of moon, 124; and bellows thought experiment, 212–13; denied extracosmic space, 66n47; fall of bodies through a hole in the earth, 184–85; fall of homogeneous mixed bodies, 134, 181; fall of mixed body in a vacuum, 218–19; fly and lance thought experiment, 206; God and faith in his questions, 109, 111, 117; and God's absolute power, 112–13; imagined God created vacuum, 158, 296; nature prevents vacuum, 217–18; on other worlds, 62–63n36; and potential theological problem, 250–51; rejected moment of rest, 185, 205, 206; on smith's wheel and top, 201n18

Albertus Magnus (Albert the Great), 7, 39, 97, 99, 107, 108, 126n24, 223, 251, 325; on Aristotle, 306; emphasized experience, 196, 198, 221; little about God and faith in commentary, 103–4; minimized theologization of natural philosophy, 102–6, 115; on surfaces of the spheres, 123–24; as theologian-natural philosopher, 105–6; weakened links between geometry and physical nature, 134–35

Alexander Hales, 25

Alexander of Aphrodisias, 18, 282, 324, 325

Alhazen (Ibn al-Haytham), 18, 21

Ambrose, St., 240

Amicus, Bartholomaeus, 89; on God moving the world rectilinearly, 55n16, 88

angel(s), 36, 38n62, 75, 80, 99n17, 155, 284, 313, 314, 316; behavior of, 242; considered by Peter Lombard, 39n62; and the empyrean heaven, 144; Floris Cohen on, 195; God and, 72n64; Gregory of Rimini on, 301; and imaginary space, 75n74; location and movement of, 80–82, 301; and the movement of orbs, 155, 156, 160; in natural philosophy, 300–301; Ockham on the place of, 82; questions on, 188–89

Anselm of Laon, 144

Apian, Peter, 141,143, 144, 145

Aquinas, Thomas. See Thomas Aquinas

Archimedes, 18

Aristotelianism: empiricism without observation, 221; longevity in Middle Ages, 312–26; never reformed from within, 325–26; reigned in universities, 47; revitalized, 325; in sixteenth century, 323–26; xii, 139; unchallenged until fifteenth century, 119

Aristotle, 1, 4, 9, 14, 16n2, 17n3, 18, 35, 38, 40, 45, 60, 61, 65, 68, 83, 86n123, 87, 95, 96, 98, 101, 104, 107, 108, 110, 111, 114, 116, 121, 130, 145, 147, 153, 155, 160, 171, 172, 173, 174, 175, 177, 178, 179, 182, 183, 192, 195, 198, 203, 205n26, 208, 209, 219, 223, 225, 232, 236, 245, 250, 257, 259, 264, 288, 289, 315n2, 317, 321n15, 322, 325; air as mover of projectiles (*antiperistasis*), 201, 202; atomization of his works, 46, 319–20; authoritative knowledge from senses, 199; basis of knowledge was perception, 163, 196; believed in secondary causation,

The Nature of Natural Philosophy in the Late Middle Ages was designed and typeset in Minion by Kachergis Book Design of Pittsboro, North Carolina. It was printed on 60-pound Natures Book Natural and bound by Thomson-Shore of Dexter, Michigan.